DEVELOPMENTAL MATHEMATICS: PREALGEBRA, ELEMENTARY ALGEBRA, AND INTERMEDIATE ALGEBRA

Tom Carson

Franklin Classical School

Bill E. Jordan

Seminole State College of Florida

PEARSON

Boston Columbus Indianapolis New York San Francisco

Amsterdam Cape Town Dubai London Madrid Milan Munich Paris Montréal Toronto

Delhi Mexico City São Paulo Sydney Hong Kong Seoul Singapore Taipei Tokyo

Reproduced by Pearson from electronic files supplied by the author.

Publishing as Pearson, 75 Arlington Street, Boston, MA 02116.

ISBN-13: 978-0-321-91520-7

ISBN-10: 0-321-91520-8

1 2 3 4 5 6 RRD-H 18 17 16 15 14

www.pearsonhighered.com

PEARSON

Table of Contents

Name:
Instructor:

Date:
Section:

Chapter 1 WHOLE NUMBERS

1.1 Introduction to Numbers, Notation, and Rounding

Learning Objectives

1. Name the digit in a specified place.
2. Write whole numbers in standard and expanded form.
3. Write the word name for a whole number.
4. Graph a whole number on a number line.
5. Use <, >, or = to write a true statement.
6. Round numbers.
7. Interpret bar graphs and line graphs.

Key Terms

Use the vocabulary terms listed below to complete each statement in Exercises 1–7.

set	**subset**	**natural numbers**	**whole numbers**
equation	**inequality**	**statistic**	

1. The number 0 is an element of the ____________________.

2. A(n) ____________________ is a collection or group of elements.

3. A number used to describe a collection of numbers is a(n) ____________________.

4. The ____________________ are the counting numbers.

5. The mathematical relationship $2 < 12$ is called a(n) ____________________.

6. The natural numbers are a(n) ____________________ of the whole numbers.

7. A statement used to show that two amounts are equal is a(n) ____________________.

GUIDED EXAMPLES AND PRACTICE

Objective 1 Name the digit in a specified place.

Review these examples for Objective 1:

1. Identify the digit in the tens place; 5,833,801,265.

5,833,801,265 6

2. Write the word name for the place value of the digit 4 in 51,468,098.

51,468,098 hundred thousands

Practice these exercises:

1. Identify the digit in the millions place; 7,838,902,215.

2. Write the word name for the place value of the digit 4 in 4,953,289,781.

Name: Date:
Instructor: Section:

Objective 2 Write whole numbers in standard and expanded form.

Review these examples for Objective 2:

3. Write the number in expanded form; 61,932.

6 ten thousands + 1 thousands + 9 hundreds + 3 tens + 2 ones

4. Write the number in standard form; 2 millions + 9 ten thousands + 7 thousands + 8 hundreds + 3 ones.

2,097,803

Practice these exercises:

3. Write the number in expanded form; 35,609,972.

4. Write the number in standard form; 9 ten thousands + 2 hundreds + 3 tens + 8 ones.

Objective 3 Write the word name for a whole number.

Review this example for Objective 3:

5. Write the word name for 51,104.

fifty-one thousand, one hundred four

Practice this exercise:

5. Write the word name for 416,995.

Objective 4 Graph a whole number on a number line.

Review this example for Objective 4:

6. Graph 3 on a number line.

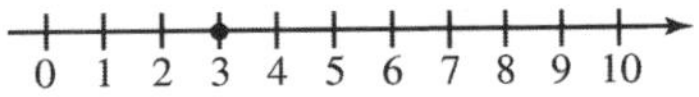

Practice this exercise:

6. Graph 5 on a number line.

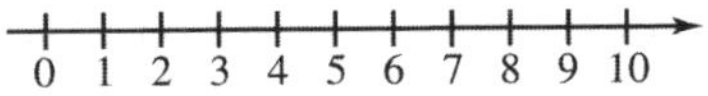

Objective 5 Use <, >, or = to write a true statement.

Review this example for Objective 5:

7. Use <, >, or = to write a true statement. 2672 ____ 1640

2672 > 1640

Practice this exercise:

7. Use <, >, or = to write a true statement. 1891 ____ 2298

Name:
Instructor:

Date:
Section:

Objective 6 Round numbers.

Review this example for Objective 6:

8. Round 456,583,127 to the ten thousands place.

A) Locate the digit in the ten thousands place, 8.
B) The digit to the right is 3, which is less then 4; so round down.
C) Change all digits to the right of 8 to zeros.

456,580,000

Practice this exercise:

8. Round 456,583,127 to the hundreds place.

Objective 7 Interpret bar graphs and line graphs.

A garden club asked its members to vote for their favorite flower. The results are represented in the following graph. Use the graph to answer the following questions.

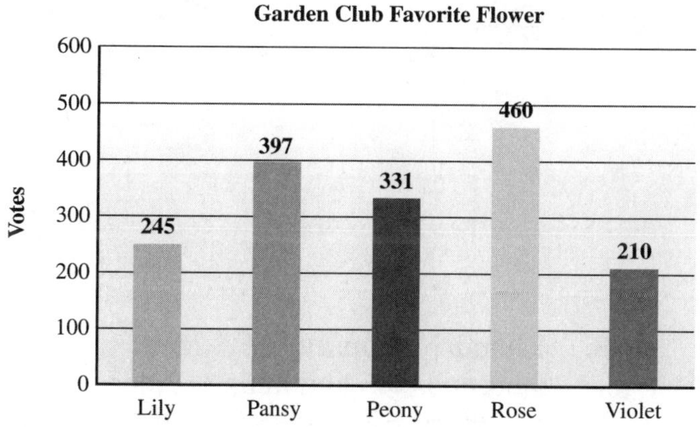

Review this example for Objective 7:

9. Which flower received the least number of votes? Which flower received the most number of votes?

The violet received the least and the rose received the most.

Practice this exercise:

9. Round the number of votes for peony to the nearest hundred.

Name: Date:
Instructor: Section:

ADDITIONAL EXERCISES

Objective 1 Name the digit in a specified place.

For extra help, see Example 1 on page 3 of your text and the Section 1.1 lecture video.

Identify the digit in the requested place.

1. 8,836,003,295; the hundred thousands place

2. 7,349,621; the thousands place

Write the word name for the place value of the digit 9 in each number.

3. 109

4. 37,492,561

Objective 2 Write whole numbers in standard and expanded form.

For extra help, see Examples 2–3 on pages 3–4 of your text and the Section 1.1 lecture video.

Write each number in expanded form.

5. 14,321

6. 720,516

Write each number in standard form.

7. 9 ten thousands + 2 hundreds + 3 tens + 8 ones

8. 5 hundred thousands + 3 thousands + 5 hundreds + 7 tens + 2 ones

Objective 3 Write the word name for a whole number.

For extra help, see Example 4 on page 4 of your text and the Section 1.1 lecture video.

Write the word name for each number.

9. 315,490

10. 2,568,157

Name: Date:
Instructor: Section:

Objective 4 Graph a whole number on a number line.
For extra help, see Example 5 on page 5 of your text and the Section 1.1 lecture video.
Graph each number on the number line.

11. 8

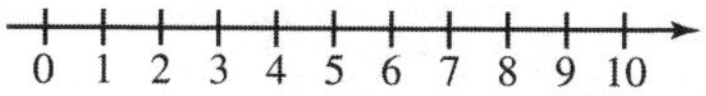

12. 1

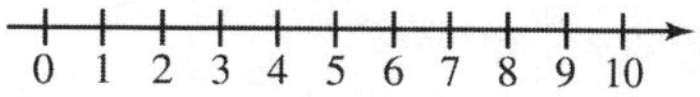

Objective 5 Use <, >, or = to write a true statement.
For extra help, see Example 6 on page 6 of your text and the Section 1.1 lecture video.
Use <, >, or = to write a true statement.

13. 562,104 ____ 562,104

14. 1,587,546 ____ 1,587,456

Objective 6 Round numbers.
For extra help, see Examples 7–8 on pages 7–8 of your text and the Section 1.1 lecture video.
Round 18,920,653 to the specified place.

15. millions

16. thousands

Objective 7 Interpret bar graphs and line graphs.
For extra help, see Examples 9–10 on pages 8–9 of your text and the Section 1.1 lecture video.

The yearly cost of tuition at a particular college, for five consecutive years, is represented in the following graph. Use the graph to answer the following questions.

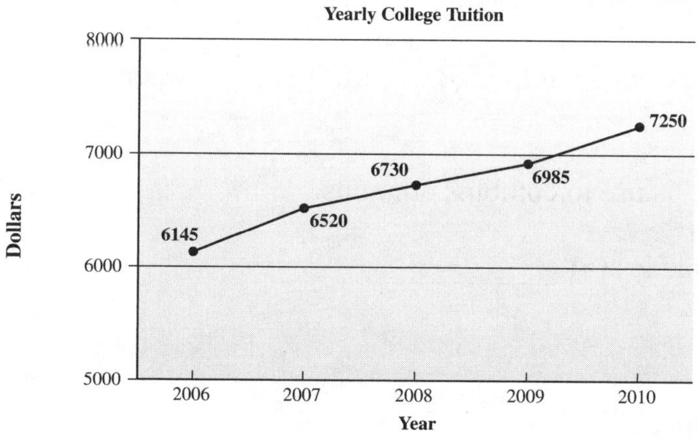

17. What was the cost of tuition in 2008? In 2006?

18. In what year was the cost of tuition the highest? What trend does the graph indicate about the cost of tuition?

Name: Date:
Instructor: Section:

Chapter 1 WHOLE NUMBERS

1.2 Adding and Subtracting Whole Numbers; Solving Equations

Learning Objectives
1. Add whole numbers.
2. Estimate sums.
3. Solve applications involving addition.
4. Subtract whole numbers.
5. Solve equations containing an unknown addend.
6. Solve applications involving subtraction.
7. Solve applications involving addition and subtraction.

Key Terms
Use the vocabulary terms listed below to complete each statement in Exercises 1–7.

addition	**perimeter**	**variable**	**constant**
subtraction	**related equation**	**solution**	

1. The ____________________ is the value that makes an equation true when it replaces the variable in the equation.

2. To find the length of fence to put around a garden, find the ____________________ of the garden.

3. A(n) ____________________ is a symbol used to represent unknown amounts.

4. Given an addition equation involving an unknown amount, write a(n) ____________________ using subtraction to solve for the amount.

5. The operation ____________________ can be interpreted as a difference between amounts.

6. The operation ____________________ is use to combine amounts.

7. A symbol that does not change its value is a(n) ____________________.

Name: Date:
Instructor: Section:

GUIDED EXAMPLES AND PRACTICE

Objective 1 Add whole numbers.

Review this example for Objective 1:

1. Add: 25,169 + 5,879 + 25

$$\begin{array}{r} {}^{1\ 1\ 1\ 2} \\ 25,169 \\ 5,879 \\ +\quad 25 \\ \hline 31,073 \end{array}$$

Practice this exercise:

1. $\begin{array}{r} 64,891 \\ +6,548 \\ \hline \end{array}$

Objective 2 Estimate sums.

Review this example for Objective 2:

2. Estimate the sum by rounding. Then find the actual sum.

$$\begin{array}{r} 7040 \\ +2666 \\ \hline \end{array} \qquad \begin{array}{r} 7000 \\ +3000 \\ \hline 10,000 \end{array} \leftarrow \text{Estimated sum.}$$

$$\begin{array}{r} {}^{1} \\ 7040 \\ +2666 \\ \hline 9706 \end{array} \leftarrow \text{Actual sum.}$$

Practice this exercise:

2. Estimate the sum by rounding. Then find the actual sum.

$$\begin{array}{r} 76,037 \\ 7,017 \\ +6,724 \\ \hline \end{array}$$

Objective 3 Solve applications involving addition.

Review these examples for Objective 3:

3. Solve. An actress waitresses part-time and spends the rest of her time trying to get an acting job. She has spent $1200 on acting classes, $55 on printing up resumes, and $150 on photos. How much has she spent trying to get an acting job?

Practice these exercises:

3. Solve. Scott and Nancy want to add an addition to their house. The cost for materials will be $16,450, the labor will cost $8000, and the necessary permits and inspections will cost $2700. What will be the total price of the addition?

Name: Date:
Instructor: Section:

$$\begin{array}{r} \overset{1}{1200} \\ 150 \\ +55 \\ \hline 1405 \end{array}$$

Notice the phrase, *how much has she spent*, which tells us to add the amounts she has spent.

She has spent \$1305.

4. Find the perimeter of a 6-sided shape with side lengths of 12 feet, 10 feet, 14 feet, 8 feet, 2 feet and 16 feet.

The perimeter is the total distance around the shape. This is found by adding the lengths of the sides.

$P = 12 + 10 + 14 + 8 + 2 + 16$

$P = 62$

The perimeter is 62 ft.

4. Jim's yard is a 13-foot-long by 12-foot-long rectangle. What is the perimeter of his yard?

Objective 4 Subtract whole numbers.

Review this example for Objective 4:

5. Subtract. $\begin{array}{r} 859 \\ -178 \\ \hline \end{array}$

$$\begin{array}{r} \overset{7}{\not{8}}\overset{15}{\not{5}}9 \\ -178 \\ \hline 681 \end{array}$$

Practice this exercise:

5. Subtract. $\begin{array}{r} 758 \\ -303 \\ \hline \end{array}$

Objective 5 Solve equations containing an unknown addend.

Review this example for Objective 5:

6. Solve and check. $8 + n = 14$

To solve for an unknown addend, write a related subtraction equation in which the known addend is subtracted from the sum.

$8 + n = 14$

$n = 14 - 8$ Check.

$n = 6$ $8 + 6 = 14$

Practice this exercise:

6. Solve and check.
$3 + r = 16$

Name: Date:
Instructor: Section:

Objective 6 Solve applications involving subtraction.

Review this example for Objective 6:

7. Solve. Jan is making a costume for her son for the school pageant. She has 5 yards of material, and the costume requires 2 yards. How much of the material will Jan have left?

The phrase, *how much of the material will Jan have left*, tells us to use subtraction. $5-2=3$

Jan will have 3 yards left.

Practice this exercise:

7. Solve. The attendance of a Thursday night baseball game is 52,366. The attendance on Friday is 53,933. How many more people attended Friday's game than Thursday's game?

Objective 7 Solve applications involving addition and subtraction.

Review this example for Objective 7:

8. Solve. The diagram shows a circuit with two currents entering the node and two currents exiting the node. Find the missing current, in Amperes (A).

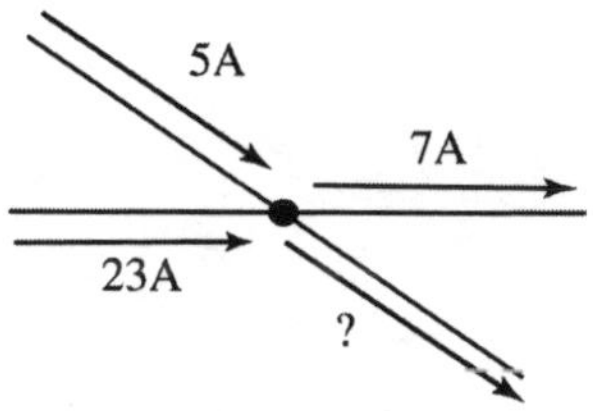

First, calculate the total current entering the node.

$23+5=28$

Then, subtract the known current exiting the node from the total entering, to find the unknown current exiting.

$28-7=21$

The unknown current is 21A.

Practice this exercise:

8. Solve. Josh has a storage capacity of 55 GB on his MP3 player. He has downloaded 9 GB of songs and 16 GB of movies. How much space does he have available?

Name: Date:
Instructor: Section:

ADDITIONAL EXERCISES

Objective 1 Add whole numbers.

For extra help, see Examples 1–2 on pages 13–14 of your text and the Section 1.2 lecture video.

Add.

1. $\begin{array}{r} 456 \\ +27 \\ \hline \end{array}$

2. $\begin{array}{r} 161{,}256 \\ 278 \\ +4{,}218 \\ \hline \end{array}$

3. $14 + 29 + 53$

4. $592 + 6375 + 1821$

Objective 2 Estimate sums.

For extra help, see Example 3 on page 15 of your text and the Section 1.2 lecture video.

Estimate each sum by rounding. Then find the actual sum.

5. $\begin{array}{r} 99{,}176 \\ 6{,}154 \\ +6{,}749 \\ \hline \end{array}$

6. $8360 + 48 + 520 + 2854$

Objective 3 Solve applications involving addition.

For extra help, see Examples 4–5 on page 16 of your text and the Section 1.2 lecture video.

Solve.

7. Oscar wants to cook dinner for a large family gathering. He estimates that he must spend $45 for the main course, $17 for side dishes, $31 for beverages, and $28 for dessert. What will be the total amount he spends on dinner?

8. Find the perimeter of a 6-sided shape with side lengths of 12 feet, 10 feet, 14 feet, 8 feet, 2 feet and 16 feet.

Name: Date:
Instructor: Section:

Objective 4 Subtract whole numbers.
For extra help, see Example 6 on page 18 of your text and the Section 1.2 lecture video.
Subtract.

9. $\begin{array}{r} 5803 \\ -426 \\ \hline \end{array}$

10. $85{,}801-12{,}424$

Objective 5 Solve equations containing an unknown addend.
For extra help, see Example 7 on page 19 of your text and the Section 1.2 lecture video.
Solve and check.

11. $66+t=122$

12. $146+x=284$

Objective 6 Solve applications involving subtraction.
For extra help, see Example 8 on page 20 of your text and the Section 1.2 lecture video.
Solve.

13. A roofer is replacing the roof on a house. He estimates it will require 240 squares of shingles to cover the whole roof. If he has 122 squares in stock, how many more squares will he need to order?

14. Emily has $200 in her savings account she has set aside for spending on new clothes. She heads to the mall and spends $55 on jeans, $38 on shoes, and $42 on a sweater. How much does she have left for accessories?

Objective 7 Solve applications involving addition and subtraction.
For extra help, see Example 9 on page 21 of your text and the Section 1.2 lecture video.
Solve.

15. A movie theater expects ticket sales of $9535 for the opening night of a new action movie. Typically, concession sales the opening night of a new film are about $1875. It will cost the theater $586 to pay employees that night, and concessions will cost the theater $265. How much will there be left from the ticket and concessions sales after these expenses are paid?

16. At midnight the outside temperature was 59 degrees Fahrenheit. The temperature rose 11 degrees over the next 8 hours, then dropped 3 degrees before noon. What was the temperature at noon?

Name: Date:
Instructor: Section:

Chapter 1 WHOLE NUMBERS

1.3 Multiplying Whole Numbers; Exponents

Learning Objectives

1. Multiply whole numbers.
2. Estimate products.
3. Solve applications involving multiplication.
4. Evaluate numbers in exponential form.
5. Write repeated factors in exponential form.
6. Solve applications.

Key Terms

Use the vocabulary terms listed below to complete each statement in Exercises 1–9.

multiplication	**base**	**exponent**	**exponential form**	**rectangular array**
square unit	**area**	**formula**	**distributive property**	

1. To find how much carpet will be needed for a living room, find the ____________________ of the room.

2. The expression 15^3 is in ____________________.

3. The desks in a classroom are arranged in a(n) ____________________ if there is the same number of desks in each row.

4. The ____________________ of an exponential expression is the value that is used as a factor a number of times.

5. The operation ____________________ can be expressed as repeated addition.

6. The ____________________ may be used if a sum or difference is multiplied by a number.

7. A(n) ____________________ uses symbols in an equation to describe a procedure.

8. In an exponential expression, the ____________________ indicates the number of times the base is used as a factor.

9. A square that measures 1 unit by 1 unit is said to measure 1 ____________________.

Name: Date:
Instructor: Section:

GUIDED EXAMPLES AND PRACTICE

Objective 1 Multiply whole numbers.

Review this example for Objective 1:

1. Multiply. $0(214)$

The product of 0 and a number is 0.

$0(214)=0$

Practice this exercise:

1. Multiply. $25\cdot 1\cdot 2$

Objective 2 Estimate products.

Review this example for Objective 2:

2. Estimate the product by rounding. Then find the actual product. $13\cdot 28$

Round each factor to the highest possible place value so that each has only one nonzero digit. Then multiply the rounded factors.

$10\cdot 30=300$ ← Estimated product.

$13\cdot 28=364$ ← Actual product.

Practice this exercise:

2. Estimate the product by rounding. Then find the actual product. 431×36

Objective 3 Solve applications involving multiplication.

Review this example for Objective 3:

3. Solve. A restaurant has a dining room that is 99 feet by 42 feet. What is the area of the dining room?

The key word, *by,* denotes multiplication.

$99\times 42=4158$

The dining room is 4158 ft^2.

Practice this exercise:

3. Solve. Lisa's car gets 20 miles per gallon of gasoline. How many miles can she drive on 53 gallons of gas?

Objective 4 Evaluate numbers in exponential form.

Review this example for Objective 4:

4. Evaluate 2^7.

To evaluate, write the base, 2, as a factor the number of times indicated by the exponent, 7, and then multiply.

$2^7=2\cdot 2\cdot 2\cdot 2\cdot 2\cdot 2\cdot 2=128$

Practice this exercise:

4. Evaluate 10^4.

Name: Date:
Instructor: Section:

Objective 5 Write repeated factors in exponential form.

Review these examples for Objective 5:

5. Write $19 \cdot 19 \cdot 19 \cdot 19 \cdot 19$ in exponential form.

Since the number 19 is repeated, it is the base. The number of times it is repeated, 5, is the exponent.

$19 \cdot 19 \cdot 19 \cdot 19 \cdot 19 = 19^5$

6. Write 17,544 in expanded form using powers of 10.

$$\begin{aligned}17{,}544 &= 1\times 10{,}000 + 7\times 1{,}000 + 5\times 100 + 4\times 10 + 4\times 1\\ &= 1\times 10^4 \quad + 7\times 10^3 \quad + 5\times 10^2 + 4\times 10 + 4\times 1\end{aligned}$$

Practice these exercises:

5. Write $3 \cdot 3 \cdot 3 \cdot 3 \cdot 3 \cdot 3 \cdot 3 \cdot 3 \cdot 3 \cdot 3 \cdot 3$ in exponential form.

6. Write 62,801,043 in expanded form using powers of 10.

Objective 6 Solve applications.

Review this example for Objective 6:

7. Solve. A multiple-choice test consists of 5 questions with each question having 2 possible answers. How many different ways are there to answer the questions?

For each question there are 2 ways to answer and there are 5 questions. So the number of ways to answer the questions is $2 \cdot 2 \cdot 2 \cdot 2 \cdot 2 = 32$.

There are 32 different ways.

Practice this exercise:

7. Solve. License plate tags in a particular state are to consist of 3 letters followed by 3 digits with repeated letters and digits allowed. How many different license plate tags can there be in this state?

ADDITIONAL EXERCISES

Objective 1 Multiply whole numbers.

For extra help, see Examples 1–2 on page 28 of your text and the Section 1.3 lecture video.

Multiply.

1. 103×608

2. 4116×692

Name: Date:
Instructor: Section:

Objective 2 Estimate products.

For extra help, see Example 3 on page 29 of your text and the Section 1.3 lecture video.

Estimate each product by rounding. Then find the actual product.

3. 439×816

4. 921×76

Objective 3 Solve applications involving multiplication.

For extra help, see Examples 4–6 on pages 29–31 of your text and the Section 1.3 lecture video.

Solve.

5. A bakery sells 120 cupcakes each day. How many cupcakes are sold in a week if the bakery is open every day?

6. If a carpenter uses 18 nails per cabinet, how many nails will he need to make 35 cabinets?

Objective 4 Evaluate numbers in exponential form.

For extra help, see Example 7 on page 33 of your text and the Section 1.3 lecture video.

Evaluate.

7. 1^9

8. 9^2

9. 7^3

10. 8^2

Objective 5 Write repeated factors in exponential form.

For extra help, see Examples 8–9 on page 34 of your text and the Section 1.3 lecture video.

Write in exponential form.

11. $21 \cdot 21 \cdot 21 \cdot 21$

12. $1 \cdot 1 \cdot 1 \cdot 1 \cdot 1 \cdot 1 \cdot 1$

13. $15 \cdot 15 \cdot 15$

14. $7 \cdot 7 \cdot 7 \cdot 7 \cdot 7 \cdot 7 \cdot 7 \cdot 7 \cdot 7 \cdot 7$

Name: Date:
Instructor: Section:

Write in expanded form using powers of 10.

15. 785,463

16. 59,201

Objective 6 Solve applications.

For extra help, see Example 10 on page 36 of your text and the Section 1.3 lecture video.

Solve.

17. A rectangular room measures 21 feet wide by 18 feet long. How many 1 foot square tiles are needed to tile the floor?

18. A man is planning on staining his deck so, he needs to know what the area of the deck is. The deck measures 20 feet wide by 30 feet long. What is the area of the deck?

Name: Date:
Instructor: Section:

Chapter 1 WHOLE NUMBERS

1.4 Dividing Whole Numbers; Solving Equations

Learning Objectives
1 Divide whole numbers.
2 Solve equations containing an unknown factor.
3 Solve applications involving division.

Key Terms
Use the vocabulary terms listed below to complete each statement in Exercises 1–4.

division **remainder** **divisor** **dividend**

1. In the equation $24 \div 4 = 6$, 4 is the __________________.

2. The operation of __________________ can be thought of as repeated subtraction.

3. In the equation $15 \div 3 = 5$, 15 is the __________________.

4. The amount left over after dividing two numbers is the __________________.

GUIDED EXAMPLES AND PRACTICE

Objective 1 Divide whole numbers.

Review these examples for Objective 1:

1. Determine the quotient and explain your answer. $55 \div 55$

$55 \div 55 = 1$ because $55 \times 1 = 55$

2. Use divisibility rules to determine whether 1230 is divisible by 2.

If a number is even, then 2 is an exact divisor. A number is even if it has 0, 2, 4, 6, or 8 in the ones place. In the ones place of 1230 is 0, so it is even and therefore divisible by 2.

Practice this exercise:

1. Determine the quotient and explain your answer.
$0 \div 8$

2. Use divisibility rules to determine whether 221 is divisible by 2.

Name: Date:
Instructor: Section:

3. Use divisibility rules to determine whether 727 is divisible by 3.

A number is divisible by 3 if the sum of the digits in the number is divisible by 3. The sum of the digits in 727 is $7+2+7=16$. Since 16 is not divisible by 3 then neither is 727.

4. Divide. $4452 \div 7$

$$\begin{array}{r} 636 \\ 7\overline{)4452} \\ \underline{-42} \\ 25 \\ \underline{-21} \\ 42 \\ \underline{-42} \\ 0 \end{array}$$

3. Use divisibility rules to determine whether 627 is divisible by 3.

4. Divide. $2960 \div 3$

Objective 2 Solve equations containing an unknown factor.

Review this example for Objective 2:

5. Solve and check. $2 \cdot y = 18$

To solve for an unknown factor, write a related division equation in which the product is divided by the known factor.

$2 \cdot y = 18$

$y = 18 \div 2$ Check.

$y = 9$ $2 \cdot 9 = 18$

Practice this exercise:

5. Solve and check. $8 \cdot x = 88$

Objective 3 Solve applications involving division.

Review this example for Objective 3:

6. Solve. A loan of $8784 will be paid off in 48 monthly payments. How much is each payment?

To solve, divide the loan amount into 48 equal payments.

Practice this exercise:

6. Solve. A bag can hold 22 kilograms of sand. How many bags can be filled with 1557 kilograms of sand? How many kilograms of sand will be left over?

Name: Date:
Instructor: Section:

$$\begin{array}{r} 183 \\ 48\overline{)8784} \\ \underline{-48} \\ 398 \\ \underline{-384} \\ 144 \\ \underline{-144} \\ 0 \end{array}$$

Each payment is $183.

ADDITIONAL EXERCISES

Objective 1 Divide whole numbers.

For extra help, see Examples 1–6 on pages 41–44 of your text and the Section 1.4 lecture video.

Determine the quotient and explain your answer.

1. $27 \div 1$

2. $15 \div 0$

Use divisibility rules to determine whether the given number is divisible by 2.

3. 357

4. 170

Use divisibility rules to determine whether the given number is divisible by 3.

5. 953

6. 642

Use divisibility rules to determine whether the given number is divisible by 5.

7. 715

8. 5251

Divide.

9. $7158 \div 49$

10. $22{,}888 \div 57$

11. $\dfrac{80{,}000}{250}$

12. $\dfrac{4253}{34}$

Name: Date:
Instructor: Section:

Objective 2 Solve equations containing an unknown factor.
For extra help, see Example 7 on page 45 of your text and the Section 1.4 lecture video.
Solve and check.

13. $7 \cdot t = 0$

14. $n \cdot 5 = 15$

15. $22 \cdot 3 \cdot x = 594$

16. $t \cdot 0 = 25$

Objective 3 Solve applications involving division.
For extra help, see Example 8 on page 46 of your text and the Section 1.4 lecture video.
Solve.

17. If the area of a table top is 8 square feet and the width is 2 feet, find the length. Use the formula for the area of rectangle.

17. Five friends go out for dinner and the bill comes to $186.25. If they want to split the bill evenly, how much should they each pay?

Name: Date:
Instructor: Section:

Chapter 1 WHOLE NUMBERS

1.5 Order of Operations; Mean, Median and Mode

Learning Objectives
1 Simplify numerical expressions by following the order of operations agreement.
2 Find the mean, median, and mode of a list of values.

Key Terms
Use the vocabulary terms listed below to complete each statement in Exercises 1–4.

average **mean** **median** **mode**

1. The ____________________ of a set of values is found by adding the values and dividing by the number of values.

2. Another word used for the mean of a set of values is ____________________.

3. The ____________________ of a set of values is the value that occurs most often.

4. In order to find the ____________________ of a set of values the first step is to arrange the values in order from least to greatest.

GUIDED EXAMPLES AND PRACTICE

Objective 1 Simplify numerical expressions by following the order of operations agreement.

Review these examples for Objective 1:

1. Simplify. $20-2^3\cdot 1^2+13$

$20-2^3\cdot 1^2+13$
$=20-8\cdot 1+13$ Evaluate the exponential form. $2^3=8, 1^2=1$
$=20-8+13$ Multiply. $8\cdot 1=8$
$=12+13$ Subtract. $20-8=12$
$=25$ Add. $12+13=25$

2. Simplify. $3^4\div(11-2)+3(6+2)-(7-5)$

Practice these exercises:

1. Simplify.
$9\cdot 2+6^2\div 3\cdot 2-5$

2. Simplify.
$[16-3(7-2)]+3^3-4$

Name: Date:
Instructor: Section:

$3^4 \div (11-2) + 3(6+2) - (7-5)$

$= 3^4 \div (9) + 3(8) - (2)$ Perform operations in parentheses.

$= 81 \div 9 + 3(8) - 2$ Evaluate the exponential form. $3^4 = 81$

$= 9 + 24 - 2$ Multiply and divide from left to right. $81 \div 9 = 9$ and $3(8) = 24$

$= 31$ Add and subtract from left to right. $9 + 24 = 33$ and $33 - 2 = 31$

Objective 2 Find the mean, median, and mode of a list of values.

Review this example for Objective 2:

3. Find the mean, median and mode. 55, 58, 62, 69

$$\frac{55+58+62+69}{4} = 61 \leftarrow \text{Mean}$$

54, 58, 62, 69

↑

median

$$\text{median} = \text{mean of 58 and 62} = \frac{58+62}{2} = 60$$

Since no value is repeated, the data has no mode.

Practice this exercise:

3. Find the mean, median and mode. 54, 26, 49, 54, 63, 78

Name: Date:
Instructor: Section:

ADDITIONAL EXERCISES

Objective 1 Simplify numerical expressions by following the order of operations agreement.

For extra help, see Examples 1–6 on pages 50–52 of your text and the Section 1.5 lecture video.

Simplify.

1. $12-5\cdot 2$

2. $16\div 4+5-2^2$

3. $(12-2)\left[8+2^5-11+7(6-1)\right]$

4. $(12-7)2+6^2-2\cdot 5$

5. $4^3-3\cdot 2^2\div 2+6(12-9)$

6. $32\div 8\cdot 2+(21\div 3)-2^3$

7. $4\times 3+\left\{7+3\left[25-4(3+1)\right]\right\}$

8. $\dfrac{16-3^2(4-3)^2}{14-5-2}$

9. $4^2\left\{6\left[7-2\left(16-2^4\right)+25\right]\right\}-3^3+5(18-15)$

10. $\dfrac{16-2^4+3(8-5)^2}{(26\div 2)+7(2)}$

Objective 2 Find the mean, median, and mode of a list of values.

For extra help, see Example 7 on page 54 of your text and the Section 1.5 lecture video.

Find the arithmetic mean, median, and mode.

11. 80, 73, 78, 80, 85, 78

12. 125, 187, 168, 187, 107, 156

13. 201, 507, 628, 578, 546, 870

14. 313, 487, 951, 574, 278, 487

Name: Date:
Instructor: Section:

Chapter 1 WHOLE NUMBERS

1.6 More with Formulas

Learning Objectives
1 Use the formula $P = 2l + 2w$ to find the perimeter of a rectangle.
2 Use the formula $A = bh$ to find the area of a parallelogram.
3 Use the formula $V = lwh$ to find the volume of a box.
4 Solve for an unknown number in a formula.
5 Use a problem-solving process to solve problems requiring more than one formula.

Key Terms
Use the vocabulary terms listed below to complete each statement in Exercises 1–5.

parallel lines **parallelogram** **right angle** **cubic unit** **volume**

1. Two lines that meet at a 90 degree angle form a ____________________.

2. ____________________ is a measure of the amount of space inside a three-dimensional object.

3. Two straight lines that never intersect are called ____________________.

4. A cube that measures 1 unit by 1 unit by 1 unit has a volume of 1 ____________________.

5. To find the area of a ____________________, multiply the measure of the base by the measure of the height.

GUIDED EXAMPLES AND PRACTICE

Objective 1 Use the formula $P = 2l + 2w$ to find the perimeter of a rectangle.

Review this example for Objective 1:
1. Find the perimeter.

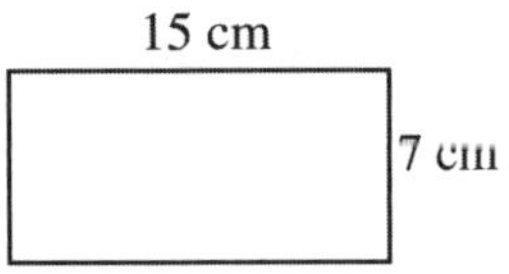

Practice this exercise:
1. Find the perimeter.

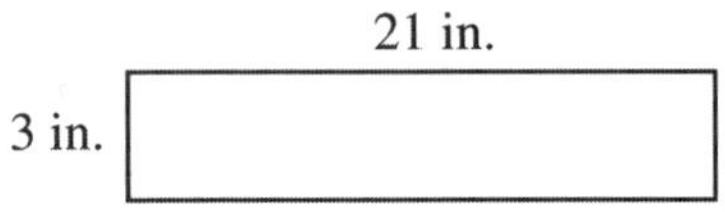

Name: Date:
Instructor: Section:

$P = 2l + 2w$

$P = 2(15) + 2(7)$ Replace l with 15 and w with 7.

$P = 30 + 14$ Multiply.

$P = 44$ Add.

The perimeter is 44 cm.

Objective 2 Use the formula $A = bh$ to find the area of a parallelogram.

Review this example for Objective 2:

2. Find the area.

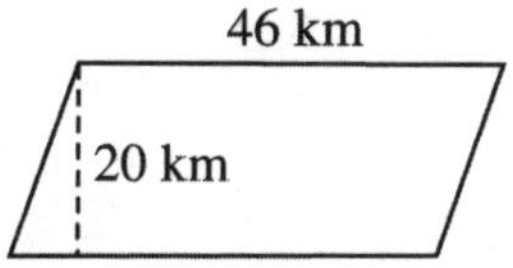

$A = bh$

$A = (46)(20)$ In $A = bh$ replace b with 46 and h with 20.

$A = 920$ Multiply.

The area is 920 km^2.

Practice this exercise:

2. Find the area.

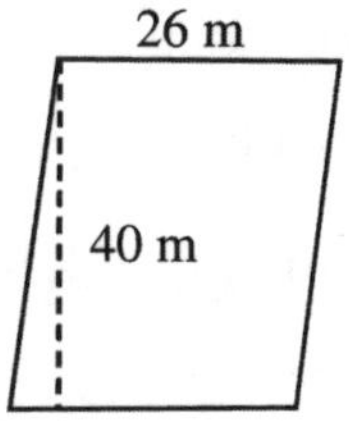

Objective 3 Use the formula $V = lwh$ to find the volume of a box.

Review this example for Objective 3:

3. Find the volume.

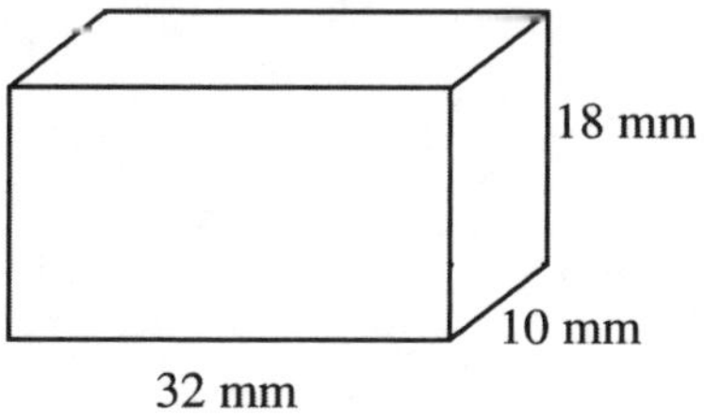

$V = lwh$

$V = (32)(10)(18)$ Replace l with 32, w with 10, and h with 18.

$V = 5760$ Multiply.

The volume is 5760 mm^3.

Practice this exercise:

3. Find the volume.

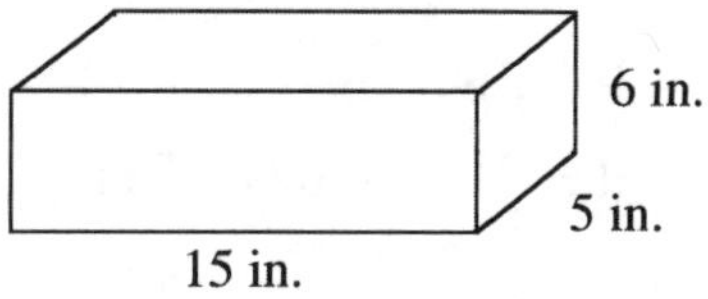

Name: Date:
Instructor: Section:

Objective 4 Solve for an unknown number in a formula.

Review this example for Objective 4:

4. Solve. The area of a parallelogram is 96 square centimeters. The base of the figure is 16 centimeters. What is the height?

Use the formula $A = bh$.

$A = bh$

$96 = 16h$ Replace A with 96 and b with 16.

$96 \div 16 = h$ Write a related division statement.

$6 = h$ Divide.

The height is 6 cm.

Practice this exercise:

4. Solve. The area of a rectangular room is 192 square feet. The width is 16 feet. What is the length?

Objective 5 Use a problem-solving process to solve problems requiring more than one formula.

Review this example for Objective 5:

5. Solve. A security fence is to be built around a 237-meter by 102-meter field. What is the perimeter of the field? Fence wire will cost \$3 per meter. What will the fence cost?

Use $P = 2l + 2w$ to find the perimeter.

$P = 2l + 2w$

$P = 2(237) + 2(102)$ Replace l with 237 and w with 102.

$P = 474 + 204$ Multiply.

$P = 678$ Add.

The perimeter is 678 m.

Multiply the length of the perimeter by the cost per meter to find the total cost.

$678 \times 3 = 2034$

The fence will cost \$2034.

Practice this exercise:

5. Solve. Dean's backyard is a rectangle 113 feet by 28 feet. Last year he put a 30-foot by 14-foot brick patio in the yard. How much of the yard is left to mow?

Name: Date:
Instructor: Section:

ADDITIONAL EXERCISES

Objective 1 Use the formula $P = 2l + 2w$ to find the perimeter of a rectangle.

For extra help, see Example 1 on page 58 of your text and the Section 1.6 lecture video.

Solve.

1. A picket fence is to be built around a 218-foot by 94-foot yard. What is the perimeter of the yard?

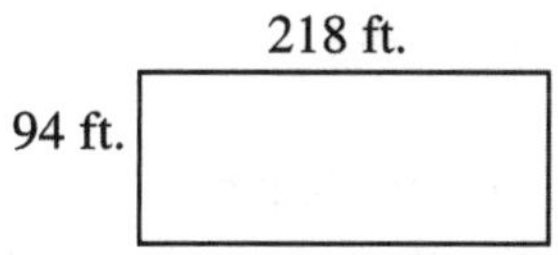

2. A rectangle has a length of 72 cm and a width of 31 cm. Find the perimeter.

Objective 2 Use the formula $A = bh$ to find the area of a parallelogram.

For extra help, see Example 2 on page 60 of your text and the Section 1.6 lecture video.

Find the area.

3.

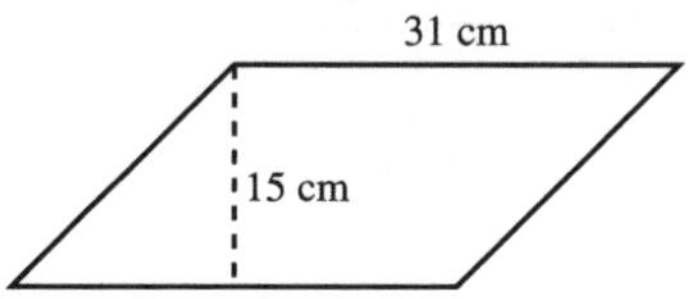

4.

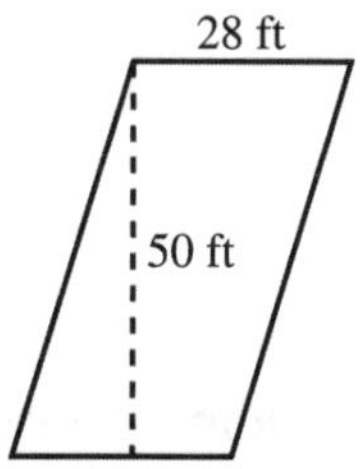

Objective 3 Use the formula $V = lwh$ to find the volume of a box.

For extra help, see Example 3 on page 61 of your text and the Section 1.6 lecture video.

Find the volume.

5.

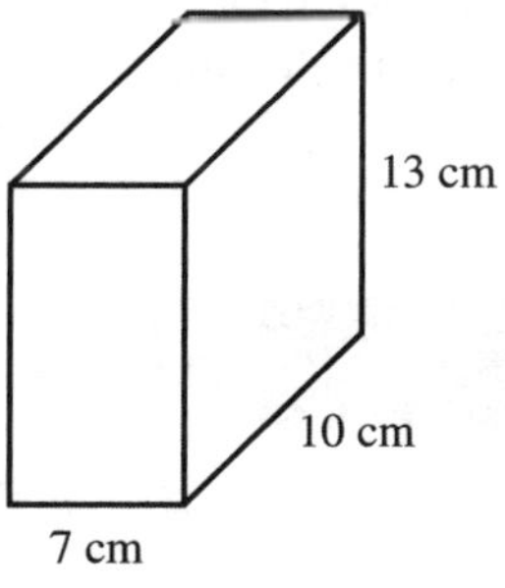

6.

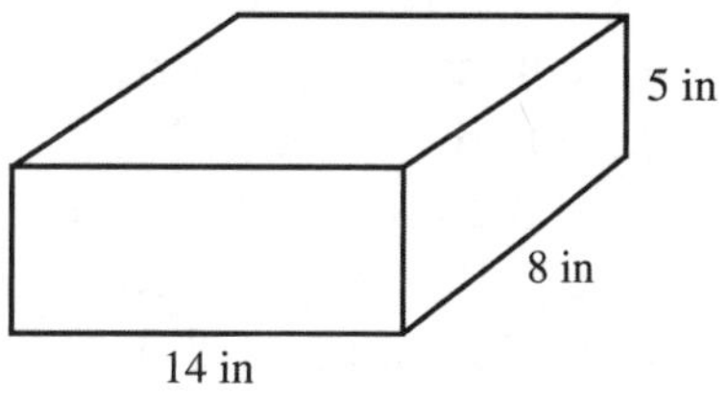

Name: Date:
Instructor: Section:

Objective 4 Solve for an unknown number in a formula.

For extra help, see Examples 4–5 on pages 61–62 of your text and the Section 1.6 lecture video.

Solve.

7. The area of a parallelogram is 294 square centimeters. The height of the figure is 14 centimeters. What is the base?

8. The area of a rectangular room is 405 square feet. The length is 27 feet. What is the width?

9. The figure shown has a volume of 144 cubic centimeters. Find the height of the figure.

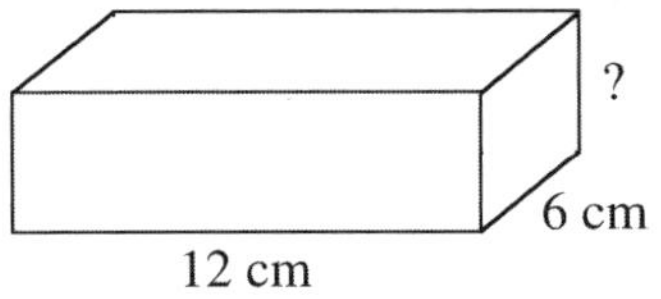

10. The figure shown has a volume of 8 cubic centimeters. Find the width of the figure.

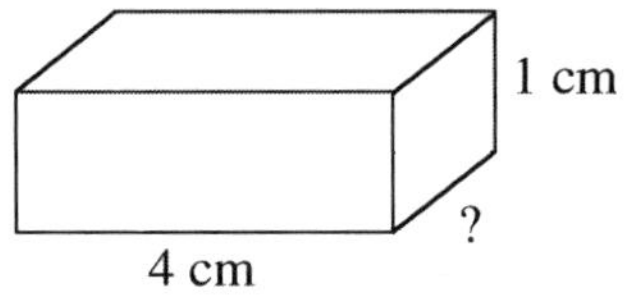

Objective 5 Use a problem-solving process to solve problems requiring more than one formula.

For extra help, see Examples 6–9 on pages 63–67 of your text and the Section 1.6 lecture video.

Solve.

11. A room is 13 feet by 21 feet. It needs to be painted. The ceiling is 8 feet above the floor. There are two windows in the room; each one is 3 feet by 4 feet. The door is 3 feet by 7 feet.

a. Find the area of the walls.

b. One gallon of paint covers 93 square feet. How many gallons are needed to paint the walls?

c. At $23 per gallon, what is the cost to paint the walls?

12. A rain gutter is to be installed around the house shown. The gutter costs $5 per foot. Find the total cost of the gutter.

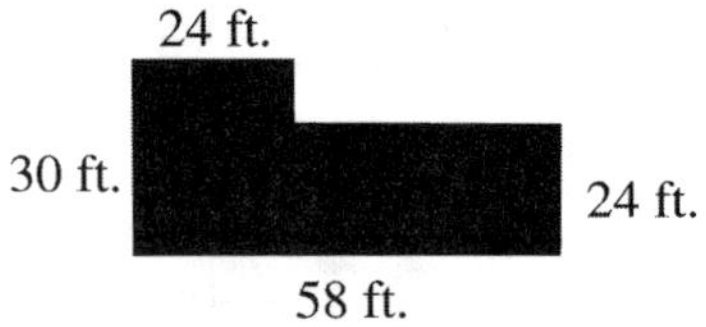

Name: Date:
Instructor: Section:

Chapter 2 INTEGERS

2.1 Introduction to Integers

Learning Objectives
1 Identify integers.
2 Interpret number lines and graphs with integers.
3 Use < or > to write a true statement.
4 Find the absolute value.

Key Terms
Use the vocabulary terms listed below to complete each statement in Exercises 1–5.

integers **absolute value** **additive inverse** **positive** **negative**

1. On the number line, the distance a number is away from 0 is the number's ____________________.

2. Distance can never be a ____________________ value.

3. The ____________________ are the whole numbers and all of their additive inverses.

4. The ____________________ of a positive number is negative.

5. The absolute value of every nonzero number is ____________________.

GUIDED EXAMPLES AND PRACTICE

Objective 1 Identify integers.

Review these examples for Objective 1:

1. Express the amount as a positive or negative integer.
A woman owes her friend $116.

Money owed is debt, which is represented by a negative number. −116

2. A geologist visits a site 2953 feet above sea level. Express this amount as a positive or negative integer.

Since 2953 feet is above sea level, it is represented by a positive number. +2953

Practice these exercises:

1. The record low temperature for a date is 21 degrees below zero. Express this amount as a positive or negative integer.

2. During a year, a person saves $1350. Express this amount as a positive or negative integer.

Name: Date:
Instructor: Section:

Objective 2 Interpret number lines and graphs with integers.

Review these examples for Objective 2:

3. Graph -4 on a number line.

Draw a dot on the mark of -4.

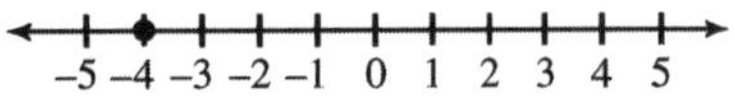

Practice these exercises:

3. Graph 3 on a number line.

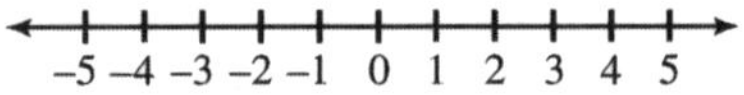

For exercise 4, use the following graph.

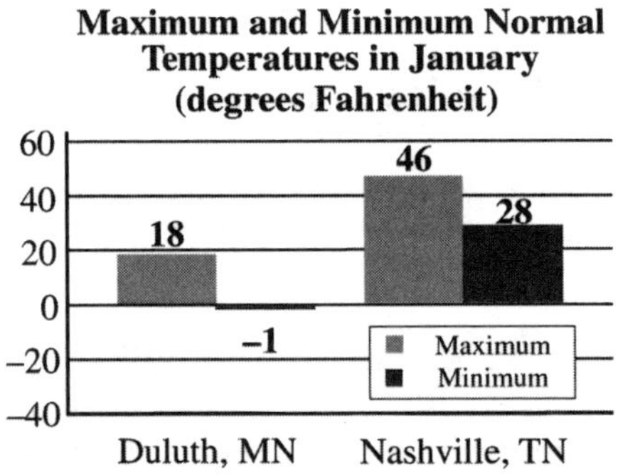

4. What is the minimum normal January temperature in Nashville, Tennessee?

28° F

4. What is the maximum normal January temperature in Duluth, Minnesota?

Objective 3 Use < or > to write a true statement.

Review this example for Objective 3:

5. Use either < or > for □ to form a true sentence.

$-3 \; \square \; 4$

Since -3 is to the left of 4 on the number line, we have $-3 < 4$.

Practice this exercise:

5. Use either < or > for □ to form a true sentence.

$-10 \; \square \; -6$

Name: Date:
Instructor: Section:

Objective 4 Find the absolute value.

Review these examples for Objective 4:

6. Find the absolute value of -21.

Because -21 is 21 steps from 0 on a number line, its absolute value is 21.

7. Find the absolute value of 10.

Because 10 is 10 steps from 0 on a number line, its absolute value is 10.

8. Find $|109|$.

Because 109 is 109 steps from 0 on a number line, its absolute value is 109.

9. Find $|-31|$.

Because -31 is 31 steps from 0 on a number line, its absolute value is 31.

Practice these exercises:

6. Find the absolute value of -59.

7. Find the absolute value of 33.

8. Find $|46|$.

9. Find $|-17|$.

ADDITIONAL EXERCISES

Objective 1 Identify integers.

For extra help, see Example 1 on page 89 of your text and the Section 2.1 lecture video.

Tell which integers correspond to each situation.

1. Express the amount as a positive or negative integer.

Steven owes \$416 on a car loan.

2. Express the amount as a positive or negative integer.

Stan receives a paycheck of \$562.

3. Express the amount as a positive or negative integer.

The owner of a diner lowered her prices on sandwiches by \$1.50.

4. Express the amount as a positive or negative integer.

The population of Hudson City decreased by 3142 people.

Name: Date:
Instructor: Section:

Objective 2 Interpret number lines and graphs with integers.
For extra help, see Examples 3–4 on page 90 of your text and the Section 2.1 lecture video.
Graph each integer on a number line.

5. 6

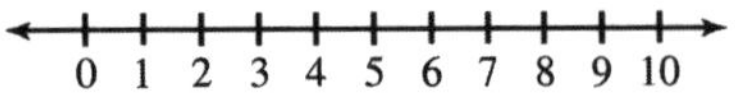

6. 7

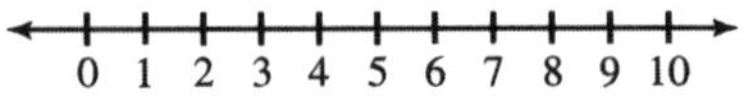

7. −7

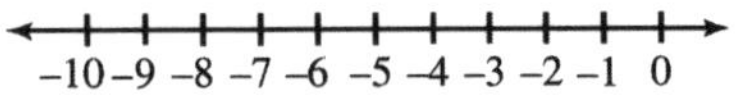

8. −10

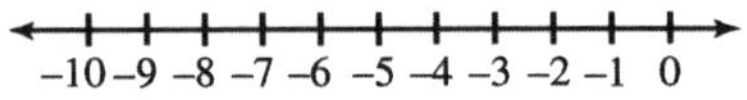

Objective 3 Use < or > to write a true statement.
For extra help, see Example 4 on page 91 of your text and the Section 2.1 lecture video.
Use either < or > for □ *to form a true sentence.*

9. 3 □ −7

10. −2 □ 0

11. −72 □ −56

12. −8 □ 3

Objective 4 Find the absolute value.
For extra help, see Examples 5–6 on page 91–92 of your text and the Section 2.1 lecture video.
Find the absolute value.

13. $|-26|$

14. $|41|$

15. $|-175|$

16. $|0|$

Name: Date:
Instructor: Section:

Chapter 2 INTEGERS

2.2 Adding Integers

Learning Objectives
1 Add integers with like signs.
2 Add integers with different signs.
3 Add integers.
4 Find the additive inverse
5 Solve applications involving addition of integers.

GUIDED EXAMPLES AND PRACTICE

Objective 1 Add integers with like signs.

Review this example for Objective 1:

1. Add without using a number line:

$28+34$ *Think*: Add the abosolute values: $28+34=62$.
Make the answer positive, 62.

Practice this exercise:

1. Add without using a number line:

$-18+(-14)$

Objective 2 Add integers with different signs.

Review this example for Objective 2:

2. Add without using a number line:

$48+(-83)$ *Think*: The absolute values are 48 and 83. The difference is 35. Since the negative number has the larger absolute value, the answer is *negative*, -35.

Practice this exercise:

2. Add without using a number line:

$62+(-72)$

Objective 3 Add integers.

Review these examples for Objective 3:

3. Explain the situation in terms of debts and credits, then add.

$52+(-12)$

A credit with a debt, so subtract and because the credit, 52, is more than the debt, the result is a *credit/positive*; 40.

Practice these exercises:

3. Explain the situation in terms of debts and credits, then add.

$41+(-63)$

Name: Date:
Instructor: Section:

4. Simplify $18+(-3)+(-7)+15$.

$18+(-3)+(-7)+15$

$15+15+(-7)$

$30+(-7)$

23

4. Simplify $-72+(-1)+36+48$.

Objective 4 Find the additive inverse.

Review this example for Objective 4:

5. Find the additive inverse of -28.

-28 and 28 are additive inverses because $-28 + 28 = 0$. Also, they have the same absolute value but opposite signs.

Practice this exercise:

5. Find the additive inverse of 83.

Objective 5 Solve applications involving addition of integers.

Review this example for Objective 5:

6. The barometric pressure at a certain city dropped 8 millibars (mb); then it rose 3 mb. After that, it dropped 6 mb and then it rose 6 mb. What was the total change in pressure?

Add the changes in pressure.

$-8+3+(-6)+6$

$-5+(-6)+6$

$-11+6$

-5

The total change in pressure is -5 mb.
The pressure dropped 5 mb.

Practice this exercise:

6. The rim of a canyon is at an altitude of 76 m. On a hike down into the canyon, a party of hikers stops for a rest 108 m below the rim. Then they descend another 60 m. What is their new altitude?

Name: Date:
Instructor: Section:

ADDITIONAL EXERCISES

Objective 1 Add integers with like signs.

For extra help, see Example 1on page 96 of your text and the Section 2.2 lecture video.
Add without using a number line.

1. $18+9$

2. $-7+(-6)$

3. $-9+(-2)$

4. $13+17$

5. $-8+(-16)+(-31)$

6. $-12+(-61)+(-21)$

Objective 2 Add integers with different signs.

For extra help, see Example 2 on page 97 of your text and the Section 2.2 lecture video.
Add.

7. $8+(-16)$

8. $-4+4$

9. $-19+7$

10. $0+(-11)$

Objective 3 Add integers.

For extra help, see Examples 3–4 on pages 97–98 of your text and the Section 2.2 lecture video.
Add.

11. Explain the situation in terms of debts and credits, then add.
$31+(-43)$

12. Explain the situation in terms of debts and credits, then add.
$-11+67$

13. $2+(-4)+13+(-7)$

14. $-6+1+(-2)+5$

Name: Date:
Instructor: Section:

Objective 4 Find the additive inverse.

For extra help, see Example 5 on page 99 of your text and the Section 2.2 lecture video.

Find the additive inverse.

15. −74

16. 8

17. 4

18. −513

Objective 5 Solve applications involving addition of integers.

For extra help, see Examples 6–8 on pages 100–101 of your text and the Section 2.2 lecture video.

19. Kyle's credit card bill is $480. Kyle sends a check to the credit card company for $53, charges another $171 in merchandise, and then pays off another $240 of the bill. How much does Kyle owe the company?

20. An airplane climbs to 35,000 feet, and then descends 2340 feet. What is its new altitude?

21. A submarine uses compressed air to launch a rocket from 35 feet below the surface of the ocean. The rocket rises 100 feet before it fires its own engine. At what height does the rocket fire its engine?

22. Charlie finds a $5 bill in the street and uses it to buy a $1 chocolate bar. How much money does he have left?

Name: Date:
Instructor: Section:

Chapter 2 INTEGERS

2.3 Subtracting Integers and Solving Equations

Learning Objectives
1 Write subtraction statements as equivalent addition statements.
2 Solve equations containing an unknown addend.
3 Solve applications involving subtraction of integers.

Key Terms
Use the vocabulary terms listed below to complete each statement in Exercises 1–4.

net **cost** **revenue** **profit** **loss**

1. ____________________ is the amount spent.

2. After the amount spent is subtracted from the amount made, the amount left is the ____________________.

3. The result when revenue is less than cost is a ____________________.

4. To obtain a profit, ____________________ must be greater than cost.

GUIDED EXAMPLES AND PRACTICE

Objective 1 Write subtraction statements as equivalent addition statements.

Review these examples for Objective 1:

1. Subtract: $6-13$
Change the operation symbol from $-$ to $+$.
Change the second number to its additive inverse.

$6+(-13)$
-7

2. Subtract: $19-(-21)$
Change the operation symbol from $-$ to $+$.
Change the second number to its additive inverse.

$19+21$
40

Practice these exercises:

1. Subtract: $15-37$

2. Subtract: $-14-(-16)$

Name: Date:
Instructor: Section:

Objective 2 Solve equations containing an unknown addend.

Review these examples for Objective 2:

3. Solve and check: $15 + p = 12$

Write a related subtraction equation in which the known addend is subtracted from the sum.

$p = 12 - 15$

$p = 12 + (-15)$

$p = -3$

Check. Verify that the equation is true.

$15 + (-3) \stackrel{?}{=} 12$

$12 = 12$

4. Solve and check: $r + (-2) = 7$

Write a related subtraction equation in which the known addend is subtracted from the sum.

$r = 7 - (-2)$

$r = 7 + 2$

$r = 9$

Check. Verify that the equation is true.

$9 + (-2) \stackrel{?}{=} 7$

$7 = 7$

Practice these exercises:

3. Solve and check:
$27 + x = 11$

4. Solve and check:
$d + (-17) = 13$

Objective 3 Solve applications involving subtraction of integers

Review this example for Objective 3:

5. One day the temperature dropped from $-1°$ F to $-14°$ F. How many degrees did it drop?

Calculate the difference.

$-14 - (-1)$

$-14 + 1$

-13

The change in temperature was -13 degrees. The temperature dropped 13 degrees.

Practice this exercise:

5. A submarine at a depth of 1881 ft ascends to a depth of 838 ft. How far did the submarine ascend?

Name: Date:
Instructor: Section:

ADDITIONAL EXERCISES

Objective 1 Write subtraction statements as equivalent addition statements.

For extra help, see Examples 1–2 on pages 106–108 of your text and the Section 2.3 lecture video.

Write as an equivalent addition, then evaluate.

1. $8-4$

2. $-15-31$

3. $21-16$

4. $-316-84$

Objective 2 Solve equations containing an unknown addend.

For extra help, see Example 3 on page 108 of your text and the Section 2.3 lecture video.

Solve and check.

5. $x+(-3)=9$

6. $-1+m=-17$

7. $f-2=81$

8. $r+(-3)=-2$

Objective 3 Solve applications involving subtraction of integers.

For extra help, see Examples 4–6 on pages 109–110 of your text and the Section 2.3 lecture video.

Find the absolute value.

9. During one year, a business had total revenues of \$2,620,760 and total costs of \$1,620,260. Calculate the net and determine if it was a profit or a loss.

10. Hikers descended a 2500 foot mountain, and set up camp 400 feet from the bottom of the mountain. Find how far the hikers descended from the top of the mountain to the camp.

Name: Date:
Instructor: Section:

Chapter 2 INTEGERS

2.4 Multiplying and Dividing Integers; Exponents; Square Roots; Solving Equations

Learning Objectives
1 Multiply integers.
2 Evaluate numbers in exponential form.
3 Divide integers.
4 Solve equations containing an unknown factor.
5 Evaluate square roots.
6 Solve applications involving multiplication or division of integers.

Key Terms

Use the vocabulary terms listed below to complete each statement in Exercises 1–6.

negative **positive** **same** **different** **even** **odd**

1. When evaluating an exponential form that has a negative base, if the exponent is ____________ the product is positive.

2. When multiplying two numbers that have different signs, the product is ____________.

3. When multiplying signed numbers, if there is an even number of negative numbers the product is ____________.

4. The product of two numbers with the ____________ sign is always positive.

5. The principal square root is the ____________ square root of a number.

6. The square root of a ____________ number is not an integer.

GUIDED EXAMPLES AND PRACTICE

Objective 1 Multiply integers.

Review these examples for Objective 1:

1. Multiply $-6 \cdot 5$
   The signs are different, so the answer is negative.
   $-6 \cdot 5 = -30$

Practice these exercises:

1. Multiply $5 \cdot (-8)$

Name: Date:
Instructor: Section:

2. Multiply: $-12\cdot(-11)$

The signs are the same, so the answer is positive.

$-12\cdot(-11)=132$

3. Multiply: $-1(-2)(-1)$

Multiply from left to right

$-1(-2)(-1)$

$2(-1)$

-2

2. Multiply: $-5\cdot(-10)$

3. Multiply: $-3(-5)(-1)$

Objective 2 Evaluate numbers in exponential form.

Review these examples for Objective 2:

4. Evaluate: $(-6)^2$

The exponent is even. The product is positive.

$(-6)^2=(-6)(-6)=36$

5. Evaluate: -2^6

The minus sign is an understood -1 multiplying 2^6.

-2^6

$=-1\cdot 2^6$

$=-1\cdot 2\cdot 2\cdot 2\cdot 2\cdot 2\cdot 2$

$=-1\cdot 64$

$=-64$

Practice these exercises:

4. Evaluate: $(-2)^3$

5. Evaluate: -5^2

Objective 3 Divide integers.

Review this example for Objective 3:

6. Divide: $14\div(-2)$

The signs are different, so the answer is negative.

$14\div(-2)=-7$

Practice this exercise:

6. Divide: $-21\div(-3)$

Name: Date:
Instructor: Section:

Objective 4 Solve equations containing an unknown factor.

Review this example for Objective 4:

7. Solve and check: $-11a = 77$

Write a related division equation in which the product is divided by the known factor.

$$a = \frac{77}{-11} = -7$$

Check. Verify that the equation is true.

$$-11(-7) \stackrel{?}{=} 77$$
$$77 = 77$$

Practice this exercise:

7. Solve and check: $-1f = -6$

Objective 5 Evaluate square roots.

Review these examples for Objective 5:

8. Find all the square roots of 9.

3 and -3, because $3^2 = 9$ and $(-3)^2 = 9$.

9. Simplify: $\sqrt{16}$

The positive square root of 16 is 4.

Practice these exercises:

8. Find all the square roots of 169.

9. Simplify: $-\sqrt{121}$

Objective 6 Solve applications involving multiplication or division of integers.

Review this example for Objective 6:

10. An estate of $4277 was divided equally among seven family members. Find the amount received by each family member.

Divide the amount of money by the number of family members.

$$4277 \div 7 = 611$$

Each family member receives $611.

Practice this exercise:

10. Kevin has three children in college. For each child, Kevin pays tuition of $13,500 per year. What is the total tuition that Kevin pays per year?

Name: Date:
Instructor: Section:

ADDITIONAL EXERCISES

Objective 1 Multiply integers.

For extra help, see Examples 1–4 on pages 114–116 of your text and the Section 2.4 lecture video.

Multiply.

1. $8\cdot(-4)$

2. $-7\cdot(-5)$

3. $-13\cdot 0$

4. $-9\cdot(-2)$

5. $(-5)(-6)(-2)(4)$

6. $-3(-18)0\cdot 15$

Objective 2 Evaluate numbers in exponential form.

For extra help, see Examples 5–6 on page 117 of your text and the Section 2.4 lecture video.

Evaluate.

7. $(-2)^4$

8. $(-2)^5$

9. -5^2

10. $(-11)^2$

Objective 3 Divide integers.

For extra help, see Examples 7–8 on pages 118–119 of your text and the Section 2.4 lecture video.

Divide.

11. $18\div(-2)$

12. $-51\div(-3)$

13. $16\div 0$

14. $-21\div(-7)$

Name: Date:
Instructor: Section:

Objective 4 Solve equations containing an unknown factor.

For extra help, see Example 8 on pages 118–119 of your text and the Section 2.4 lecture video.

Solve and check.

15. $-2y=-6$

16. $f \cdot 7=-63$

17. $4r=-72$

18. $(-5)(2)m=130$

Objective 5 Evaluate square roots.

For extra help, see Examples 9–10 on pages 119–120 of your text and the Section 2.4 lecture video.

Find all the square roots of each number.

19. 144

20. 64

Simplify.

21. $\sqrt{25}$

22. $-\sqrt{49}$

Objective 6 Solve applications involving multiplication or division of integers.

For extra help, see Example 11 on page 121 of your text and the Section 2.4 lecture video.

Solve.

23. A family of four bought tickets to a baseball game. If the tickets were $34 each, how much did it cost for the family to go to the game?

Name: Date:
Instructor: Section:

Chapter 2 INTEGERS

2.5 Order of Operations

Learning Objectives
1 Simplify numerical expressions by following the order of operations agreement.

GUIDED EXAMPLES AND PRACTICE

Objective 1 Simplify numerical expressions by following the order of operations agreement.

Review these examples for Objective 1:

1. Simplify:
$-16 \div 4 \cdot 2$
$= -4 \cdot 2$ Divide $-16 \div 4 = -4$
$= -8$ Multiply $-4 \cdot 2 = -8$

2. Simplify:
$\sqrt{16} + \sqrt{9}$
$= 4 + 3$ Evaluate the square roots. $\sqrt{16} = 4, \sqrt{9} = 3$
$= 7$ add $4 + 3 = 7$

3. Simplify:
$-24 \div 4\left[5-(6+8) \div 7\right]+|-7|$
$-24 \div 4\left[5-(14) \div 7\right]+|-7|$ Add in the parentheses $6 + 8 = 14$
$-24 \div 4[5-2]+|-7|$ Divide $14 \div 7=2$
$-24 \div 4[3]+|-7|$ Subtract $5 - 2 = 3$
$-24 \div 4[3]+7$ Find the absolute value of -7
$-6[3]+7$ Divide $-24 \div 4 = -6$
$-18+7$ Multiply $-6 \cdot 3 = -18$
-11 Add $-18 + 7$

Practice these exercises:

1. Simplify: $18 - 2 \cdot 16$

2. Simplify: $\sqrt{16+9}$

3. Simplify:
$10-8\left[-15-2 \div (-2)\right]+3^3$

Name: Date:
Instructor: Section:

4. Simplify:

$2|7-2\cdot 4|+18-2^3$

$2|7-8|+18-2^3$ Multiply in the absolute value symbols $2\cdot 4=8$

$2|-1|+18-2^3$ Subtract in the absolute value symbols $7-8=-1$

$2(1)+18-2^3$ Find the absolute value of -1

$2(1)+18-8$ Evaluate the exponential form $2^3=8$

$2+18-8$ Multiply $2\cdot 1=2$

$20-8$ Add 2+18=20

12 Subtract $20-8=12$

4. Simplify:

$-|-40\div(-2)+6|+7\sqrt{4}$

5. Simplify:

$\dfrac{14-6(4)}{(8-4)+6}$

$=\dfrac{14-24}{(8-4)+6}$ Multiply in the numerator $6\cdot 4=24$

$=\dfrac{14-24}{(4)+6}$ Subtract in the denominator $8-4=4$

$=\dfrac{-10}{4+6}$ Subtract in the numerator $14-24=-10$

$=\dfrac{-10}{10}$ Add in the numerator $4+6=10$

$=-1$ Divide $-10\div 10=-1$

5. Simplify:

$\dfrac{-3^2+2(7-12)^2+3}{16-20+(8-4)^2-1}$

Name: Date:
Instructor: Section:

ADDITIONAL EXERCISES

Objective 1 Simplify numerical expressions by following the order of operations agreement.

For extra help, see Examples 1–10 on pages 124–127 of your text and the Section 2.5 lecture video.

Simplify.

1. $-32 \div 8 \cdot 3$

2. $(9+5) \div 7 \cdot 3$

3. $14 - 2 \cdot 8$

4. $63 \div (-9) \cdot (-2)$

5. $-2^3(-14+17)^2 + \sqrt{36}$

6. $-(12-1)^2 + \sqrt{49}$

7. $(-3)^2 + 2(14-20) + 6^2$

8. $-5^2 + 12 \div 6 + 6^2$

9. $\sqrt{100} - \sqrt{36}$

10. $\sqrt{100-36}$

11. $2|7-2\cdot 4| + 18 - 2^3$

12. $-|-40 \div (-2) + 6| + 7\sqrt{4}$

13. $-6\sqrt{25} + 15 \div (-5) - 7(-2)$

14. $2\sqrt{50-5\cdot 5} + 7(-3)^2 - 2^5$

15. $\dfrac{-|15(16-12)| - 6(-10)}{5\{7-3[2(6-10)+8]\} - 2^4}$

16. $\dfrac{-96+5\left[6(4-2)^2+7\right]-12(-7)}{-6\left|16 \div [2(-4)]\right|^3 + 48}$

Name: Date:
Instructor: Section:

Chapter 2 INTEGERS

2.6 Additional Applications and Problem Solving

Learning Objectives
1 Solve problems involving net.
2 Solve problems involving voltage.
3 Solve problems involving average rate.

GUIDED EXAMPLES AND PRACTICE

Objective 1 Solve problems involving net.

Review this example for Objective 1

1. Tomas put $1000 down when he bought a new car. He made 36 payments of $326 and spent $750 in maintenance and repairs. Three years after paying off the car he sold it for $5200. What was his net? Was it a profit or loss?

Use the formula $N = R - C$ to calculate the net.
First find the total cost for the car.
Cost = amount down + total of all payments+ maintenance cost.

$\text{Cost} = 1000 + 36(326) + 750$

$\text{Cost} = 1000 + 11,736 + 750$

$\text{Cost} = 13,486$

Now use $N = R - C$ to calculate the net.

$N = 5200 - 13486$

$N = -8286$

Tomas' net is $-\$8286$, which is a loss of $\$8286$.

Practice this exercise:

1. Tanya takes out a loan to buy a fixer-upper house. She then spends $4900 in repairs and improvements. She sells the house for $80,880. The payoff amount for the loan that she took out to buy the house is $70,974. What is her net? Is it a profit or loss?

Name: Date:
Instructor: Section:

Objective 2 Solve problems involving voltage.

Review this example for Objective 2

2. The voltage in an electrical circuit measures -252 volts. If the resistance is 7 ohms, find the current.

Use the formula $V = ir$, where V is the voltage, and r is the resistance.

$$V = ir$$
$$-252 = i \cdot 7$$
$$\frac{-252}{7} = i$$
$$-36 = i$$

The current is -36 amps.

Practice this exercise:

2. An electrical circuit has a resistance of 45 ohms and a current of -11 amps. Find the voltage.

Objective 3 Solve problems involving average rate.

Review this example for Objective 3

3. A car traveled 268 miles in 4 hours. How far did it travel in 1 hour?

Use the formula $d = rt$, where d is the distance, and t is the time.

$$d = rt$$
$$268 = r \cdot 4$$
$$\frac{268}{4} = r$$
$$67 = r$$

The car traveled 67 miles in 1 hour.

Practice this exercise:

3. A plane flies 151 kilometers per hour for 3 hours. How far does it travel?

Name: Date:
Instructor: Section:

ADDITIONAL EXERCISES

Objective 1 Solve problems involving net.

For extra help, see Example 1 on page 131 of your text and the Section 2.6 lecture video.

Solve.

1. In 1995, Priscilla bought a collection of four ceramic village pieces. Two of the buildings cost \$125 each and the other two cost \$175 each. In 2007 she sold the entire collection in an auction for \$725. What was her net? Was it a profit or loss?

2. Jason bought 60 shares of stock at a price of \$72 per share. One day, he sold 20 shares for \$66 per share. A week later, he sold the rest of his shares for \$58 per share. What was his net? Was it a profit or loss?

Objective 2 Solve problems involving voltage.

For extra help, see Example 2 on page 132 of your text and the Section 2.6 lecture video.

Solve.

3. The voltage in an electrical circuit measures -512 volts. If the resistance is 8 ohms, find the current.

4. An electrical circuit has a resistance of 18 ohms and a current of -13 amps. Find the voltage.

Objective 3 Solve problems involving average rate.

For extra help, see Examples 3–4 on page 133 of your text and the Section 2.6 lecture video.

Solve.

5. On a scenic drive, Ava left at 9 A.M. and drove 75 miles, then took a 15-minute break. She then traveled another 50 miles and stopped for 1 hour to eat lunch. Finally she traveled 122 miles and arrived at her destination at 3 P.M. What was her average speed?

6. A small airplane flew 648 miles in 8 hours. How far did it fly in 1 hour?

Name: Date:
Instructor: Section:

Chapter 3 EXPRESSIONS AND EQUATIONS

3.1 Translating and Evaluating Expressions

Learning Objectives
1 Differentiate between an expression and an equation.
2 Translate word phrases to expressions
3 Evaluate expressions.

Key Terms
Use the vocabulary terms listed below to complete each statement in Exercises 1–5.

equation **expression** **evaluate** **undefined** **indeterminate**

1. To ________________ an expression, replace the variables with the corresponding given values.
2. An ________________ is a mathematical relationship that has an equal sign.
3. The quotient $\frac{0}{0}$ is said to be ________________.
4. An ________________ may be a combination of constants, variables, and arithmetic symbols.
5. For any number $n \neq 0$, the quotient $\frac{n}{0}$ is ________________.

GUIDED EXAMPLES AND PRACTICE

Objective 1 Differentiate between an expression and an equation.

Review these examples for Objective 1:

1. Determine whether the following is an expression or an equation.
   $3x+5(x-1)$

   $3x+5(x-1)$ is an expression because it does not contain an equals sign.

2. Determine whether the following is an expression or an equation.
   $4x+13=6(3-x)$

   $4x+13=6(3-x)$ is an equation because it contains an equals sign.

Practice these exercises:

1. Determine whether the following is an expression or an equation.
   $1+\sqrt{7}=8x$

2. Determine whether the following is an expression or an equation.
   $\frac{2x^2+1}{x-7}$

Name: Date:
Instructor: Section:

Objective 2 Translate word phrases to expressions.

Review these examples for Objective 2:

3. Translate to a variable expression.
Eight subtracted from the product of five and x.

The product of five and x translates to $5x$.
Eight subtracted from $5x$ translates to $5x-8$.

4. Translate the words to a variable expression.
The product of nine and the sum of some number and two.

"The sum of some number and two" translates to $n+2$ when n is the unknown number.
Nine times this sum translates to $9(n+2)$

Practice these exercises:

3. Translate to a variable expression.
The product of five and the square of n.

4. Translate the words to a variable expression.
The quotient of the difference of x and y and the sum of x and four.

Objective 3 Evaluate expressions.

Review these examples for Objective 3:

5. Evaluate $1+y; y=12$.

$1+(12)$	Replace y with 12 using parentheses
13	Simplify by adding

6. Evaluate $x^2-6y+9; x=-4, y=3$.

$(-4)-6(3)+9$	Replace x with -4 and y with -3.
$-4-18+9$	Multiply 6 and 3.
-13	Simplify by adding.

Practice these exercises:

5. Evaluate.
$4x-2y; x=1, y=-7$

6. Evaluate.
$\frac{-4mn+2}{6n}; m=-2, n=-1$

Name:
Instructor:

Date:
Section:

ADDITIONAL EXERCISES

Objective 1 Differentiate between an expression and an equation.

For extra help, see Example 1 on page 144 of your text and the Section 3.1 lecture video.

Determine whether each of the following is an expression or an equation.

1. $|14x-9|=6$

2. $\sqrt{3-\frac{x}{c}}$

Objective 2 Translate word phrases to expressions.

For extra help, see Examples 2–3 on pages 146–147 of your text and the Section 3.1 lecture video.

Translate to a variable expression.

3. Negative three multiplies by the difference of some number and four.

4. Four minus the quotient of two and x.

Objective 3 Evaluate expressions.

For extra help, see Example 4 on page 147 of your text and the Section 3.1 lecture video.

Evaluate.

5. $8x-y;\ x=5,\ y=40$

6. $x^2-7y+9;\ x=-1,\ y=-2$

7. $b^2-4ac;\ a=-3,\ b=9,\ c=-4$

8. $2t+1+4t-5t;\ t=-1$

9. $-|-x-y|;\ x=-3,\ y=18$

10. $\sqrt{c^2-b^2};\ c=15,\ b=9$

11. $\sqrt{m^2-n};\ m=-5,\ n=9$

12. $\sqrt{\frac{7x-3}{3x+1}};\ x=1$

13. $\frac{-m+26n}{s+5};\ m=5,\ n=-5,\ s=0$

14. $\frac{x+h}{h};\ x=2,\ h=1$

Name: Date:
Instructor: Section:

Complete each table by evaluating the expression using the given values for the variable.

15.

x	$-3x+5$
-2	
-1	
0	
1	
2	

16.

y	$\lvert y^2-4\rvert$
-4	
-2	
0	
2	
4	

Find all values for the variable that cause the expression to be undefined or indeterminate.

17. $\dfrac{x+1}{x-4}$

18. $\dfrac{12}{(x+2)(x-3)}$

Name: Date:
Instructor: Section:

Chapter 3 EXPRESSIONS AND EQUATIONS

3.2 The Distributive Property; Combining Like Terms

Learning Objectives
1 Rewrite an expression using the distributive property.
2 Rewrite an expression by combining like terms.
3 Solve problems involving perimeter.
4 Solve problems involving surface area.

Key Terms
Use the vocabulary terms listed below to complete each statement in Exercises 1–3.

terms **coefficient** **like terms**

1. __________________ are the expressions that are the addends in an expression that is a sum.

2. A(n) __________________ is a numerical factor in a term.

3. Constant terms or variable terms that have the same variable raised to the same exponents are called __________________.

GUIDED EXAMPLES AND PRACTICE

Objective 1 Rewrite an expression using the distributive property.

Review this example for Objective 1:

1. Use the distributive property to write an equivalent expression and simplify.
$2(x+5)$

$2(x+5)$
$2\cdot x+2\cdot 5$ Distributive property.
$2x+10$ Simplify.

Practice this exercise:

1. Use the distributive property to write an equivalent expression and simplify.
$4(c+7)$

Name: Date:
Instructor: Section:

Objective 2 Rewrite an expression by combining like terms.

Review this example for Objective 2:

2. Simplify by combining like terms.
$2p+4q-18p+2q$

$2p+4q-18p+2q$
Collect the like terms.
$2p-18p+4q+2q$
Combine the like terms.
$-16p+6q$

Practice this exercise:

2. Simplify by combining like terms.
$6a+3d-18d+4a$

Objective 3 Solve problems involving perimeter.

Review this example for Objective 3:

3. Write an expression in simplest form for the perimeter. Then use the given value for the variable to calculate the value of the perimeter.

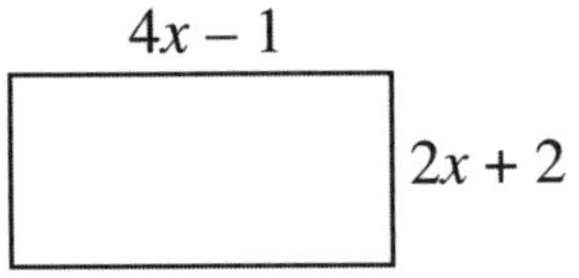

$x = 7$ centimeters

Because perimeter is the sum of the side lengths, we add the lengths of the sides together.

$$\text{Perimeter} = 2(4x-1)+2(2x+2)$$
$$= 8x-2+4x+4$$
$$= 12x+2$$

Now evaluate $12x+2$ when $x=7$.
$12(7)+2=84+2=86$ centimeters

Practice this exercise:

3. Write an expression in simplest form for the perimeter. Then use the given value for the variable to calculate the value of the perimeter.

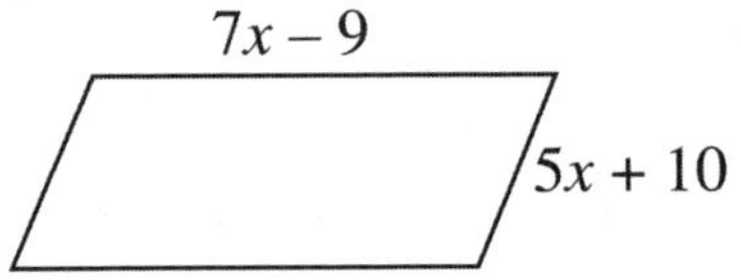

$x = 15$ feet

Name:
Instructor:

Date:
Section:

Objective 4 Solve problems involving surface area.

Review this example for Objective 2:

4. A gift box 18 centimeters long by 9 centimeters wide by 2 centimeters high is to be wrapped in paper. What is the total area of paper that will be needed if there is no overlapping?

The surface area of a box is $SA = 2(lw + lh + wh)$ where l is length, w is width, and h is height. In this equation replace l with 18, w with 9, and h with 2.

$$SA = 2(18 \cdot 9 + 18 \cdot 2 + 9 \cdot 2)$$
$$= 2(162 + 36 + 18)$$
$$= 2(216)$$
$$= 432$$

The total area of paper needed is 432 cm^2.

Practice this exercise:

4. A large cube of ice is 3 feet along each side. What is the total surface area of the cube?

ADDITIONAL EXERCISES

Objective 1 Rewrite an expression using the distributive property.

For extra help, see Example 1 on page 151 of your text and the Section 3.2 lecture video.

Use the distributive property to write an equivalent expression and simplify.

1. $9(3m + 2)$

2. $-2(b - 5)$

Objective 2 Rewrite an expression by combining like terms.

For extra help, see Examples 2–3 on pages 153–154 of your text and the Section 3.2 lecture video.

Identify the coefficient of each term.

3. $8w^5$

4. $-x^7$

Simplify by combining like terms.

5. $47x - 89x$

6. $3p + 13 - 20p + 34$

7. $-36-4x+3y-93-y+8x$

8. $20r+3s-5r+16s$

Objective 3 Solve problems involving perimeter.

For extra help, see Example 4 on pages 154–155 of your text and the Section 3.2 lecture video.

Write an expression in simplest form for the perimeter. Then use the given value for the variable to calculate the value of the perimeter.

9.

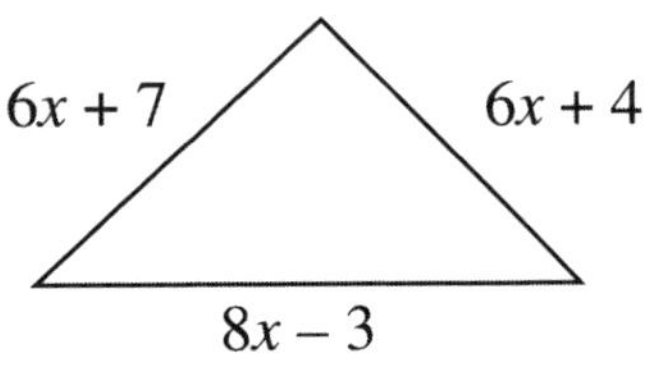

$x = 3$ yards

10.

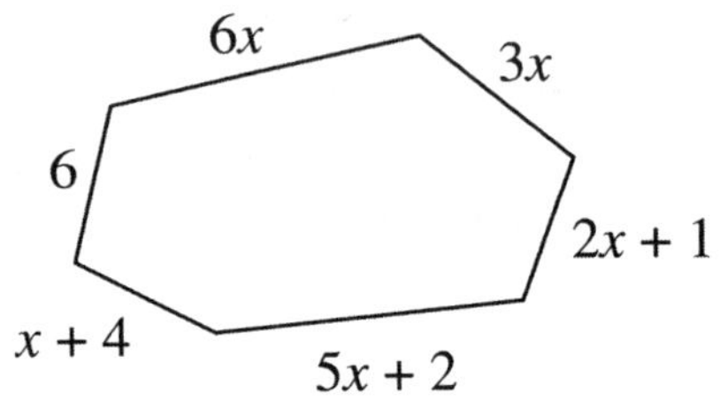

$x = 3$ meters

Objective 4 Solve problems involving surface area.

For extra help, see Example 5 on page 156 of your text and the Section 3.2 lecture video.

Solve.

11. A brick 12 centimeters long by 6 centimeters wide by 5 centimeters high is to be painted on every side. What is the total area to be painted?

12. A cereal box is printed on all sides with color and type. If the box is 8 inches long by 2 inches wide by 10 inches high, what is the total area that is printed?

Name: Date:
Instructor: Section:

Chapter 3 EXPRESSIONS AND EQUATIONS

3.3 Equations and Their Solutions

Learning Objectives
1 Differentiate between an expression and an equation.
2 Check to see whether a number is a solution for an equation.

Key Terms
Use the vocabulary terms listed below to complete each statement in Exercises 1–4.

equation **expression** **equate** **solve** **solution**

1. To check a solution, simplify each side of the __________________ separately to determine if it is true or false.

2. The goal with an equation is to __________________ it.

3. An equation has a(n) __________________ on each side of the equal sign.

4. To check a(n) __________________ of an equation, replace the variable(s) with the number and simplify both sides of the equation.

GUIDED EXAMPLES AND PRACTICE

Objective 1 Differentiate between an expression and an equation.

Review these examples for Objective 1:

1. Determine whether the following is an expression or an equation.
$7(x+5)-6(2x-5)+13$

It is an expression because it does not contain an equal sign.

2. Determine whether the following is an expression or an equation.
$9(x+3)-5(9x-5)=17$

It is an equation because it does contain an equal sign.

Practice these exercises:

1. Determine whether the following is an expression or an equation.
$5x^2+17=2x$

2. Determine whether the following is an expression or an equation.
$2x^2+3xy-7y$

Name: Date:
Instructor: Section:

Objective 2 Check to see whether a number is a solution for an equation.

Review these examples for Objective 2:

3. Check to see if the given number is a solution for the given equation.
$x+17=40$; check $x=16$

$$x+17=40$$

$$16+17\overset{?}{=}40 \quad \text{Replace.}$$

$$33=40 \quad \text{Simplify.}$$

False; so 16 is not a solution.

4. Check to see if the given number is a solution for the given equation.
$6x^2-60=36$; check $x=4$

$$6x^2-60\overset{?}{=}36$$

$$6\left(4^2\right)-60\overset{?}{=}36 \quad \text{Replace.}$$

$$6(16)-60\overset{?}{=}36 \quad \text{Simplify.}$$

$$36=36 \quad \text{True; so 4 is a solution.}$$

Practice these exercises:

3. Check to see if the given number is a solution for the given equation.
$x+14=35$; check $x=21$

4. Check to see if the given number is a solution for the given equation.
$32x^2-76=54$; check $x=2$

ADDITIONAL EXERCISES

Objective 1 Differentiate between an expression and an equation.

For extra help, see Example 1 on page 159 of your text and the Section 3.3 lecture video.

Indicate whether each of the following is an expression or an equation.

1. $8x^2-4x+1=0$

2. $9(x-2)+7(x+5)-3$

3. $9+2(x-5)+x^2$

4. $6y^2-3(5+y)+11=42$

Name: Date:
Instructor: Section:

Objective 2 Check to see whether a number is a solution for an equation.
For extra help, see Examples 2-3 on page 160-161 of your text and the Section 3.3 lecture video.

Check to see if the given number is a solution for the given equation.

5. $6x = -15$
check $x = -6$

6. $7x + 5 = 40$
check $x = 5$

7. $7x + 7 = 56$
check $x = 7$

8. $5x + 7 = 92$
check $x = 17$

9. $5x + 5 = 56$
check $x = 10$

10. $9x + 7 = 56$
check $x = 5$

11. $9x + 7 = 10x$
check $x = 9$

12. $2x^2 - 75 = 53$
check $x = 8$

13. $7x + 7 = 8x$
check $x = 8$

14. $6x^2 - 135 = 15$
check $x = 5$

15. $9(y - 4) = -2y + 52$
check $y = 8$

16. $7(x - 3) = -2x + 60$
check $x = 9$

17. $9(z - 4) = -3z + 52$
check $z = 9$

18. $2(x - 4) = -6x + 56$
check $x = 5$

Name: Date:
Instructor: Section:

Chapter 3 EXPRESSIONS AND EQUATIONS

3.4 The Addition Principle of Equality

Learning Objectives
1 Determine whether a given equation is linear.
2 Solve linear equations in one variable using the addition principle of equality.
3 Solve application problems.

Key Terms
Use the vocabulary terms listed below to complete each statement in Exercises 1–4.

linear	**degree**	**monomial**	**distributive**	**solution**
nonlinear	**multiply**	**inequality**	**subtract**	**add**

1. Adding the same amount to both sides of an equation does not change the equation's ________________.

2. If an equation in one variable has degree greater than 1, it is a(n) ________________ equation.

3. To clear a term that is added to one side of an equation, ________________ the additive inverse of the term to both sides of the equation.

4. If an equation has parentheses, use the ________________ property to clear the parentheses.

GUIDED EXAMPLES AND PRACTICE

Objective 1 Determine whether a given equation is linear.

Review this example for Objective 1:

1. Determine whether the given equation is linear.
$4x^2+2(x-3)=11$

$4x^2+2(x-3)=11$ is nonlinear because it has a variable term, $4x^2$, with a degree other than 1. The degree of $4x^2$ is 2.

Practice this exercise:

1. $3x+5(x-2)=7$

Name: Date:
Instructor: Section:

Objective 2 Solve linear equations in one variable using the addition principle of equality.

Review these examples for Objective 2:

2. Solve and check
$x+13=15$

$x+13=15$
$-13 \quad -13$ Subtract 13 from both sides
$x+0=2$ The sum of x and 0 is x.
$x=2$

Check:
$x+13=15$
$2+13=15$ Replace x with 2.
$15=15$ True; so 2 is the solution.

3. Solve and check
$11+7=4(y+3)-3y$

$11+7=4(y+3)-3y$
$11+7=4y+12-3y$ Distribute to clear parentheses.
$18=y+12$ Simplify each side.
$-12 \quad -12$ Subtract 12.
$6=y$

Check:
$11+7=4(y+3)-3y$
$11+7=4((6)+3)-3(6)$ Replace y with 6 in the original equation.
$11+7=4(9)-3(6)$ Simplify each side.
$11+7=36-18$
$18=18$ True; so 6 is the solution.

Practice these exercises:

2. Solve and check
$y+12=-8$

3. Solve and check
$2(h-3)-h=21-18$

Name: Date:
Instructor: Section:

Objective 3 Solve application problems.

Review this example for Objective 3:

4. Translate to an equation, then solve. Suppose Aida is planning to buy a new car. The down payment is \$4444. She has \$238 saved toward that amount. How much more does she need?

Adding the amount needed to her current amount, \$238, should equal the total needed. Let x represent the unknown addend.

$$\text{Current amount} + \text{needed amount} = 4444$$
$$238 \quad + \quad x \quad = 4444$$

Solve:

$$238 + x = 4444$$
$$\underline{-238} \quad \underline{-238}$$ Subtract 238 from both sides.
$$0 + x = 4206$$
$$x = 4206$$

Practice this exercise:

4. A golfer's average annual expense for greens fees, cart rentals, and golf balls is \$346. If he spends \$168 for greens fees and \$148 for cart rentals, how much does he spend for golf balls?

ADDITIONAL EXERCISES

Objective 1 Determine whether a given equation is linear.

For extra help, see Example 1 on page 163 of your text and the Section 3.4 lecture video.

Determine whether the given equation is linear.

1. $5x + 2(x - 1) = 9 - x$

2. $n^3 + 9 = y - 3$

Objective 2 Solve linear equations in one variable using the addition principle of equality.

For extra help, see Examples 2–4 on pages 165–166 of your text and the Section 3.4 lecture video.

Solve and check.

3. $-9 = n - 21$

4. $15 = q - 19$

Name: Date:
Instructor: Section:

5. $4t+12-3t=18$

6. $8m+12-7m=14-3$

7. $17+8=3w-5-2w$

8. $10x-3(3x+2)=-15+14$

Objective 3 Solve application problems.

For extra help, see Example 5 on page 166 of your text and the Section 3.4 lecture video.

Translate to an equation, then solve.

9. The perimeter of the triangle is 86 millimeters. Find the length of the missing side.

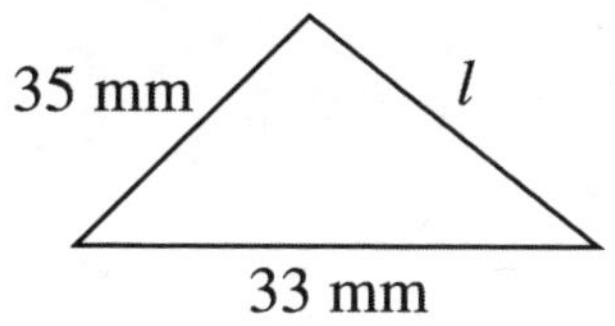

10. Marc is tiling his dining room with tiles that measure 1 foot by 1 foot. The room is 180 square feet. He has already purchased 125 tiles. How many more tiles does he need?

Name: Date:
Instructor: Section:

Chapter 3 EXPRESSIONS AND EQUATIONS

3.5 The Multiplication Principle of Equality

Learning Objectives

1. Solve equations using the multiplication principle of equality.
2. Solve equations using both the addition and multiplication principles of equality.
3. Solve application problems.

GUIDED EXAMPLES AND PRACTICE

Objective 1 Solve equations using the multiplication principle of equality.

Review this example for Objective 1:

1. Solve using the multiplication principle, then check.

$7y=-21$

$\frac{7y}{7}=\frac{-21}{7}$ Divide both sides by 7.

$y=-3$

Check:

$7y=-21$

$7(-3)\stackrel{?}{=}-21$ In the original equation, replace y with -3.

$-21=-21$ True; so -3 is the solution.

Practice this exercise:

1. Solve using the multiplication principle, then check.

$8y=16$

Objective 2 Solve equations using both the addition and multiplication principles of equality.

Review these examples for Objective 2:

2. Solve using the multiplication and addition principles of equality, then check.

$4x+8=20$

Practice these exercises:

2. Solve using the multiplication and addition principles of equality, then check.

$5m-7=-12$

Name: Date:
Instructor: Section:

$4x+8=20$

$\underline{-8} \quad \underline{-8}$ Subtract 8 from both sides.

$4x=12$

$\frac{4x}{4}=\frac{12}{4}$ Divide both sides by 4.

$x=3$

Check:

$4x+8=20$

$4(3)+8\stackrel{?}{=}20$ Replace x with 3.

$12+8\stackrel{?}{=}20$ Simplify.

$20=20$ True; so 3 is the solution.

3. Solve using the multiplication and addition principles of equality, then check.
$6x-15=4x-13$

$6x-15=4x-13$

$\underline{-4x} \quad \underline{-4x}$ Subtract $4x$ from both sides.

$2x-15=0-13$

$2x-15=-13$

$\underline{+15} \quad \underline{+15}$ Add 15 to both sides.

$x2+0=2$

$\frac{x2}{2}=\frac{2}{2}$ Divide both sides by 2.

$x=1$

Check:

$6x-15=4x-13$

$6(1)-15\stackrel{?}{=}4(1)-13$ Replace x with 1.

$6-15\stackrel{?}{=}4-13$ Simplify.

$-9=-9$ True; so 1 is the solution.

3. Solve using the multiplication and addition principles of equality, then check.
$8c+15=-19-9c$

Name: Date:
Instructor: Section:

4. Solve using the multiplication and addition principles of equality, then check.

$3(u-2)=5(u-1)+1$

$3(u-2)=5(u-1)+1$

$3u-6=5u-4$ Distribute 3 and 5.

$\underline{-3u}$ $\underline{-3u}$ Subtract $3u$ from both sides.

$0-6=2u-4$

$-6=2u-4$

$\underline{+4}$ $\underline{+4}$ Add 4 to both sides.

$-2=2u-0$

$-2=2u$

$\frac{-2}{2}=\frac{2u}{2}$ Divide by 2.

$-1=u$

Check:

$3(u-2)=5(u-1)+1$

$3((-1)-2)\overset{?}{=}5((-1)-1)+1$ Replace u with -1.

$3(-3)\overset{?}{=}5(-2)+1$ Simplify.

$-9\overset{?}{=}-10+1$

$-9=-9$ True; so -1 is the solution.

4. Solve using the multiplication and addition principles of equality, then check.

$-9(v+1)-3=-2v+11-3(v+1)$

Name: Date:
Instructor: Section:

Objective 3 Solve application problems.

Review this example for Objective 3:

5. Solve for the unknown amount. The area of a rectangular room is 156 square feet. The width is 13 feet. What is the length?

$A = lw$

$156 = l \cdot 13$ Replace A with 156 and w with 13.

$156 = 13l$

$\frac{156}{13} = \frac{13l}{13}$ Divide by 13.

$12 = l$

The length is 12 feet.

Practice this exercise:

5. The Klein family is planning a 300 mile trip. If they travel at an average speed of 60 miles per hour, what will be their travel time?

ADDITIONAL EXERCISES

Objective 1 Solve equations using the multiplication principle of equality.

For extra help, see Example 1 on page 171 of your text and the Section 3.5 lecture video.

Solve using the multiplication principle, then check.

1. $-5a = 25$

2. $7g = -35$

3. $-13r = -39$

4. $-26y = 104$

Objective 2 Solve equations using both the addition and multiplication principles of equality.

For extra help, see Examples 2–5 on pages 172–175 of your text and the Section 3.5 lecture video.

Solve using the multiplication and addition principles of equality, then check.

5. $18 = 2s + 14$

6. $-24 = 3y + 21$

7. $17 = -3r - 7$

8. $81 = 30 - 17n$

Name: Date:
Instructor: Section:

9. $-6q+13-3q=17-q-20$

10. $13y-4-14y=16+3y-12$

11. $-3p-19+7p+12=17p+16-p+1$

12. $14h+4-9h-1=h+7-2h+2$

13. $7(x+4)-2(x-1)=-4(x-3)+36$

14. $-3(m-10)-2(m+4)=-2+6m-(m+16)$

Objective 3 Solve application problems.

For extra help, see Examples 6–7 on page 176 of your text and the Section 3.5 lecture video.

Solve for the unknown amount.

15. The perimeter of a rectangle is 94 centimeters and the length is 27 centimeters. Find the width of the rectangle.

16. The voltage of a circuit is -144 volts. If the current is -6 amps, find the resistance in ohms. (Use $V = ir$.)

Name: Date:
Instructor: Section:

Chapter 3 EXPRESSIONS AND EQUATIONS

3.6 Translating Word Sentences to Equations

Learning Objectives
1 Translate sentences to equations using key words; then solve.

GUIDED EXAMPLES AND PRACTICE

Objective 1 Translate sentences to equations using key words; then solve.

Review these examples for Objective 1:

1. Translate to an equation; then solve. Ten more than a number is equal to negative six.

Translate:

Ten	more than	a number	is equal to	negative six.
10	+	n	=	−6

Solve:

$$10+n=-6$$
$$-10 \qquad -10$$ Subtract 10 from both sides.
$$0+n=-16$$
$$n=-16$$

2. Translate to an equation; then solve. Six more than the product of five and x yields forty-six.

Translate:

Six	more than	the product	of five	and	x	yields	forty-six.
6	+		5	·	x	=	46

Solve:

$$6+5x=46$$
$$-6 \qquad -6$$ Subtract 6 from both sides.
$$0+5x=40$$
$$5x=40$$
$$\frac{5x}{5}=\frac{40}{5}$$ Divide by 5.
$$x=8$$

Practice these exercises:

1. Translate to an equation; then solve. The difference of a number and seven is equal to fourteen.

2. Translate to an equation; then solve. Fourteen subtracted from the product of negative two and y is sixteen.

Name: Date:
Instructor: Section:

3. Translate to an equation; then solve. Four times the difference of b and six is equal to eighteen plus seven times b.

Translate:

$$\underbrace{\text{Four}}_{4}\ \underbrace{\text{times}}_{\cdot}\ \underbrace{\text{the difference of b and six}}_{b-6}$$

$$\underbrace{\text{is equal to}}_{=}\ \underbrace{\text{eighteen}}_{18}\ \underbrace{\text{plus}}_{+}\ \underbrace{\text{seven}}_{7}\ \underbrace{\text{times}}_{\cdot}\ \underbrace{\text{b.}}_{b}$$

Solve:

$4(b-6)=18+7b$

$4b-24=18+7b$ Distribute the 4.

$\underline{-4b} \quad \underline{-4b}$ Subtract $4b$.

$0-24=18+3b$

$-24=18+3b$

$\underline{-18} \quad \underline{-18}$ Subtract 18.

$-42=0+3b$

$-42=3b$

$\frac{-42}{3}=\frac{3b}{3}$ Divide by 3.

$-14=b$

3. Translate to an equation; then solve. Four more than the product of six and n is the same as four times the sum of n and twelve.

ADDITIONAL EXERCISES

Objective 1 Translate sentences to equations using key words; then solve.

For extra help, see Examples 1–3 on pages 181–182 of your text and the Section 3.6 lecture video.

Translate to an equation, then solve.

1. Fourteen less than a number is ten.

2. The sum of a number and eight is negative twelve.

3. A number increased by eleven is negative nineteen.

4. A number decreased by six is negative twenty.

5. The product of negative five and a number is thirty-five.

6. Negative one times a number is equal to twenty-six.

Name: Date:
Instructor: Section:

7. A number multiplied by eight is negative sixteen.

8. Negative three times a number is equal to negative eighteen.

9. The sum of eleven and four times t is the same as the difference of six times t and thirteen.

10. Six times x plus three times the difference of x and twenty is equal to sixty-two minus the sum of x and two.

11. Twenty less than negative ten times the difference of y and fifteen is the same as the product of negative four and y minus the difference of y and five.

12. Four times the sum of b and eight is equal to negative forty-eight.

13. Six times m subtracted from seven times the sum of m and three is equal to three times the sum of m and three.

Name: Date:
Instructor: Section:

Chapter 4 MULTIPLYING AND DIVIDING FRACTIONS

4.1 Prime Numbers and GCF

Learning Objectives

1 Determine whether a number is prime, composite, or neither.
2 Find the prime factorization of a given number.
3 Find all factors of a given number.
4 Find the greatest common factor of a given set of numbers by listing.
5 Find the greatest common factor of a given set of numbers using prime factorization.

Key Terms

Use the vocabulary terms listed below to complete each statement in Exercises 1–4.

natural	**prime**	**composite**	**factor**
greatest	**quotient**	**least**	

1. A prime factorization is written with only __________________ factors.

2. A __________________ number has exactly one other factor than itself.

3. The GCF is the __________________ number that divides all given numbers with no remainder.

4. A __________________ number has factors other than one and itself.

GUIDED EXAMPLES AND PRACTICE

Objective 1 Determine whether a number is prime, composite, or neither.

Review these examples for Objective 1:

1. Determine if the number is prime, composite, or neither.
6

Divide by the list of primes.
Is 6 divisible by 2? Yes, since $2 \cdot 3 = 6$. Since 2 is a factor of 6 which is neither 1 nor 6, 6 is not prime.

Practice these exercises:

1. Determine if the number is prime, composite, or neither.
39

Name: Date:
Instructor: Section:

2. Determine if the number is prime, composite, or neither.
17

Divide by the list of primes.
Is 17 divisible by 2? No, since 17 is odd.
Is 17 divisible by 3?

$$\begin{array}{r} 5 \\ 3\overline{)17} \\ -15 \\ \hline 2 \end{array}$$

No, 17 is not divisible by 3.
Because the quotient, 5 is greater than the divisor, 3, we must go on to the next prime

Is 17 divisible by 5?

$$\begin{array}{r} 3 \\ 5\overline{)17} \\ -15 \\ \hline 2 \end{array}$$

No, 17 is not divisible by 5.
Because the quotient, 3 is less than the divisor, 5, we can stop and conclude that 17 is a prime number.

2. Determine if the number is prime, composite, or neither.
23

Objective 2 Find the prime factorization of a given number.

Review this example for Objective 2:

3. Find the prime factorization. Write the answer in exponential form.
405

Use a factor tree.

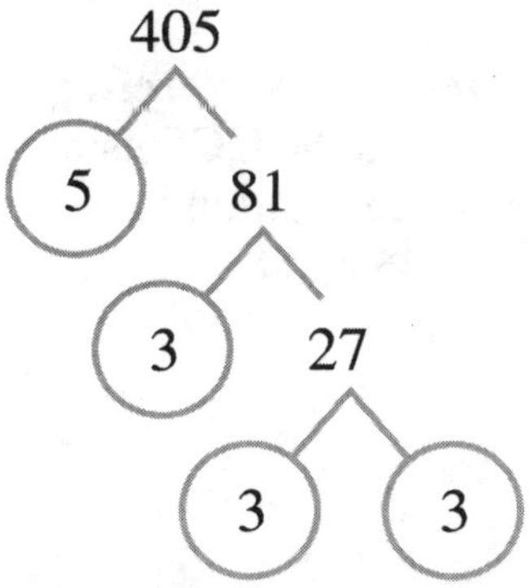

The prime factorization is $405 = 3^4 \cdot 5$.

Practice this exercise:

3. Find the prime factorization. Write the answer in exponential form.
200

Name: Date:
Instructor: Section:

Objective 3 Find all factors of a given number.

Review this example for Objective 3:

4. List all factors of each number.
36

Divide by the natural numbers $1, 2, 3, \ldots$ until we have all of the factors

$1 \cdot 36$
$2 \cdot 18$
$3 \cdot 12$
$4 \cdot 9$
$6 \cdot 6$

Because 36 is not divisible by 5, we have found all factors. The list of factors is
1, 2, 3, 4, 6, 9, 12, 18, 36.

Practice this exercise:

4. List all factors of each number.
54

Objective 4 Find the greatest common factor of a given set of numbers by listing.

Review this example for Objective 4:

5. Find the GCF by listing.
189 and 36

List the factors of each number.
Factors of 189: 1, 3, 7, 9, 21, 27, 63, 189
Factors of 36: 1, 3, 4, 6, 9, 12, 18, 36
Since 9 is the greatest number in both of these lists, $\text{GCF} = 9$.

Practice this exercise:

5. Find the GCF by listing.
27 and 26

Objective 5 Find the greatest common factor of a given set of numbers using prime factorization.

Review this example for Objective 5:

6. Find the GCF using prime factorization.
135 and 225

First, find the prime factorizations of 135 and 225.

$135 = 3^3 \cdot 5$
$225 = 3^2 \cdot 5^2$

The common prime factors are 3 and 5. For 3, the least exponent is 2. For 5, the least exponent is 1. The GCF is the product
$\text{GCF} = 3^2 \cdot 5 = 9 \cdot 5 = 45$.

Practice this exercise:

6. Find the GCF using prime factorization.
343 and 8

Name: Date:
Instructor: Section:

ADDITIONAL EXERCISES

Objective 1 Determine whether a number is prime, composite, or neither.

For extra help, see Example 1 on page 194 of your text and the Section 4.1 lecture video.

Determine if the number is prime, composite, or neither.

1. 37

2. 1

3. 125

4. 243

Objective 2 Find the prime factorization of a given number.

For extra help, see Example 2 on page 195 of your text and the Section 4.1 lecture video.

Find the prime factorization. Write the answer in exponential form.

5. 512

6. 675

7. 2205

8. 138

Objective 3 Find all factors of a given number.

For extra help, see Example 3 on page 195–196 of your text and the Section 4.1 lecture video.

List all factors of each number.

9. 375

10. 225

Objective 4 Find the greatest common factor of a given set of numbers by listing.

For extra help, see Example 4 on page 197–198 of your text and the Section 4.1 lecture video.

Find the GCF by listing.

11. 22 and 29

12. 24 and 225

Objective 5 Find the greatest common factor of a given set of numbers using prime factorization.

For extra help, see Examples 5–6 on pages 199–200 of your text and the Section 4.1 lecture video.

Find the GCF using prime factorization.

13. 40, 100, and 20

14. 20, 28, and 24

Name: Date:
Instructor: Section:

Chapter 4 MULTIPLYING AND DIVIDING FRACTIONS

4.2 Introduction to Fractions

Learning Objectives
1 Name a fraction represented by a shaded region.
2 Write improper fractions as mixed numbers.
3 Write mixed numbers as improper fractions.
4 Graph fractions or mixed numbers on a number line.
5 Write equivalent fractions.
6 Use <, >, or = to write a true statement.

Key Terms
Use the vocabulary terms listed below to complete each statement in Exercises 1–5.

fraction	**numerator**	**rational**	**simplify**	**improper**
equivalent	**greatest**	**multiple**	**dividend**	**mixed**
denominator	**fewest**	**simplest form**		

1. An expression is in simplest form if it is written with the ___________________ symbols possible.

2. In a fraction, the number written in the top position is the ___________________.

3. If a number is evenly divisible by a given number, that number is a ___________________ of the given number.

4. A(n) ___________________ number can be written as a ratio of integers.

5. An integer combined with a fraction is a(n) ___________________ number.

GUIDED EXAMPLES AND PRACTICE

Objective 1 Name a fraction represented by a shaded region.

Review these examples for Objective 1:

1. Name the fraction represented by the shaded region.

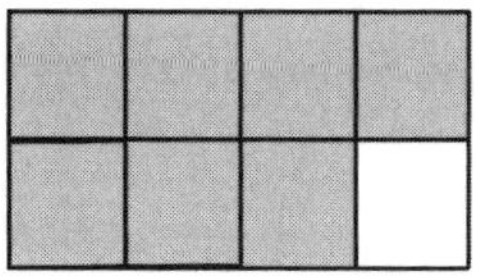

Practice these exercises:

1. Name the fraction represented by the shaded region.

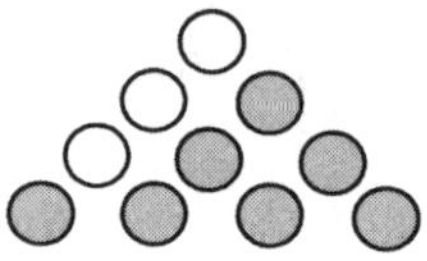

Name: Date:
Instructor: Section:

The figure has been divided into eight regions of equal size. Seven regions have been shaded, so the fraction $\frac{7}{8}$ describes the shaded region.

2. In an American Sign Language (A.S.L.) class of 19 students, 5 are hearing impaired. What fraction of the students are hearing impaired?

The fraction of the students that are hearing impaired is $\frac{5}{19}$.

2. Of 61 cars making up a freight train, 30 are boxcars. What fraction of the cars are not boxcars?

Objective 2 Write improper fractions as mixed numbers.

Review these examples for Objective 2:

3. Simplify $\frac{-31}{-31}$.

$$\frac{-31}{-31} = 1$$

4. Write $\frac{10}{7}$ as a mixed number.

$$\begin{array}{r} 1 \\ 7\overline{)10} \\ -7 \\ \hline 3 \end{array} \qquad \frac{10}{7} = 1\frac{3}{7}$$

The quotient becomes the integer, the remainder becomes the new numerator, and the denominator stays the same.

5. Divide $23 \div 6$ and write the quotient as a mixed number.

$$\begin{array}{r} 3 \\ 6\overline{)23} \\ -18 \\ \hline 5 \end{array} \qquad 23 \div 6 = 3\frac{5}{6}$$

The quotient becomes the integer, the remainder becomes the new numerator, and the denominator stays the same.

Practice these exercises:

3. Simplify $\frac{0}{7}$.

4. Write $\frac{23}{4}$ as a mixed number.

5. Divide $-61 \div 7$ and write the quotient as a mixed number.

Name: Date:
Instructor: Section:

Objective 3 Write mixed numbers as improper fractions.

Review this example for Objective 3:

6. Write $6\frac{7}{8}$ as an improper fraction.

$6\frac{7}{8}=\frac{8\cdot 6+7}{8}$ Rewrite the numerator.

$=\frac{48+7}{8}$ Multiply.

$=\frac{55}{8}$ Add.

Practice this exercise:

6. Write $-10\frac{2}{3}$ as an improper fraction.

Objective 4 Graph fractions or mixed numbers on a number line.

Review these examples for Objective 4:

7. Graph $\frac{3}{10}$ on a number line.

The denominator is 10; so draw 10 evenly spaced marks to divide the segment between 0 and 1 into 10 equal spaces. The numerator is 3; so draw a dot on the third mark to the right of 0.

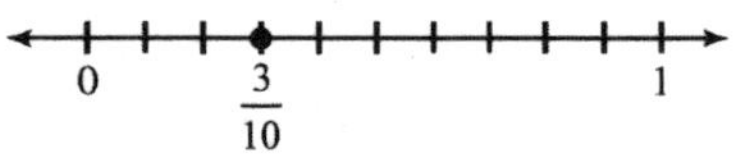

8. Graph $2\frac{3}{4}$ on a number line.

The denominator is 4; so draw 4 evenly spaced marks to divide the segment between 2 and 3 into 4 equal spaces. The numerator is 3; so draw a dot on the third mark to the right of 2.

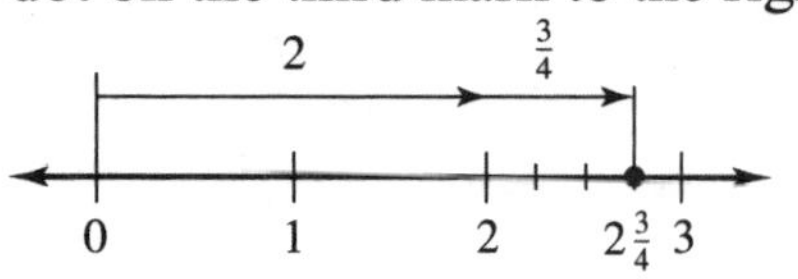

Practice these exercises:

7. Graph $-\frac{1}{3}$ on a number line.

8. Graph $-1\frac{1}{3}$ on a number line.

Name: Date:
Instructor: Section:

Objective 5 Write equivalent fractions.

Review these examples for Objective 5:

9. Fill in the blank so that the fractions are equivalent.

$\frac{9}{10} = \frac{?}{20}$

$\frac{9}{10} = \frac{9 \cdot 2}{10 \cdot 2}$ Multiplying 10 by 2 produces 20; so multiplying 9 by 2 will produce the unknown numerator.

$= \frac{18}{20}$ Multiply the numerator and the denominator by 2.

10. Fill in the blank so that the fractions are equivalent.

$\frac{16}{40} = \frac{2}{?}$

$\frac{16}{40} = \frac{16 \div 8}{40 \div 8}$ Dividing 16 by 8 gives 2; so dividing 40 by 8 will give the unknown denominator.

$= \frac{2}{5}$ Divide the numerator and denominator by 8.

Practice these exercises:

9. Fill in the blank so that the fractions are equivalent.

$\frac{-7}{10} = \frac{?}{40}$

10. Fill in the blank so that the fractions are equivalent.

$\frac{-8}{12} = \frac{-2}{?}$

Objective 6 Use <, >, or = to write a true statement.

Review this example for Objective 3:

11. Use <, >, or = to write a true statement.

$\frac{1}{3} \ ? \ \frac{6}{25}$

Notice that a common multiple of 3 and 25 is 75, which can be found by multiplying 25 times 3.

Practice this exercises:

11. Use <, >, or = to write a true statement.

$\frac{3}{5} \ ? \ \frac{5}{7}$

Name: Date:
Instructor: Section:

To write $\frac{1}{3}$ with a denominator of 75, multiply its numerator and denominator by 25.

$$\frac{1}{3} = \frac{1 \cdot 25}{3 \cdot 25} = \frac{25}{75}$$

To write $\frac{6}{25}$ with a denominator of 75, multiply its numerator and denominator by 3.

$$\frac{6}{25} = \frac{6 \cdot 3}{25 \cdot 3} = \frac{18}{75}$$

$$\frac{25}{75} > \frac{18}{75}, \text{ so } \frac{1}{3} > \frac{6}{25}$$

ADDITIONAL EXERCISES

Objective 1 Name a fraction represented by a shaded region.

For extra help, see Examples 1–2 on pages 203–204 of your text and the Section 4.2 lecture video.

Name a fraction represented by the shaded region.

1.

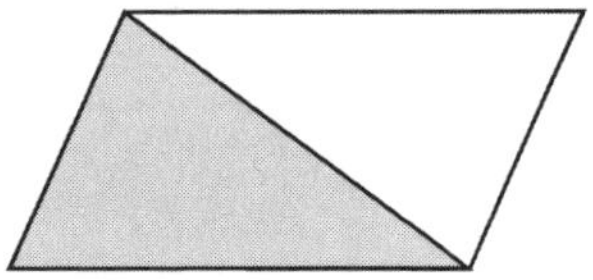

2.

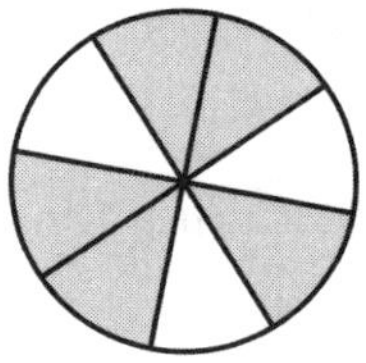

Write the fraction for each situation.

3. Twelve students board a school bus in the morning, and 11 students return on the bus in the afternoon. What fraction of the students that board the bus in the morning do not return on the bus in the afternoon?

4. Four of 13 friends do not have siblings. What fraction of the friends have siblings?

Name: Date:
Instructor: Section:

Objective 2 Write improper fractions as mixed numbers.
For extra help, see Examples 3–5 on pages 204–206 of your text and the Section 4.2 lecture video.
Simplify.

5. $\dfrac{-17}{1}$

6. $\dfrac{960}{0}$

Write the improper fraction as a mixed number.

7. $\dfrac{47}{6}$

8. $-\dfrac{75}{8}$

Divide and write the quotient as a mixed number.

9. $179 \div 14$

10. $-59 \div 4$

Objective 3 Write mixed numbers as improper fractions.
For extra help, see Example 6 on page 207 of your text and the Section 4.2 lecture video.
Write the mixed number as an improper fraction.

11. $-2\dfrac{3}{7}$

12. $6\dfrac{3}{5}$

Objective 4 Graph fractions or mixed numbers on a number line.
For extra help, see Example 7 on page 208 of your text and the Section 4.2 lecture video.
Graph the fraction on a number line.

13. $\dfrac{4}{5}$

14. $\dfrac{-3}{4}$

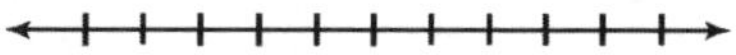

15. $3\dfrac{1}{2}$

16. $-4\dfrac{3}{4}$

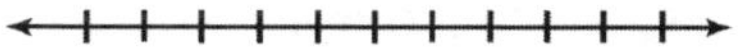

Name: Date:
Instructor: Section:

Objective 5 Write equivalent fractions.
For extra help, see Example 8 on page 209 of your text and the Section 4.2 lecture video.
Fill in the blank so that the fractions are equivalent.

17. $\frac{5}{7}=\frac{?}{56}$

18. $\frac{54}{21}=\frac{18}{?}$

Objective 6 Use <, >, or = to write a true statement.
For extra help, see Example 9 on pages 210–211 of your text and the Section 4.2 lecture video.
Use <, >, or = to write a true statement.

19. $\frac{2}{4}\ ?\ \frac{4}{5}$

20. $-\frac{1}{5}\ ?\ -\frac{2}{10}$

21. $-\frac{2}{3}\ ?\ -\frac{11}{15}$

22. $\frac{4}{6}\ ?\ \frac{10}{15}$

Name: Date:
Instructor: Section:

Chapter 4 MULTIPLYING AND DIVIDING FRACTIONS

4.3 Simplifying Fractions

Learning Objectives
1 Simplify fractions to lowest terms.
2 Simplify improper fractions or fractions within mixed numbers.

Key Terms
Use the vocabulary terms listed below to complete each statement in Exercises 1–4.

lowest **greatest** **numerator** **denominator** **least** **rational**

1. A ____________________ expression is a ratio of monomials or polynomials.

2. A fraction is in ____________________ terms if the greatest common factor of its numerator and denominator is 1.

3. To simplify a fraction to lowest terms, divide the numerator and denominator by their ____________________ common factor.

4. To begin simplifying a fraction to lowest terms, replace the ____________________ and ____________________ with their prime factorizations.

GUIDED EXAMPLES AND PRACTICE

Objective 1 Simplify fractions to lowest terms.

Review these examples for Objective 1:

1. Simplify $\frac{24}{42}$ to lowest terms.

$\frac{24}{42}=\frac{24\div 6}{42\div 6}$ Divide both 24 and 42 by their GCF, 6.

$=\frac{4}{7}$ Simplify.

2. Simplify $-\frac{48}{72}$ to lowest terms.
The prime factorization of 48 is $2\cdot 2\cdot 2\cdot 2\cdot 3$.
The prime factorization of 72 is $2\cdot 2\cdot 2\cdot 3\cdot 3$.

Practice these exercises:

1. Simplify $\frac{21}{35}$ to lowest terms.

2. Simplify $-\frac{96}{120}$ to lowest terms.

Name: Date:
Instructor: Section:

$$-\frac{48}{72} = -\frac{\overset{1}{\not{2}} \cdot \overset{1}{\not{2}} \cdot \overset{1}{\not{2}} \cdot 2 \cdot \overset{1}{\not{3}}}{\underset{1}{\not{2}} \cdot \underset{1}{\not{2}} \cdot \underset{1}{\not{2}} \cdot \underset{1}{\not{3}} \cdot 3}$$

Dividing out the common factors divides out the GCF.

$= -\frac{2}{3}$ Multiply.

3. A 30-foot picket fence is being painted white. So far, 18 feet of the fence has been painted. What is the fraction, in lowest terms, of how much of the fence is not painted yet?

If 18 feet of fence has been painted, then $30 - 18 = 12$ feet of the fence has not been painted.

$\frac{12}{30} = \frac{12 \div 6}{30 \div 6}$ Divide both 12 and 30 by their GCF, 6.

$= \frac{2}{5}$ Simplify.

3. A baseball player has hit 24 home runs so far this season. Nine of those home runs have come in games against division rivals. What is the fraction, in lowest terms, of the number of home runs not hit against division rivals?

Objective 2 Simplify improper fractions or fractions within mixed numbers.

Review these examples for Objective 2:

4. Write $\frac{10}{4}$ as a mixed number.

$\frac{10}{4} = \frac{5 \cdot \overset{1}{\not{2}}}{2 \cdot \underset{1}{\not{2}}} = \frac{5}{2}$ Simplify to lowest terms.

$= 2\frac{1}{2}$ Write the mixed number. Its fraction is in lowest terms.

5. Write $\frac{63}{12}$ as a mixed number.

Practice these exercises:

4. Write $\frac{20}{6}$ as a mixed number.

5. Write $\frac{38}{8}$ as a mixed number.

Name: Date:
Instructor: Section:

$$\frac{63}{12} = 5\frac{3}{12}$$ Write as a mixed number.

$$= 5\frac{\overset{1}{\cancel{3}}}{2 \cdot 2 \cdot \underset{1}{\cancel{3}}}$$ Simplify the mixed number's fraction.

$$= 5\frac{1}{4}$$

ADDITIONAL EXERCISES

Objective 1 Simplify fractions to lowest terms.

For extra help, see Examples 1–3 on pages 215–217 of your text and the Section 4.3 lecture video.

Simplify to lowest terms.

1. $\frac{9}{15}$

2. $\frac{19}{57}$

3. $-\frac{28}{36}$

4. $\frac{48}{132}$

5. $\frac{60}{108}$

6. $\frac{205}{287}$

Write each requested fraction in lowest terms.

7. What fraction of a minute is 51 seconds?

8. What fraction of September is 18 days?

9. A survey is conducted in front of a home improvement store. The person conducting the survey asks 132 men if they use a certain cleansing agent and 120 respond that they do use the product. What is the fraction, in lowest terms, of the men surveyed that said they used the product?

10. A cubic meter of concrete mix contains 400 kilograms of cement, 160 kilograms of stone, and 110 kilograms of sand. What is the total weight of the cubic meter of concrete mix, and what part is stone?

Name: Date:
Instructor: Section:

Objective 2 Simplify improper fractions or fractions within mixed numbers.
For extra help, see Example 4 on page 217 of your text and the Section 4.3 lecture video.
Write each fraction as a mixed number and simplify.

11. $\frac{90}{55}$

12. $-\frac{85}{70}$

13. $-\frac{90}{65}$

14. $-\frac{111}{9}$

Name: Date:
Instructor: Section:

Chapter 4 MULTIPLYING AND DIVIDING FRACTIONS

4.4 Multiplying Fractions and Mixed Numbers

Learning Objectives
1 Multiply fractions.
2 Multiply and simplify fractions.
3 Multiply mixed numbers.
4 Simplify fractions raised to a power.
5 Solve applications involving multiplying fractions.
6 Find the area of a triangle.
7 Find the radius and diameter of a circle.
8 Find the circumference of a circle.

Key Terms
Use the vocabulary terms listed below to complete each statement in Exercises 1–4.

circle	**center**	**radius**	**diameter**
ratio	**irrational**	**circumference**	

1. A number that cannot be expressed exactly as a fraction is a(n) ____________________ number.

2. A collection of points that are equally distant from a central point is called a(n) ____________________.

3. The distance around a circle is its ____________________.

4. The distance across a circle along a straight line through the center is called the ____________________.

GUIDED EXAMPLES AND PRACTICE

Objective 1 Multiply fractions.

Review these examples for Objective 1:

1. Multiply $\frac{1}{3}\cdot\frac{2}{5}$.

Multiply numerator by numerator and denominator by denominator.

$$\frac{1}{3}\cdot\frac{2}{5}=\frac{1\cdot 2}{3\cdot 5}=\frac{2}{15}$$

Practice these exercises:

1. Multiply $\frac{3}{4}\cdot\frac{3}{7}$.

Name: Date:
Instructor: Section:

2. Multiply $-\frac{1}{2}\cdot\frac{5}{9}$.

Multiply numerator by numerator and denominator by denominator. The product is negative because the two fractions have different signs.

$$-\frac{1}{2}\cdot\frac{5}{9}=-\frac{1\cdot5}{2\cdot9}=-\frac{5}{18}$$

3. Multiply $-\frac{5}{11}\cdot\left(-\frac{4}{13}\right)$.

Multiply numerator by numerator and denominator by denominator. The product is positive because the two fractions have the same signs.

$$-\frac{5}{11}\cdot\left(-\frac{4}{13}\right)=\frac{5\cdot4}{11\cdot13}=\frac{20}{143}$$

2. Multiply $\frac{2}{5}\cdot\left(-\frac{4}{7}\right)$.

3. Multiply $-\frac{3}{5}\cdot\left(-\frac{7}{8}\right)$.

Objective 2 Multiply and simplify fractions.

Review these examples for Objective 2:

4. Multiply $\frac{4}{6}\cdot\frac{9}{15}$. Write the product in lowest terms.

$\frac{4}{6}\cdot\frac{9}{15}=\frac{36}{90}$ Multiply.

$$=\frac{\overset{1}{\cancel{2}}\cdot2\cdot\overset{1}{\cancel{3}}\cdot\overset{1}{\cancel{3}}}{\underset{1}{\cancel{2}}\cdot\underset{1}{\cancel{3}}\cdot\underset{1}{\cancel{3}}\cdot5}$$

Simplify to lowest terms by dividing out the common factors.

$$=\frac{2}{5}$$

Practice these exercises:

4. Multiply $\frac{2}{9}\cdot\frac{3}{8}$. Write the product in lowest terms.

Name: Date:
Instructor: Section:

5. Multiply $\frac{10}{15}\cdot\frac{12}{30}$. Write the product in lowest terms.

$\frac{10}{15}\cdot\frac{12}{30}=\frac{2\cdot 5}{3\cdot 5}\cdot\frac{2\cdot 2\cdot 3}{2\cdot 3\cdot 5}$

Replace each numerator and denominator with its prime factorization.

$=\frac{2\cdot \cancel{5}^{1}}{3\cdot \cancel{5}_{1}}\cdot\frac{\cancel{2}^{1}\cdot 2\cdot \cancel{3}^{1}}{\cancel{2}_{1}\cdot \cancel{3}_{1}\cdot 5}$

Divide out the common factors.

$=\frac{4}{15}$

5. Multiply $\frac{9}{21}\cdot\frac{14}{33}$. Write the product in lowest terms.

Objective 3 Multiply mixed numbers.

Review these examples for Objective 3:

6. Estimate $2\frac{2}{3}\cdot 4\frac{1}{5}$, then find the actual product expressed as a mixed number in lowest terms.

$2\frac{2}{3}$ rounds up to 3. $4\frac{1}{5}$ rounds down to 4. The actual product should be around $3\cdot 4=12$.

$2\frac{2}{3}\cdot 4\frac{1}{5}=\frac{8}{3}\cdot\frac{21}{5}$ Write as improper fractions.

$=\frac{2\cdot 2\cdot 2}{\cancel{3}_{1}}\cdot\frac{\cancel{3}^{1}\cdot 7}{5}$ Divide out a common factor, 3.

$=\frac{56}{5}$ Multiply.

$=11\frac{1}{5}$ Write as a mixed number.

The actual product agrees with the estimate.

Practice these exercises:

6. Estimate $4\frac{1}{2}\cdot 2\frac{1}{6}$, then find the actual product expressed as a mixed number in lowest terms.

Name: Date:
Instructor: Section:

7. Estimate $12 \cdot \frac{4}{9}$, then find the actual product expressed as a mixed number in lowest terms.

Rounding $\frac{4}{9}$ down to 0 is not helpful, so instead round $\frac{4}{9}$ to $\frac{1}{2}$. The actual product should be around $12 \cdot \frac{1}{2} = 6$.

$12 \cdot \frac{4}{9} = \frac{12}{1} \cdot \frac{4}{9}$ Write as improper fractions.

$= \frac{2 \cdot 2 \cdot \overset{1}{\cancel{3}}}{1} \cdot \frac{2 \cdot 2}{\underset{1}{\cancel{3}} \cdot 3}$ Divide out a common factor, 3.

$= \frac{16}{3}$ Multiply.

$= 5\frac{1}{3}$ Write as a mixed number.

The actual product agrees with the estimate.

7. Estimate $10 \cdot \frac{5}{8}$, then find the actual product expressed as a mixed number in lowest terms.

8. Multiply $-4\frac{2}{9} \cdot 3\frac{3}{4}$. Write the product as a mixed number in lowest terms.

$-4\frac{2}{9} \cdot 3\frac{3}{4} = -\frac{38}{9} \cdot \frac{15}{4}$

Write as improper fractions.

$= -\frac{19 \cdot \overset{1}{\cancel{2}}}{3 \cdot \underset{1}{\cancel{3}}} \cdot \frac{\overset{1}{\cancel{3}} \cdot 5}{\underset{1}{\cancel{2}} \cdot 2}$

Divide out the common factors, which are 2 and 3.

8. Multiply $2\frac{1}{2} \cdot 4\frac{3}{5}$. Write the product as a mixed number in lowest terms.

Name: Date:
Instructor: Section:

$= -\frac{95}{6}$

Multiply. The product is negative because the two numbers have different signs.

$= -15\frac{5}{6}$

Write the improper fractions as a mixed number.

Objective 4 Simplify fractions raised to a power.

Review these examples for Objective 4:

9. Simplify $\left(\frac{3}{5}\right)^3$.

Write the base $\frac{3}{5}$ as a factor three times; then multiply.

$$\left(\frac{3}{5}\right)^3 = \frac{3}{5} \cdot \frac{3}{5} \cdot \frac{3}{5}$$
$$= \frac{27}{125}$$

10. Simplify $\left(-\frac{4}{7}\right)^3$.

Write the base $-\frac{4}{7}$ as a factor three times; then multiply.

$$\left(-\frac{4}{7}\right)^3 = \left(-\frac{4}{7}\right)\left(-\frac{4}{7}\right)\left(-\frac{4}{7}\right)$$
$$= -\left(\frac{64}{343}\right)$$

Practice these exercises:

9. Simplify $\left(\frac{2}{9}\right)^4$.

10. Simplify $\left(-\frac{5}{6}\right)^3$.

Name: Date:
Instructor: Section:

Objective 5 Solve applications involving multiplying fractions.

Review these examples for Objective 5:

11. Three of every 7 pies at a bake sale are apple pies. If there are 84 pies at the bake sale, how many are apple pies?

3 of every 7 is the fraction $\frac{3}{7}$. This means $\frac{3}{7}$ of the 84 pies are apple.

$\frac{3}{7}\cdot 84=\frac{3}{7}\cdot\frac{84}{1}$ Write as improper fractions.

$=\frac{3}{\cancel{7}_1}\cdot\frac{2\cdot 2\cdot 3\cdot \cancel{7}^1}{1}$ Divide out a common factor, 7.

$=36$ Multiply.

There are 36 apple pies.

12. A billiards trick shot artist makes $\frac{5}{6}$ of her shots in competition. Of those shots, she makes $\frac{3}{5}$ on her first attempt. What fraction of her shots does she make on her first attempt?

$\frac{3}{5}$ of the $\frac{5}{6}$ successful shots were made on the artist's first attempt.

$\frac{3}{5}\cdot\frac{5}{6}=\frac{\cancel{3}^1}{\cancel{5}_1}\cdot\frac{\cancel{5}^1}{2\cdot\cancel{3}_1}$ Divide out common factors.

$=\frac{1}{2}$ Multiply.

She makes $\frac{1}{2}$ of her shots on her first attempt.

Practice these exercises:

11. A mayoral candidate is confident he will win 7 of every 12 votes in his town. If 372 people vote in the first two hours after polls open, how many of those votes does the candidate expect he will get?

12. A group of elementary school students are at a lakeside summer camp. Nine-sixteenths of the group go swimming in a part of the lake designated for swimming. Of the swimmers, $\frac{2}{3}$ pass the test allowing them to swim in the deeper parts of the designated swimming area. What fraction of the group of students are allowed to swim in the deeper parts of the designated swimming area?

Name: Date:
Instructor: Section:

Objective 6 Find the area of a triangle.

Review this example for Objective 6:

13. A child's triangular building blocks are $3\frac{2}{5}$ cm along the base and $2\frac{1}{3}$ cm high. Find the area of the triangular face of the block.

$$A=\frac{1}{2}bh=\frac{1}{2}\left(3\frac{2}{5}\right)\left(2\frac{1}{3}\right)$$

Replace b with $3\frac{2}{5}$ and h with $2\frac{1}{3}$.

$$=\frac{1}{2}\left(\frac{17}{5}\right)\left(\frac{7}{3}\right)$$

Write the mixed numbers as improper fractions.

$$=\frac{119}{30}$$ Multiply.

$$=3\frac{29}{30}$$ Write the result as an improper fraction.

The area is $3\frac{29}{30}$ cm^2.

Practice this exercise:

13. Find the area of a triangle with height $7\frac{1}{2}$ in. and base $2\frac{1}{2}$ in.

Objective 7 Find the radius and diameter of a circle.

Review these examples for Objective 7:

14. Find the diameter of a circle with radius $1\frac{7}{8}$ ft.

Practice these exercises:

14. Find the diameter of a circle with radius $4\frac{5}{6}$ in.

Name: Date:
Instructor: Section:

$d = 2r = 2\left(1\frac{7}{8}\right)$ Replace r with $1\frac{7}{8}$

$= \frac{\overset{1}{\cancel{2}}}{1} \cdot \frac{15}{\underset{4}{\cancel{8}}}$

Write the mixed number as an improper fraction. Then divide out the common factor, 2.

$= \frac{15}{4}$ Multiply.

$= 3\frac{3}{4}$ Write the result as a mixed number.

The diameter is $3\frac{3}{4}$ ft.

15. Find the radius of a circle with diameter 13 cm.

$r = \frac{1}{2}d = \frac{1}{2}(13)$ Replace d with 13.

$= \frac{1}{2} \cdot \frac{13}{1}$ Write as an improper fraction.

$= \frac{13}{2}$ Multiply.

$= 6\frac{1}{2}$ Write as a mixed number.

The radius is $6\frac{1}{2}$ cm.

15. Find the radius of a circle with diameter $3\frac{1}{7}$ m.

Objective 8 Find the circumference of a circle.

Review this example for Objective 8:

16. Find the circumference of a circle with a radius of $3\frac{3}{8}$ in. Use $\frac{22}{7}$ for π.

Practice this exercise:

16. Find the circumference of a circle with a radius of $2\frac{5}{8}$ ft. Use $\frac{22}{7}$ for π.

Name: Date:
Instructor: Section:

$C = 2\pi r = 2\left(\frac{22}{7}\right)\left(3\frac{3}{8}\right)$

Replace π with $\frac{22}{7}$ and r with $3\frac{3}{8}$.

$= \frac{\overset{1}{\cancel{2}}}{1}\cdot\frac{\overset{1}{\cancel{2}}\cdot 11}{7}\cdot\frac{3\cdot 3\cdot 3}{\cancel{2}\cdot\cancel{2}\cdot 2}$

Write as improper fractions and divide out the common factors.

$= \frac{297}{14}$ Multiply.

$= 21\frac{3}{14}$ Write as a mixed number.

The circumference is $21\frac{3}{14}$ in.

ADDITIONAL EXERCISES

Objective 1 Multiply fractions.

For extra help, see Example 1 on page 221 of your text and the Section 4.4 lecture video.

Multiply.

1. $-\frac{1}{8}\cdot\frac{1}{6}$

2. $\frac{5}{7}\cdot\frac{6}{13}$

3. $\frac{3}{10}\cdot\frac{3}{100}$

4. $-\frac{7}{9}\cdot\left(-\frac{2}{9}\right)$

Objective 2 Multiply and simplify fractions.

For extra help, see Example 2 on page 222 of your text and the Section 4.4 lecture video.

Multiply. Write the product in lowest terms.

5. $\frac{6}{7}\cdot\frac{5}{9}$

6. $\frac{14}{39}\cdot\frac{3}{4}$

7. $-\frac{10}{39}\cdot\frac{3}{8}$

8. $\frac{16}{51}\cdot\frac{3}{10}$

9. $-\frac{16}{27}\cdot\left(-\frac{9}{32}\right)$

10. $\frac{-7}{15}\cdot\left(\frac{10}{-14}\right)$

Name: Date:
Instructor: Section:

Objective 3 Multiply mixed numbers.
For extra help, see Examples 3–4 on pages 223–224 of your text and the Section 4.4 lecture video.
Estimate and then find the actual product expressed as a mixed number in simplest form.

11. $3\frac{2}{5} \cdot 5\frac{3}{4}$

12. $45\left(\frac{4}{5}\right)$

Multiply. Write the product as a mixed number in simplest form.

13. $2\frac{3}{8} \cdot 1\frac{5}{8}$

14. $21 \cdot \frac{2}{3}$

Objective 4 Simplify fractions raised to a power.
For extra help, see Example 5 on page 224 of your text and the Section 4.4 lecture video.
Simplify.

15. $\left(\frac{5}{4}\right)^2$

16. $\left(-\frac{2}{7}\right)^3$

Objective 5 Solve applications involving multiplying fractions.
For extra help, see Examples 6–7 on pages 225–226 of your text and the Section 4.4 lecture video.
Solve. Write all answers in lowest terms.

17. Cheryl receives \$32 for working a full day doing inventory at a hardware store. How much can she get for working $\frac{3}{8}$ of a day?

18. A recipe asks for $\frac{1}{7}$ cup of granola. How much is needed to make $\frac{6}{7}$ of the recipe?

Objective 6 Find the area of a triangle.
For extra help, see Example 8 on page 227 of your text and the Section 4.4 lecture video.
Find the area of each triangle.

19.

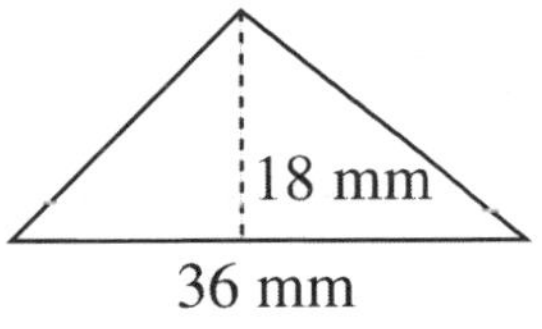

20.

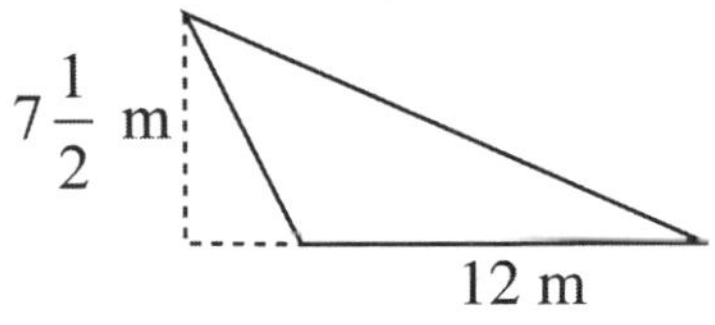

Name: Date:
Instructor: Section:

Objective 7 Find the radius and diameter of a circle.
For extra help, see Examples 9–10 on pages 228–229 of your text and the Section 4.4 lecture video.
Solve.

21. The radius of a circle is $\frac{5}{8}$ inch. What is its diameter?

22. The diameter of a circle is 15 centimeters. What is its radius?

Objective 8 Find the circumference of a circle.
For extra help, see Example 11 on page 230 of your text and the Section 4.4 lecture video.
Solve. Use $\frac{22}{7}$ *for* π.

23. The radius of a circle is $\frac{7}{10}$ inch. What is its circumference?

24. The radius of a circle is $1\frac{19}{44}$ m. What is its circumference?

Name: Date:
Instructor: Section:

Chapter 4 MULTIPLYING AND DIVIDING FRACTIONS

4.5 Dividing Fractions and Mixed Numbers

Learning Objectives
1 Divide fractions.
2 Divide mixed numbers.
3 Find the square root of a fraction.
4 Solve equations involving fractions.
5 Solve applications involving division of fractions.

Key Terms
Use the vocabulary terms listed below to complete each statement in Exercises 1–4.

reciprocal **complex** **numerator** **denominator** **improper**

1. A __________________ fraction is a fractional expression with a fraction in its numerator and/or denominator.

2. The __________________ of a fraction is the same things as its multiplicative inverse.

3. To simplify before multiplying fractions, divide out any numerator factors with any like __________________ factors.

4. To divide mixed numbers, begin by writing the mixed numbers as __________________ fractions.

GUIDED EXAMPLES AND PRACTICE

Objective 1 Divide fractions.

Review these examples for Objective 1:

1. Find the reciprocal of $\frac{2}{13}$.

Invert the numerator and the denominator to get $\frac{13}{2}$.

$\frac{2}{13} \cdot \frac{13}{2} = 1$

Practice these exercises:

1. Find the reciprocal of 4.

Name: Date:
Instructor: Section:

2. Divide $\frac{8}{15} \div \frac{4}{5}$. Write the quotient in lowest terms.

$\frac{8}{15} \div \frac{4}{5} = \frac{8}{15} \cdot \frac{5}{4}$ Write an equivalent multiplication statement

$= \frac{2 \cdot \overset{1}{\cancel{2}} \cdot \overset{1}{\cancel{2}}}{3 \cdot \underset{1}{\cancel{5}}} \cdot \frac{\overset{1}{\cancel{5}}}{\underset{1}{\cancel{2}} \cdot \underset{1}{\cancel{2}}}$ Divide out the common factors.

$= \frac{2}{3}$ Multiply.

2. Divide $-\frac{6}{11} \div 4$. Write the quotient in lowest terms.

Objective 2 Divide mixed numbers.

Review this example for Objective 2:

3. Estimate $5\frac{3}{5} \div 2\frac{1}{3}$ and then find the actual quotient expressed as a mixed number in simplest form.

$5\frac{3}{5}$ rounds up to 6, and $2\frac{1}{3}$ rounds down to 2. The actual quotient should be around $6 \div 2 = 3$.

$5\frac{3}{5} \div 2\frac{1}{3} = \frac{28}{5} \div \frac{7}{3}$ Write the mixed numbers as improper fractions.

$= \frac{28}{5} \cdot \frac{3}{7}$ Write an equivalent multiplication statement.

$= \frac{2 \cdot 2 \cdot \overset{1}{\cancel{7}}}{5} \cdot \frac{3}{\underset{1}{\cancel{7}}}$ Divide out the common factor, 7.

$= \frac{12}{5}$ Multiply.

$= 2\frac{2}{5}$ Write as a mixed number.

Practice this exercise:

3. Estimate $7\frac{1}{3} \div \frac{4}{9}$ and then find the actual quotient expressed as a mixed number in simplest form.

Name: Date:
Instructor: Section:

Objective 3 Find the square root of a fraction.

Review these examples for Objective 4:

4. Simplify $\sqrt{\frac{9}{4}}$.

Find the square root of the numerator and the denominator separately; then simplify.

$\sqrt{\frac{9}{4}} = \frac{\sqrt{9}}{\sqrt{4}} = \frac{3}{2}$

5. Simplify $\sqrt{\frac{28}{7}}$.

Simplify the fraction; then find the square root of the quotient.

$\sqrt{\frac{28}{7}} = \sqrt{4} = 2$

Practice these exercises:

4. Simplify $\sqrt{\frac{16}{81}}$.

5. Simplify $\sqrt{\frac{48}{3}}$.

Objective 4 Solve equations involving fractions.

Review these examples for Objective 4:

6. Solve and check. $\frac{5}{6}x = 2\frac{2}{9}$

$\frac{6}{5} \cdot \frac{5}{6}x = \frac{20}{9} \cdot \frac{6}{5}$ Write the mixed number as an improper fraction, then multiply both sides by $\frac{6}{5}$.

$\frac{\overset{1}{\cancel{6}}}{\underset{1}{\cancel{5}}} \cdot \frac{\overset{1}{\cancel{5}}}{\underset{1}{\cancel{6}}}x = \frac{2 \cdot 2 \cdot \overset{1}{\cancel{5}}}{3 \cdot \underset{1}{\cancel{3}}} \cdot \frac{2 \cdot \overset{1}{\cancel{3}}}{\underset{1}{\cancel{5}}}$ Divide out the common factors.

$1x = \frac{8}{3}$ Multiply.

$x = \frac{8}{3}$, or $2\frac{2}{3}$

Practice these exercises:

6. Solve and check. $\frac{4}{7}y = 1\frac{3}{5}$

Name: Date:
Instructor: Section:

Check: $\frac{5}{6}x = 2\frac{2}{9}$

$\frac{5}{6} \cdot \frac{8}{3} \stackrel{?}{=} 2\frac{2}{9}$ Replace x with $\frac{8}{3}$ and verify that the equation is true.

$\frac{40}{18} \stackrel{?}{=} 2\frac{2}{9}$ Multiply.

$\frac{20}{9} \stackrel{?}{=} 2\frac{2}{9}$ Simplify.

$2\frac{2}{9} = 2\frac{2}{9}$ True; so $\frac{8}{3}$ is the solution.

7. Solve and check. $\frac{j}{4} = -2\frac{3}{8}$

$\frac{4}{1} \cdot \frac{j}{4} = -\frac{19}{8} \cdot \frac{4}{1}$

Write the mixed number as an improper fraction, then multiply both sides by $\frac{4}{1}$.

$\frac{\not{4}}{1} \cdot \frac{j}{\not{4}} = -\frac{19}{2 \cdot \not{4}} \cdot \frac{\not{4}}{1}$ Divide out the common factors.

$j = -\frac{19}{2}$, or $-9\frac{1}{2}$ Multiply.

Check: $\frac{j}{4} = -2\frac{3}{8}$

Replace j with $-\frac{19}{2}$ and verify that the equation is true.

7. Solve and check. $5\frac{2}{9} = \frac{k}{6}$

Name: Date:
Instructor: Section:

$$\frac{-\frac{19}{2}}{4} \stackrel{?}{=} -2\frac{3}{8}$$

$$-\frac{19}{2} \div 4 \stackrel{?}{=} -2\frac{3}{8}$$

$$-\frac{19}{2} \cdot \frac{1}{4} \stackrel{?}{=} -2\frac{3}{8}$$ Write an equivalent multiplication statement.

$$-\frac{19}{8} \stackrel{?}{=} -2\frac{3}{8}$$ Multiply.

$$-2\frac{3}{8} = -2\frac{3}{8}$$ True; so $-\frac{19}{2}$ is the solution.

Objective 5 Solve applications using division of fractions.

Review this example for Objective 5:

8. A breakfast cereal box contains 49 servings of cereal. A certain bowl can hold $2\frac{1}{3}$ servings of cereal. How many bowls of cereal does the box contain?

$$2\frac{1}{3} \cdot b = 49$$

$$\frac{3}{7} \cdot \frac{7}{3} \cdot b = \frac{49}{1} \cdot \frac{3}{7}$$

Write the mixed numbers as improper fractions, then multiply both sides by $\frac{3}{7}$.

$$\frac{\cancel{3}}{\cancel{7}} \cdot \frac{\cancel{7}}{\cancel{3}} \cdot b = \frac{7 \cdot \cancel{7}}{1} \cdot \frac{3}{7}$$ Divide out the common factors.

$$b = 21$$ Multiply.

The box contains 21 bowls of cereal.

Practice this exercise:

8. The circumference of a circle is $10\frac{3}{14}$ ft. What is the radius of the circle? Use $\frac{22}{7}$ for π.

Name: Date:
Instructor: Section:

ADDITIONAL EXERCISES

Objective 1 Divide fractions.

For extra help, see Examples 1–2 on pages 235–237 of your text and the Section 4.5 lecture video.

Find the reciprocal.

1. $\frac{4}{9}$

2. $\frac{1}{8}$

3. -11

4. $-\frac{4}{7}$

Divide. Write the quotient in lowest terms

5. $\frac{1}{5} \div \frac{1}{15}$

6. $\frac{49}{45} \div \frac{7}{5}$

7. $-\frac{18}{13} \div 6$

8. $\frac{12}{7} \div \frac{15}{26}$

9. $\dfrac{\frac{16}{3}}{\frac{12}{13}}$

10. $\dfrac{14}{\frac{2}{5}}$

Objective 2 Divide mixed numbers.

For extra help, see Example 3 on pages 237–239 of your text and the Section 4.5 lecture video.

Estimate, then find the actual quotient expressed as a mixed number.

11. $2\frac{1}{2} \div 1\frac{8}{9}$

12. $3\frac{3}{4} \div 2\frac{8}{9}$

13. $4\frac{1}{4} \div 1\frac{4}{7}$

14. $\dfrac{-10\frac{1}{4}}{-2\frac{3}{5}}$

Name: Date:
Instructor: Section:

Objective 3 Find the square root of a fraction.
For extra help, see Example 4 on page 239 of your text and the Section 4.5 lecture video.
Simplify.

15. $\sqrt{\dfrac{25}{36}}$

16. $\sqrt{\dfrac{25}{49}}$

17. $\sqrt{\dfrac{18}{2}}$

18. $\sqrt{\dfrac{45}{5}}$

Objective 4 Solve equations involving fractions.
For extra help, see Examples 5–6 on pages 240–241 of your text and the Section 4.5 lecture video.
Solve and check.

19. $\dfrac{-5}{7}z = -20$

20. $\dfrac{y}{7} = \dfrac{-3}{2}$

Objective 5 Solve applications involving division of fractions.
For extra help, see Example 7 on page 242 of your text and the Section 4.5 lecture video.
Solve.

21. A cream is sold in a 34-gram container. The average amount of cream used per application is $3\dfrac{2}{5}$ grams. How many applications can be made with the container?

22. A sporting goods manufacturer requires $\dfrac{5}{3}$ yards of fabric to make a pair of soccer shorts. How many shorts can be made from 15 yards of fabric?

Name: Date:
Instructor: Section:

Chapter 5 ADDING AND SUBTRACTING FRACTIONS

5.1 Least Common Multiple

Learning Objectives
1. Find the least common multiple (LCM) by listing.
2. Find the LCM using prime factorization.
3. Write fractions as equivalent fractions with the least common denominator (LCD).

Key Terms
Use the vocabulary terms listed below to complete each statement in exercises 1-3.

least	**greatest**	**multiple**	**factor**
numerator	**denominator**	**divisor**	**quotient**

1. The LCD is the least common ____________________ of the denominators.

2. The ____________________ common multiple of a set of numbers is divisible by all the numbers in the set.

3. To find the LCM using prime factorization, find the prime factorization of each given number then write a factorization that contains each prime factor the ____________________ number of times it occurs in the factorizations.

GUIDED EXAMPLES AND PRACTICE

Objective 1 Find the least common multiple (LCM) by listing.

Review this example for Objective 1:

1. Find the least common multiple of 18 and 24 by listing.

24 is not evenly divisible by 18, so go to the next multiple of 24.
48 is not evenly divisible by 18, so go to the next multiple of 24
72 is evenly divisible by 18, so it is the LCM.

Practice this exercise:

1. Find the least common multiple of 21 and 35 by listing.

Name: Date:
Instructor: Section:

Objective 2 Find the LCM using prime factorization.

Review this example for Objective 2:

2. Find the least common multiple of 15, 24 and 28 by prime factorization.

$15 = 3 \cdot 5$

$24 = 2^3 \cdot 3$

$28 = 2^2 \cdot 7$

$\text{LCM}(15, 24, 28) = 2^3 \cdot 3 \cdot 5 \cdot 7$

2^3 has the greatest exponent for 2, 3 has the greatest exponent for 3, 5 has the greatest exponent for 5, and 7 has the greatest exponent for 7.

$= 8 \cdot 3 \cdot 5 \cdot 7$

$= 840$

Practice this exercise:

2. Find the least common multiple of 56, 72, and 80 by prime factorization.

Objective 3 Write fractions as equivalent fractions with the least common denominator (LCD).

Review this example for Objective 4:

3. Write $\frac{3}{8}$ and $\frac{7}{12}$ as equivalent fractions with the least common denominator.

12 is not evenly divisible by 8, so go to the next multiple of 12.
24 is evenly divisible by 8. Write equivalent fractions using 24 as the denominator.

To find the factor that multiplies by 8 to equal 24, divide 24 by 8.

$24 \div 8 = 3$

To write the equivalent fraction, multiply the numerator and denominator by 3.

$$\frac{3}{8} = \frac{3 \cdot 3}{8 \cdot 3} = \frac{9}{24}$$

Practice this exercise:

3. Write $\frac{5}{6}$ and $\frac{13}{15}$ as equivalent fractions with the least common denominator.

Name: Date:
Instructor: Section:

To find the factor that multiplies by 12 to equal 24, divide 24 by 12.

$24 \div 12 = 2$

To write the equivalent fraction, multiply the numerator and denominator by 2.

$$\frac{7}{12} = \frac{7 \cdot 2}{12 \cdot 2} = \frac{14}{24}$$

ADDITIONAL EXERCISES

Objective 1 Find the least common multiple (LCM) by listing.

For extra help, see Example 1 on page 256 of your text and the Section 5.1 lecture video.

Find the LCM by listing.

1. 12 and 15

2. 21 and 63

3. 9, 15, and 25

4. 4, 14, and 49

Objective 2 Find the LCM using prime factorization.

For extra help, see Example 2 on page 258 of your text and the Section 5.1 lecture video.

Find the LCM using prime factorization.

5. 18 and 12

6. 98 and 28

7. 147 and 63

8. 105 and 189

9. 27, 14, and 63

10. 8, 15, and 20

Name: Date:
Instructor: Section:

Objective 3 Write fractions as equivalent fractions with the least common denominator (LCD).

For extra help, see Example 3 on page 259 of your text and the Section 5.1 lecture video.

Write as equivalent fractions with the LCD.

11. $\frac{11}{21}$ and $\frac{7}{18}$

12. $\frac{5}{6}$ and $\frac{3}{4}$

13. $\frac{13}{15}$ and $\frac{41}{45}$

14. $\frac{3}{4}$, $\frac{5}{14}$ and $\frac{43}{49}$

Name: Date:
Instructor: Section:

Chapter 5 ADDING AND SUBTRACTING FRACTIONS

5.2 Adding and Subtracting Fractions and Mixed Numbers

Learning Objectives

1. Add and subtract fractions with the same denominator.
2. Add and subtract fractions with different denominators.
3. Add and subtract mixed numbers.
4. Add and subtract negative mixed numbers.
5. Solve equations.
6. Solve applications.

GUIDED EXAMPLES AND PRACTICE

Objective 1 Add and subtract fractions with the same denominator.

Review these examples for Objective 1:

1. Add $\frac{1}{12}+\frac{7}{12}$.

$\frac{1}{12}+\frac{7}{12}=\frac{1+7}{12}$ Add the numerators and keep the same denominator.

$=\frac{8}{12}$

$=\frac{2}{3}$ Simplify.

2. Subtract $\frac{3}{4}-\frac{1}{4}$.

$\frac{3}{4}-\frac{1}{4}=\frac{3-1}{4}$ Subtract the numerators and keep the same denominator.

$=\frac{2}{4}$ Simplify.

$=\frac{1}{2}$

Practice these exercises:

1. Add $\frac{-3}{8}+\frac{5}{8}$.

2. Subtract $\frac{8}{9}-\frac{4}{9}$.

Name: Date:
Instructor: Section:

Objective 2 Add and subtract fractions with different denominators.

Review this example for Objective 2:

3. Add $\frac{5}{12}+\frac{4}{9}$.

The LCD for 9 and 12 is 36. To write equivalent fractions with 36 as the denominator, rewrite $\frac{5}{12}$ by multiplying its numerator and denominator by 3, and rewrite $\frac{4}{9}$ by multiplying its numerator and denominator by 4.

$$\begin{aligned}\frac{5}{12}+\frac{4}{9}&=\frac{5(3)}{12(3)}+\frac{4(4)}{9(4)}\\&=\frac{15}{36}+\frac{16}{36}\\&=\frac{15+16}{36}\\&=\frac{31}{36}\end{aligned}$$

Practice this exercise:

3. Subtract $\frac{5}{6}-\frac{7}{15}$.

Objective 3 Add and subtract mixed numbers.

Review these examples for Objective 3:

4. Estimate $2\frac{1}{4}+5\frac{4}{5}$; then find the sum expressed as a mixed number in lowest terms.

$2\frac{1}{4}$ rounds down to 2, and $5\frac{4}{5}$ rounds up to 6, so the actual sum should be around $2+6=8$.
Notice that the LCD is 20.

Practice these exercises:

4. Estimate $4\frac{1}{4}+4\frac{2}{3}$; then find the sum expressed as a mixed number in lowest terms.

Name: Date:
Instructor: Section:

$$2\frac{1}{4}+5\frac{4}{5}=\frac{9}{4}+\frac{29}{5}$$ Write as improper fractions.

$$=\frac{9(5)}{4(5)}+\frac{29(4)}{5(4)}$$

$$=\frac{45}{20}+\frac{116}{20}$$

$$=\frac{161}{20}$$ Add the numerators.

$$=8\frac{1}{20}$$ Write the result as a mixed number.

The actual sum agrees with our estimate.

5. Estimate $4\frac{1}{6}-3\frac{2}{9}$; then find the difference expressed as a mixed number in lowest terms.

$4\frac{1}{6}$ rounds down to 4, and $3\frac{2}{9}$ rounds down to 3, so the actual difference should be around $4-3=1$.

Notice that the LCD is 18.

$$4\frac{1}{6}-3\frac{2}{9}=\frac{25}{6}-\frac{29}{9}$$ Write as improper fractions.

$$=\frac{25(3)}{6(3)}-\frac{29(2)}{9(2)}$$

$$=\frac{75}{18}-\frac{58}{18}$$

$$=\frac{17}{18}$$ Subtract.

The actual difference agrees with our estimate.

5. Estimate $3\frac{5}{6}-\frac{7}{8}$; then find the difference expressed as a mixed number in lowest terms.

Name: Date:
Instructor: Section:

Objective 4 Add and subtract negative mixed numbers.

Review this example for Objective 4:

6. Add $3\frac{1}{3}+\left(-4\frac{2}{5}\right)$.

Notice that the LCD is 15.

$$3\frac{1}{3}+\left(-4\frac{2}{5}\right)=\frac{10}{3}+\left(-\frac{22}{5}\right)$$

Write the mixed numbers as improper fractions.

$$=\frac{10(5)}{3(5)}+\left[-\frac{22(3)}{5(3)}\right]$$

$$=\frac{50}{15}+\left(-\frac{66}{15}\right)$$

$$=\frac{50+(-66)}{15}$$

$$=\frac{-16}{15}$$ Add.

$$=-1\frac{1}{15}$$ Write the result as a mixed number.

Practice this exercise:

6. Subtract $-2\frac{7}{8}-\left(-5\frac{1}{4}\right)$.

Objective 5 Solve equations.

Review this example for Objective 7:

7. Solve and check. $r+\frac{1}{5}=\frac{2}{7}$.

Notice that the LCD is 35.

Practice this exercise:

7. Solve and check. $f-\frac{5}{8}=\frac{1}{12}$.

Name: Date:
Instructor: Section:

$$r+\frac{1}{5}-\frac{1}{5}=\frac{2}{7}-\frac{1}{5}$$ Subtract $\frac{1}{5}$ from both sides to isolate r.

$$r+0=\frac{2(5)}{7(5)}-\frac{1(7)}{5(7)}$$ Write equivalent fractions with the LCD, 35.

$$r=\frac{10}{35}-\frac{7}{35}$$

$$r=\frac{3}{35}$$ Subtract.

Check: $r+\frac{1}{5}=\frac{2}{7}$

$$\frac{3}{35}+\frac{1}{5}\stackrel{?}{=}\frac{2}{7}$$ Replace x with $\frac{3}{35}$.

$$\frac{3}{35}+\frac{1(7)}{5(7)}\stackrel{?}{=}\frac{2}{7}$$ Write equivalent fractions with the LCD, 35.

$$\frac{3}{35}+\frac{7}{35}\stackrel{?}{=}\frac{2}{7}$$

$$\frac{10}{35}\stackrel{?}{=}\frac{2}{7}$$ Add the fractions.

$$\frac{2}{7}=\frac{2}{7}$$ Simplify. The equation is true; so $\frac{3}{35}$ is a solution.

Objective 6 Solve applications.

Review these examples for Objective 6:

8. A computer help company is taking job applications, where $\frac{5}{9}$ of the applicants have less than 2 years experience doing computer help, and $\frac{4}{15}$ of the applicants have more than 5 years experience doing computer help. What fraction of the applicants have between 2 and 5 years of experience?

Practice these exercises:

8. One car gets $29\frac{3}{10}$ miles per gallon on the highway. A second car gets $35\frac{4}{15}$ miles per gallon on the highway. How many more miles per gallon does the second car get on the highway than the first car?

There are three distinct categories into which an applicant can fall: less than 2 years experience, between 2 and 5 years experience, and more than 5 years experience. Let a represent the number of applicants with between 2 and 5 years of experience.

$$\frac{5}{9}+\frac{4}{15}+a=1$$

$$\frac{5(5)}{9(5)}+\frac{4(3)}{15(3)}+a=\frac{1(45)}{1(45)}$$

Write equivalent fractions with the LCD, 45.

$$\frac{25}{45}+\frac{12}{45}+a=\frac{45}{45}$$

$$\frac{37}{45}+a=\frac{45}{45}$$

Combine fractions on the left side.

$$\frac{37}{45}+a-\frac{37}{45}=\frac{45}{45}-\frac{37}{45}$$

Subtract $\frac{37}{45}$ from both sides to isolate a.

$$a=\frac{8}{45}$$

$\frac{8}{45}$ of the applicants have between 2 and 5 years of experience.

ADDITIONAL EXERCISES

Objective 1 Add and subtract fractions with the same denominator.

For extra help, see Examples 1–3 on pages 261–262 of your text and the Section 5.2 lecture video.

Add or subtract.

1. $\frac{3}{5}+\frac{1}{5}$

2. $\frac{4}{15}+\frac{9}{15}$

Name: Date:
Instructor: Section:

3. $-\frac{3}{14}+\left(-\frac{8}{14}\right)$

4. $\frac{17}{20}-\frac{4}{20}$

Objective 2 Add and subtract fractions with different denominators.
For extra help, see Example 4 on page 263 of your text and the Section 5.6 lecture video.
Add or subtract.

5. $\frac{2}{11}+\frac{1}{33}$

6. $\frac{7}{12}-\frac{2}{9}$

7. $\frac{1}{12}-\frac{5}{24}$

8. $\frac{2}{3}+\frac{1}{5}+\frac{3}{8}$

Objective 3 Add and subtract mixed numbers.
For extra help, see Examples 5–6 on pages 264–262 of your text and the Section 5.6 lecture video.
Add or subtract.

9. $6\frac{2}{3}+8$

10. $9\frac{5}{7}-4$

11. $7\frac{1}{8}-\frac{3}{8}$

12. $9\frac{5}{11}-5\frac{3}{5}$

13. $1\frac{1}{9}+5\frac{1}{2}$

14. $4\frac{1}{2}-7\frac{3}{4}$

15. $3\frac{4}{5}-8\frac{5}{6}$

16. $-4\frac{2}{5}-9\frac{1}{2}$

Objective 4 Add and subtract negative mixed numbers.
For extra help, see Example 7 on page 266 of your text and the Section 5.2 lecture video.
Add or subtract.

17. $7\frac{6}{7}+\left(-4\frac{1}{3}\right)$

18. $-3\frac{8}{15}+2\frac{4}{5}$

19. $-7\frac{4}{9}-8\frac{5}{18}$

20. $-1\frac{3}{5}-\left(-4\frac{2}{3}\right)$

Name: Date:
Instructor: Section:

Objective 5 Solve equations.

For extra help, see Example 8 on page 267 of your text and the Section 5.6 lecture video.

Solve and check.

21. $y+\frac{4}{5}=\frac{11}{20}$

22. $\frac{7}{9}+m=\frac{4}{21}$

23. $\frac{5}{11}=q-\frac{3}{4}$

24. $-\frac{9}{13}+p=-\frac{3}{26}$

Objective 6 Solve applications.

For extra help, see Example 9 on page 268 of your text and the Section 5.2 lecture video.

Solve.

25. A new long-life tire has a tread depth of $\frac{7}{16}$ inch, instead of the more typical $\frac{13}{32}$ inch. How much deeper is the new tire tread?

26. A parent died and left an estate to four children. One inherited $\frac{1}{9}$ of the estate, the second inherited $\frac{15}{81}$, and the third inherited $\frac{16}{27}$. How much did the fourth child inherit?

27. A book cover is $6\frac{4}{5}$ inches by $8\frac{19}{25}$ inches. What is the total distance around (perimeter) the cover of the book?

28. Phil biked $2\frac{3}{4}$ miles from his house to the town library. He then biked $3\frac{1}{6}$ miles from the library to the local market. His house is $4\frac{7}{10}$ miles away from the local market. If Phil had biked straight from his house to the market, how many fewer miles would he have biked?

Name: Date:
Instructor: Section:

Chapter 5 ADDING AND SUBTRACTING FRACTIONS

5.3 Order of Operations; Evaluating and Simplifying Expressions

Learning Objectives
1. Use the order of operations agreement to simplify expressions containing fractions and mixed numbers.
2. Find the mean, median and mode of a set of values.
3. Evaluate expressions.
4. Find the area of a trapezoid.
5. Find the area of a circle.
6. Rewrite expressions containing fractions.

GUIDED EXERCISES AND PRACTICE

Objective 1 Use the order of operations agreement to simplify expressions containing fractions and mixed numbers.

Review these examples for Objective 1:

1. Simplify $3\frac{1}{5}-1\frac{2}{3}\cdot\left(-\frac{4}{5}\right)$.

$3\frac{1}{5}-1\frac{2}{3}\cdot\left(-\frac{4}{5}\right)=\frac{16}{5}-\frac{5}{3}\cdot\left(-\frac{4}{5}\right)$

Write the mixed numbers as improper fractions.

$=\frac{16}{5}-\frac{\overset{1}{\cancel{5}}}{3}\cdot\left(-\frac{4}{\underset{1}{\cancel{5}}}\right)$

Divide out the common factor of 5.

$=\frac{16}{5}-\left(-\frac{4}{3}\right)$

$=\frac{16(3)}{5(3)}-\left(-\frac{4(5)}{3(5)}\right)$

Write equivalent fractions with the LCD, 15.

$=\frac{48}{15}-\left(-\frac{20}{15}\right)$

$=\frac{68}{15}$ Subtract.

$=4\frac{8}{15}$ Write as a mixed number.

Practice these exercises:

1. Simplify $-\frac{5}{12}+1\frac{3}{4}\cdot\frac{2}{3}$.

Name: Date:
Instructor: Section:

2. Simplify $3\left(\frac{1}{6}+\frac{1}{9}\right)-\left(\frac{3}{4}\right)^2$.

$$3\left(\frac{1}{6}+\frac{1}{9}\right)-\left(\frac{3}{4}\right)^2=3\left(\frac{1(3)}{6(3)}+\frac{1(2)}{9(2)}\right)-\left(\frac{3}{4}\right)^2$$

Write equivalent fractions with the LCD, 18.

$$=3\left(\frac{3}{18}+\frac{2}{18}\right)-\left(\frac{3}{4}\right)^2$$

$$=3\cdot\frac{5}{18}-\frac{9}{16}$$

Add and evalute the expression with the exponent.

$$=\frac{\overset{1}{\cancel{3}}}{1}\cdot\frac{5}{\underset{1}{\cancel{3}}\cdot 6}-\frac{9}{16}$$

Divide out the common factors.

$$=\frac{5}{6}-\frac{9}{16} \qquad \text{Multiply.}$$

$$=\frac{5(8)}{6(8)}-\frac{9(3)}{16(3)}$$

Write equivalent fractions with the LCD, 48.

$$=\frac{40}{48}-\frac{27}{48}$$

$$=\frac{13}{48} \qquad \text{Subtract.}$$

2. Simplify $\sqrt{\frac{25}{4}}-3\left(\frac{2}{5}+\frac{1}{2}\right)$.

Objective 2 Find the mean, median and mode of a set of values.

Review this example for Objective 2:

3. A practicing shot putter puts her shot $10\frac{1}{3}$ m, $15\frac{5}{6}$ m, $12\frac{1}{12}$ m, $11\frac{1}{3}$ m, and $15\frac{5}{6}$ m. Find the mean, median, and mode distance.

Practice this exercise:

3. A teacher times his students taking a short test. Seven students took $7\frac{1}{3}$ minutes, $5\frac{1}{3}$ minutes, $4\frac{5}{6}$ minutes, $6\frac{1}{6}$ minutes, $7\frac{2}{3}$ minutes,

Name: Date:
Instructor: Section:

$$\text{mean} = \left(10\frac{1}{3}+15\frac{5}{6}+12\frac{1}{12}+11\frac{1}{3}+15\frac{5}{6}\right)\div 5$$

Divide the sum of the distances by the number of distances, 5.

$$= \left(10\frac{4}{12}+15\frac{10}{12}+12\frac{1}{12}+11\frac{4}{12}+15\frac{10}{12}\right)\div 5$$

Write each fraction with the LCD, 12.

$$= 63\frac{29}{12}\div 5$$ Add the mixed numbers.

$$= \frac{785}{12}\cdot\frac{1}{5}$$

Write the mixed number as an improper fraction and write the division as multiplication.

$$= \frac{157\cdot \overset{1}{\cancel{5}}}{12}\cdot\frac{1}{\underset{1}{\cancel{5}}}$$ Divide out the common factor, 5.

$$= \frac{157}{12}, \text{ or } 13\frac{1}{12}$$ Multiply.

The mean distance is $13\frac{1}{2}$ m.

Median and mode:

$$10\frac{1}{3}, 11\frac{1}{3}, 12\frac{1}{12}, 15\frac{5}{6}, 15\frac{5}{6}$$

This list has an odd number of values, so the middle value, $12\frac{1}{12}$ m, is the median.

The mode is the number that occurs most often, which is $15\frac{5}{6}$ m.

minutes, $5\frac{1}{3}$ minutes, and $6\frac{1}{2}$ minutes to finish the test.

a. Find the mean test time of these students.

b. Find the median test time of these students.

c. Find the mode test time of these students.

Name: Date:
Instructor: Section:

Objective 3 Evaluate expressions.

Review this example for Objective 3:

4. Evaluate $7x-yz$ when $x=\frac{3}{4}$, $y=1\frac{1}{3}$, and $z=2\frac{1}{5}$.

$7\left(\frac{3}{4}\right)-\left(1\frac{1}{3}\right)\left(2\frac{1}{5}\right)=\frac{7}{1}\left(\frac{3}{4}\right)-\left(\frac{4}{3}\right)\left(\frac{11}{5}\right)$

Replace x with $\frac{3}{4}$, y with $1\frac{1}{3}$, and z with $2\frac{1}{5}$, then rewrite mixed numbers as improper fractions.

$=\frac{21}{4}-\frac{44}{15}$ Multiply.

$=\frac{21(15)}{4(15)}-\frac{44(4)}{15(4)}$

Write equivalent fractions with the LCD, 60.

$=\frac{315}{60}-\frac{176}{60}$

$=\frac{139}{60}$, or $2\frac{19}{60}$ Subtract.

Practice this exercise:

4. Evaluate b^2-4ac when $a=1\frac{1}{3}$, $b=2\frac{2}{3}$, and $c=\frac{5}{8}$.

Objective 4 Find the area of a trapezoid.

Review this example for Objective 4:

5. Find the area of a trapezoid with height $3\frac{2}{5}$ cm, smaller base length 8 cm, and larger base length $9\frac{1}{2}$ cm.

Use the formula $A=\frac{1}{2}h(a+b)$.

Practice this exercise:

5. Find the area of a trapezoid with height $5\frac{1}{4}$ ft, smaller base length 12 ft, and larger base length 15 ft.

$A=\frac{1}{2}\left(3\frac{2}{5}\right)\left(8+9\frac{1}{2}\right)$ Replace h with $3\frac{2}{5}$, a with 8, and b with $9\frac{1}{2}$.

$A=\frac{1}{2}\left(3\frac{2}{5}\right)\left(17\frac{1}{2}\right)$ Add inside the parentheses.

$A=\frac{1}{2}\left(\frac{17}{5}\right)\left(\frac{35}{2}\right)$ Write as improper fractions.

$A=\frac{1}{2}\left(\frac{17}{\not{5}_1}\right)\left(\frac{\not{5}^1\cdot 7}{2}\right)$ Divide out the common factor.

$A=\frac{119}{4}$ Multiply.

$A=29\frac{3}{4}$

The area is $29\frac{3}{4}$ cm^2.

Objective 5 Find the area of a circle.

Review this example for Objective 5:

6. Find the area of a circle with diameter 13 cm. Use $\frac{22}{7}$ for π.

$r=\frac{1}{2}d$

$r=\frac{1}{2}(13)$ Replace d with 13.

$r=\frac{1}{2}\cdot\frac{13}{1}$

$r=\frac{13}{2}$

Practice this exercise:

6. Find the area of a circle with radius $\frac{3}{4}$ m. Use $\frac{22}{7}$ for π.

$A = \pi r^2$

$A \approx \frac{22}{7} \cdot \left(\frac{13}{2}\right)^2$ Replace π with $\frac{22}{7}$ and r with $\frac{13}{2}$.

$A \approx \frac{22}{7} \cdot \frac{169}{4}$ Square $\frac{13}{2}$ to get $\frac{169}{4}$.

$A \approx \frac{\overset{11}{\cancel{22}}}{7} \cdot \frac{169}{\underset{2}{\cancel{4}}}$ Divide out the common factor of 2.

$A \approx \frac{1859}{14}$ Multiply.

$A \approx 132\frac{11}{14}$ Write the result as a mixed number.

The area is $132\frac{11}{14}$ cm^2.

Objective 6 Rewrite expressions containing fractions.

Review these examples for Objective 6:

7. Distribute. $\frac{4}{7}\left(\frac{21}{22}b + 35\right)$

Use the distributive property.

$\frac{4}{7}\left(\frac{21}{22}b + 35\right)$

$= \frac{4}{7} \cdot \frac{21}{22}b + \frac{4}{7} \cdot \frac{35}{1}$ Distribute the $\frac{4}{7}$.

$= \frac{2 \cdot \overset{2}{\cancel{2}}}{\underset{1}{\cancel{7}}} \cdot \frac{\overset{1}{\cancel{7}} \cdot 3}{\underset{1}{\cancel{2}} \cdot 11}b + \frac{4}{\underset{1}{\cancel{7}}} \cdot \frac{\overset{1}{\cancel{7}} \cdot 5}{1}$ Divide out the common factors.

$= \frac{6}{11}b + 20$ Multiply.

Practice these exercises:

7. Distribute.
$-\frac{3}{5}\left(\frac{4}{9n} - \frac{5}{12}\right)$

Name:
Instructor:

Date:
Section:

8. Combine like terms. $\frac{4}{9}w+\frac{11}{12}w^2-\frac{1}{8}w^2+\frac{1}{6}w$

$\frac{4}{9}w+\frac{11}{12}w^2-\frac{1}{8}w^2+\frac{1}{6}w$

$=\frac{11}{12}w^2-\frac{1}{8}w^2+\frac{4}{9}w+\frac{1}{6}w$

Use the commutative property to group the like terms together. Note that this step is optional.

$=\frac{11(2)}{12(2)}w^2-\frac{1(3)}{8(3)}w^2+\frac{4(2)}{9(2)}w+\frac{1(3)}{6(3)}w$

Write equivalent fractions. The LCD for the w^2 coefficients is 24. The LCD for the w coefficients is 18.

$=\frac{22}{24}w^2-\frac{3}{24}w^2+\frac{8}{18}w+\frac{3}{18}w$

$=\frac{19}{24}w^2+\frac{11}{18}w$ Add or subtract coefficients.

8. Combine like terms.
$\frac{1}{2}j^3+\frac{3}{4}-\frac{2}{5}j^2+\frac{3}{7}j^3-\frac{3}{10}j^2$

ADDITIONAL EXERCISES

Objective 1 Use the order of operations agreement to simplify expressions containing fractions and mixed numbers.

For extra help, see Examples 1–2 on pages 272–273 of your text and the Section 5.3 lecture video.

Simplify.

1. $\frac{3}{5}\div\left(-\frac{6}{5}\right)-3\frac{3}{5}$

2. $\frac{4}{5}\div\left(-\frac{4}{3}\right)-1\frac{3}{4}$

3. $2\frac{1}{2}-\left(\frac{3}{5}\right)^2$

4. $4\frac{1}{2}-3\sqrt{\frac{50}{2}}$

5. $3\frac{1}{4}-3\sqrt{\frac{100}{4}}$

6. $\left(\frac{2}{3}\right)^2\div\frac{1}{5}\left(-4\frac{3}{4}+\frac{3}{7}\right)$

Name: Date:
Instructor: Section:

Objective 2 Find the mean, median and mode of a set of values.
For extra help, see Example 3 on pages 273–274 of your text and the Section 5.3 lecture video.

7. A first grade teacher asks her students about their height and weight. The table lists the information she collected from six of her students.
a. Find the mean height and weight.
b. Find the median height and weight.
c. Find the mode height and weight.

Height (inches)	Weight (pounds)
$46\frac{1}{4}$	$42\frac{1}{4}$
48	$60\frac{1}{2}$
$44\frac{3}{4}$	$42\frac{1}{4}$
$47\frac{1}{2}$	$52\frac{1}{2}$
$45\frac{3}{4}$	$48\frac{1}{2}$
$47\frac{1}{2}$	$51\frac{1}{4}$

8. During a particular half-hour, a radio station plays eight songs and eight ads. The table lists the duration, in minutes, of each song and ad played.
a. Find the mean duration of the songs and ads.
b. Find the median duration of the songs and ads.
c. Find the mode duration of the songs and ads.

Songs	Ads
$2\frac{1}{2}$	$\frac{1}{6}$
$3\frac{1}{6}$	$1\frac{1}{4}$
$2\frac{5}{8}$	$\frac{1}{2}$
$3\frac{1}{10}$	$\frac{3}{4}$
$2\frac{8}{9}$	$\frac{5}{6}$
$2\frac{1}{4}$	$\frac{1}{2}$
$3\frac{3}{8}$	$\frac{1}{3}$
$3\frac{1}{4}$	$1\frac{1}{4}$

Name: Date:
Instructor: Section:

Objective 3 Evaluate expressions.
For extra help, see Example 4 on page 274 of your text and the Section 5.3 lecture video.
Evaluate each expression using the given values.

9. $r + ps$; $r = 2\frac{1}{2}$, $p = 3$, $s = 1\frac{1}{4}$

10. hr^2; $h = 3\frac{3}{5}$, $r = \frac{2}{9}$

11. $\frac{1}{2}ap^2$; $a = -10\frac{5}{6}$, $p = \frac{1}{2}$

12. $3lmn$; $l = \frac{4}{7}$, $m = -3\frac{1}{2}$, $n = -\frac{7}{11}$

Objective 4 Find the area of a trapezoid.
For extra help, see Example 5 on page 275 of your text and the Section 5.3 lecture video.
Find the area of the trapezoids shown.

13.

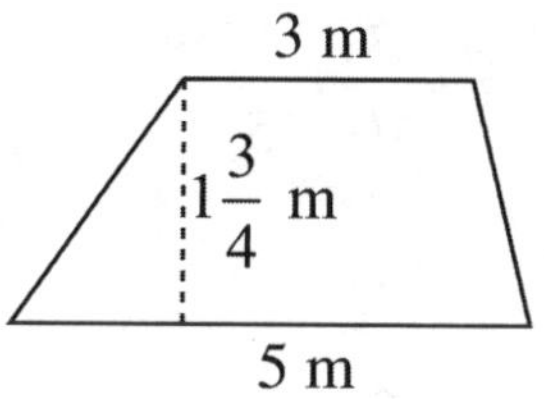

14.

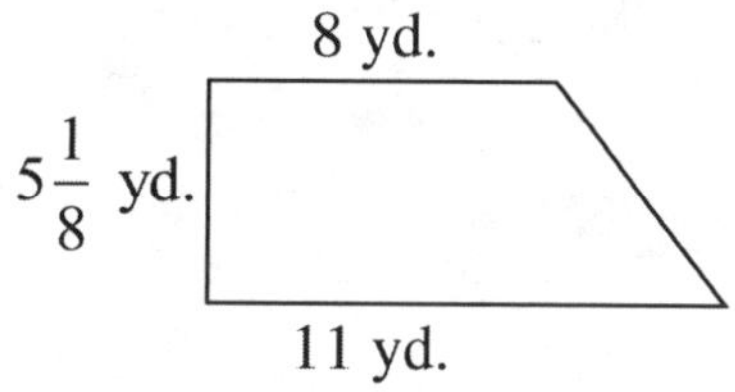

Objective 5 Find the area of a circle.
For extra help, see Example 6 on pages 276–277 of your text and the Section 5.3 lecture video.

Find each area. Use $\frac{22}{7}$ for π.

15. Find the area of a circle with radius of 9 centimeters.

16. Find the area of a circle with a diameter of $5\frac{1}{4}$ feet.

Objective 6 Simplify polynomials containing fractions.
For extra help, see Examples 7–8 on pages 277–278 of your text and the Section 5.3 lecture video.
Combine like terms.

17. $\frac{5}{6}\left(3n - \frac{4}{7}\right)$

18. $-\frac{7}{12}\left(\frac{2}{3y} + 4\right)$

19. $\frac{5}{8}y - \frac{1}{6}y^3 + 7 - \frac{3}{4}y^3 - \frac{1}{4}$

20. $\frac{2}{3}n^3 - \frac{3}{5}n + \frac{5}{8}n^2 - \frac{1}{4}n - \frac{1}{2} + \frac{1}{6}n^3$

Name: Date:
Instructor: Section:

Chapter 5 ADDING AND SUBTRACTING FRACTIONS

5.4 Solving Equations

Learning Objectives
1 Use the LCD to eliminate fractions from equations.
2 Translate sentences to equations; then solve.
3 Solve applications involving one unknown.

GUIDED EXAMPLES AND PRACTICE

Objective 1 Use the LCD to eliminate fractions from equations.

Review this example for Objective 1:

1. Solve and check. $\frac{3}{5}(r-5)=\frac{2}{7}r-\frac{1}{2}$

$$\frac{3}{5}(r-5)=\frac{2}{7}r-\frac{1}{2}$$

$$\frac{3}{5}r-\frac{3}{\cancel{5}_1}\cdot\frac{\cancel{5}^1}{1}=\frac{2}{7}r-\frac{1}{2}$$

Distribute $\frac{3}{5}$ to eliminate the parentheses. Then divide out common factors.

$$\frac{3}{5}r-3=\frac{2}{7}r-\frac{1}{2} \quad \text{Multiply.}$$

$$70\left(\frac{3}{5}r-3\right)=\left(\frac{2}{7}r-\frac{1}{2}\right)70$$

Eliminate both fractions by multiplying both sides by the LCD, 70.

$$\frac{14\cdot\cancel{5}^1}{1}\cdot\frac{3}{\cancel{5}_1}r-70\cdot3=\frac{10\cdot\cancel{7}^1}{1}\cdot\frac{2}{\cancel{7}_1}r-\frac{35\cdot\cancel{2}^1}{1}\cdot\frac{1}{\cancel{2}_1}$$

Distribute 70 and divide out the common factors.

$42r-210=20r-35$ Multiply.

$\underline{-20r} \qquad \underline{-20r}$ Subtract $20r$ from both sides.

$22r-210=0-35$

Practice this exercise:

1. Solve and check.

$$\frac{4}{7}k+\frac{3}{10}=\frac{4}{5}$$

$$22r - 210 = -35$$
$$\underline{+210} \quad \underline{+210}$$

Add 210 to both sides to isolate the $22r$ term.

$$22r - 0 = 175$$

$$\frac{22r}{22} = \frac{175}{22}$$

Divide both sides by 22 to isolate r.

$$1r = \frac{175}{22}$$

Simplify.

$$r = 7\frac{21}{22}$$

Write the solution as a mixed number.

Check:

$$\frac{3}{5}(r-5) = \frac{2}{7}r - \frac{1}{2}$$

$$\frac{3}{5}\left(7\frac{21}{22} - 5\right) \stackrel{?}{=} \frac{2}{7}\left(7\frac{21}{22}\right) - \frac{1}{2}$$

In the original equation, replace r with $7\frac{21}{22}$.

$$\frac{3}{5}\left(\frac{175}{22} - 5\right) \stackrel{?}{=} \frac{2}{7} \cdot \frac{175}{22} - \frac{1}{2}$$

Write $7\frac{21}{22}$ as an improper fraction.

$$\frac{3}{5}\left(\frac{175}{22} - \frac{5(22)}{1(22)}\right) \stackrel{?}{=} \frac{2}{7} \cdot \frac{175}{22} - \frac{1}{2}$$

Write equivalent fractions using the LCD, 22.

$$\frac{3}{5}\left(\frac{175}{22} - \frac{110}{22}\right) \stackrel{?}{=} \frac{2}{7} \cdot \frac{175}{22} - \frac{1}{2}$$

$$\frac{3}{5} \cdot \frac{65}{22} \stackrel{?}{=} \frac{2}{7} \cdot \frac{175}{22} - \frac{1}{2}$$

Subtract.

$$\frac{3}{\cancel{5}_1} \cdot \frac{\cancel{5}^1 \cdot 13}{22} \stackrel{?}{=} \frac{\cancel{2}^1}{\cancel{7}_1} \cdot \frac{\cancel{7}^1 \cdot 25}{\cancel{2}_1 \cdot 11} - \frac{1}{2}$$

Divide out common factors.

$$\frac{39}{22} \stackrel{?}{=} \frac{25}{11} - \frac{1}{2}$$

Multiply.

$$\frac{39}{22} \stackrel{?}{=} \frac{25(2)}{11(2)} - \frac{1(11)}{2(11)}$$

Write equivalent fractions with the LCD, 22.

$$\frac{39}{22} \stackrel{?}{=} \frac{50}{22} - \frac{11}{22}$$

$$\frac{39}{22} = \frac{39}{22}$$

Subtract. True; so $7\frac{21}{22}$ is the solution.

Objective 2 Translate sentences to equations; then solve.

Review this example for Objective 2:

2. Two-ninths less than the product of $1\frac{2}{5}$ and b is 1 less than $2\frac{1}{2}$ of b. Translate to an equation; then solve for b.

Translate:

Two-ninths less than the product of $1\frac{2}{5}$ and b

$$1\frac{2}{5}b - \frac{2}{9}$$

is 1 less than $2\frac{1}{2}$ of b.

$$= 2\frac{1}{2}b - 1$$

$$1\frac{2}{5}b - \frac{2}{9} = 2\frac{1}{2}b - 1$$

$$90\left(\frac{7}{5}b - \frac{2}{9}\right) = \left(\frac{5}{2}b - 1\right)90$$

$$\frac{\overset{18}{\cancel{90}}}{1} \cdot \frac{7}{\underset{1}{\cancel{5}}}b - \frac{\overset{10}{\cancel{90}}}{1} \cdot \frac{2}{\underset{1}{\cancel{9}}} = \frac{\overset{45}{\cancel{90}}}{1} \cdot \frac{5}{\underset{1}{\cancel{2}}}b - 90 \cdot 1$$

$$126b - 20 - 225b - 90$$

Practice this exercise:

2. The sum of $3\frac{1}{3}$ of n and $\frac{3}{8}$ is the same as $1\frac{1}{6}$ less than $\frac{7}{9}$ of n. Translate to an equation; then solve for n.

Name: Date:
Instructor: Section:

$$225b - 90 = 126b - 20$$
$$\underline{-126b} \qquad \underline{-126b}$$
$$99b - 90 = 0 - 20$$
$$\underline{+90 \qquad +90}$$
$$99b + 0 = 70$$
$$\frac{99b}{99} = \frac{70}{99}$$
$$b = \frac{70}{99}$$

Objective 3 Solve applications involving one unknown.

Review these examples for Objective 3:

3. Find the height h of a triangle with a base of $2\frac{1}{4}$ cm and an area of $3\frac{3}{4}\text{ cm}^2$.

The given area is for a triangle; so use $A = \frac{1}{2}bh$.

$$3\frac{3}{4} = \frac{1}{2}\left(2\frac{1}{4}\right)h$$

In $A = \frac{1}{2}bh$, replace A with $3\frac{3}{4}$ and b with $2\frac{1}{4}$.

$\frac{15}{4} = \frac{1}{2} \cdot \frac{9}{4}h$ Write the mixed numbers as improper fractions.

$\frac{15}{4} = \frac{9}{8}h$ Multiply.

$\frac{\overset{5}{\cancel{15}}}{\underset{1}{\cancel{4}}} \cdot \frac{\overset{2}{\cancel{8}}}{\underset{3}{\cancel{9}}} = \frac{\overset{1}{\cancel{8}}}{\underset{1}{\cancel{9}}} \cdot \frac{\overset{1}{\cancel{9}}}{\underset{1}{\cancel{8}}}h$ Multiply both sides by $\frac{8}{9}$ and divide out the common factors.

$\frac{10}{3} = 1h$ Multiply.

$3\frac{1}{3} = h$ Write the improper fraction as a mixed number.

The height of the triangle is $3\frac{1}{3}$ cm.

Practice these exercises:

3. Find the length a of a trapezoid with a second length b of $4\frac{3}{8}$ ft, a height h of 5 ft, and an area A of $16\frac{9}{16}\text{ ft}^2$.

Name: Date:
Instructor: Section:

4. For Janet to win a bowling tournament, she must have an average (mean) score of 248 after all five games. In her first four games, she has bowled a 233, a 260, a 257, and a 229. What is the minimum score Janet needs in her final game to win the tournament?

To calculate the mean, divide the sum of the scores by the number of scores. Let g represent the unknown score.

$$\frac{233+260+257+229+g}{5}=248$$

$$\frac{979+g}{5}=248$$

Add in the numerator.

$$\cancel{5}\cdot\frac{979+g}{\cancel{5}}=248\cdot 5$$

Eliminate the denominator by multiplying both sides by 5.

$$\begin{aligned}979+g&=1240\\ -979\quad&\quad -979\\ 0+g&=261\end{aligned}$$

Subtract 979 from both sides.

$$g=261$$

If Janet bowls a 261 in her final game, the mean of her five games will be exactly 248 and she will win the tournament.

4. A new study suggests that students who do an average of at least 35 algebra problems per weekday every week see a full letter grade improvement in their algebra class over those students that do not. If you have done 41, 20, 39, and 37 problems in the first four weekdays of this week, what is the minimum number of problems you should do on the final weekday to keep pace for the grade improvement?

ADDITIONAL EXERCISES

Objective 1 Use the LCD to eliminate fractions from equations.

For extra help, see Examples 1–2 on pages 282–285 of your text and the Section 5.4 lecture video.

Use the LCD to simplify the equation, then solve and check.

1. $t+\frac{2}{7}=\frac{5}{14}$

2. $x-\frac{1}{6}=-\frac{2}{3}$

Name: Date:
Instructor: Section:

3. $-\frac{4}{7}x=\frac{12}{35}$

4. $\frac{4}{7}x=\frac{8}{35}$

5. $\frac{1}{10}+n=\frac{5}{6}n-\frac{4}{5}$

6. $\frac{1}{9}+x=\frac{2}{3}x-\frac{1}{2}$

7. $\frac{3}{4}(y-2)=\frac{1}{6}y-\frac{3}{8}$

8. $\frac{7}{12}(f-24)=\frac{11}{30}f-\frac{5}{18}$

Objective 2 Translate sentences to equations; then solve.

For extra help, see Example 3 on pages 285–286 of your text and the Section 5.4 lecture video.

Translate to an equation, then solve.

9. $\frac{1}{3}$ of a number is $25\frac{2}{3}$.

10. The product of $5\frac{1}{4}$ and a number is $1\frac{1}{2}$.

11. $2\frac{7}{10}$ less than h is $-4\frac{3}{5}$.

12. $6\frac{1}{5}$ more than twice k is $-\frac{1}{3}$.

13. $\frac{1}{6}$ of the sum of a and 3 is equal to $1\frac{1}{9}$ added to a.

14. The sum of $\frac{2}{5}$ of z and 3 is half the difference of $5z$ and 3.

Name: Date:
Instructor: Section:

Objective 3 Solve applications involving one unknown.
For extra help, see Example 4 on pages 287–287 of your text and the Section 5.4 lecture video.
Solve.

15. Find the height of the shape if the area is $7\frac{7}{8}$ square feet.

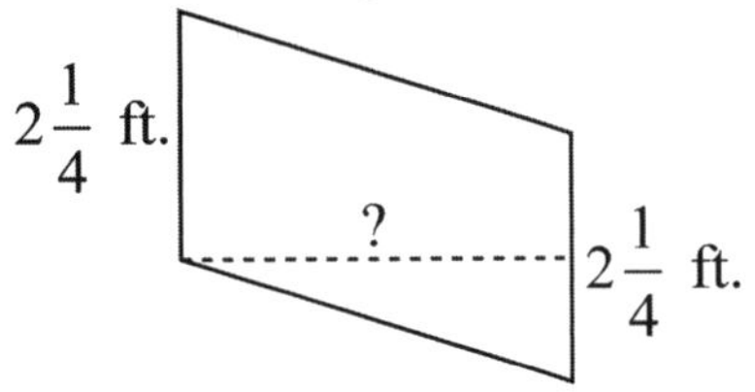

16. Warren's grade in a course is determined by the average (mean) of five tests. His scores on the first four tests are 79, 82, 84, and 80. To get a B he must have a final average of 85 or higher. What is the minimum score he needs on the fifth test to receive a B?

Name: Date:
Instructor: Section:

Chapter 6 DECIMALS

6.1 Introduction to Decimal Numbers

Learning Objectives

1 Write decimals as fractions or mixed numbers.
2 Write a word name for a decimal number.
3 Graph decimals on a number line.
4 Use < or > to write a true statement.
5 Round decimal numbers to a specified place.

Key Terms

Use the terms listed below to complete each statement in Exercises 1–4.

right **fractions** **hundredths** **left** **4** **5** **ten-thousandths**

1. Decimal notation is a base-10 notation for expressing ___________________.

2. The name of the next place value to the right of the thousands place is the ___________________ place.

3. To determine which of two positive decimal numbers is greater, compare the digits in the corresponding places from ___________________ to ___________________.

4. To round a number to a given place value, round down if the digit to the right is ___________________or less.

GUIDED EXAMPLES AND PRACTICE

Objective 1 Write decimals as fractions or mixed numbers.

Review this example for Objective 1:

1. Write as a fraction or mixed number in lowest terms.
0.4

$0.4 = \frac{4}{10}$ Write the decimal in the numberator.

$= \frac{4 \div 2}{10 \div 2}$ Simplify to lowest terms.

$= \frac{2}{5}$

Practice this exercise:

1. Write as a fraction or mixed number in lowest terms.
0.35

Name: Date:
Instructor: Section:

Objective 2 Write a word name for a decimal number.

Review these examples for Objective 2:

2. Write the word name.
0.204

To write the word name for a decimal number with no integer part (no digits other than 0 to the left of the decimal point):
1. Write the word name for digits as if they represented a whole number.
2. Write the name of the last place value.

So .204 is written as two hundred four thousandths.

3. Write the word name.
10.285

To write a word name for a decimal number with both integer and fractional parts:
1. Write the name of the integer number part.
2. Write the word *and* for the decimal point.
3. Write the name of the fractional part.

So 10.285 is written as ten and two hundred eighty-five thousandths.

Practice these exercises:

2. Write the word name.
0.105

3. Write the word name.
125.37

Objective 3 Graph decimals on a number line.

Review this example for Objective 3:

4. Graph on a number line.
0.6

Because 0.6 means $\frac{6}{10}$, we divide the distance between 0 and 1 into 10 equal-size divisions and draw a dot on the sixth division mark.

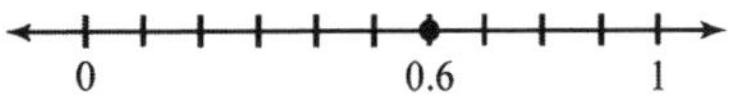

Practice this exercise:

4. Graph on a number line.
2.3

Name: Date:
Instructor: Section:

Objective 4 Use < or > to write a true statement.

Review this example for Objective 4:

5. Use < or > to write a true statement.
0.54 ? 0.5

To determine which of two positive decimal numbers is greater, compare the digits in the corresponding place values from left to right until you find two different digits in the same place. The greater number contains the greater of those two digits.

The digits match until the hundredths digit, where $4 > 0$ so $0.54 > 0.5$.

Practice this exercise:

5. Use < or > to write a true statement.
3.457 ? 3.4569

Objective 5 Round decimal numbers to a specified place.

Review this example for Objective 5:

6. Round 21.06281 to the tenths place.

The nearest tenths are 21.0 and 21.1. Rounding to the nearest tenth means that we consider the digit in the hundredths place, which is 6. Because 6 is greater than 5, we round up to 21.1.

Practice this exercise:

6. Round 21.06281 to the thousandths place.

ADDITIONAL EXERCISES

Objective 1 Write decimals as fractions or mixed numbers.

For extra help, see Example 1 on page 300 of your text and the Section 6.1 lecture video.

Write as a fraction or mixed number in lowest terms.

1. 0.875

2. 0.06

3. 12.2

4. –0.006

Name: Date:
Instructor: Section:

Objective 2 Write a word name for a decimal number.
For extra help, see Example 2 on page 301 of your text and the Section 6.1 lecture video.
Write the word name.

5. .00789

6. 50.0117

7. −4.83

8. −526.3247

Objective 3 Graph decimals on a number line.
For extra help, see Example 3 on page 302 of your text and the Section 6.1 lecture video.
Graph on a number line.

9. 6.25

10. 7.08

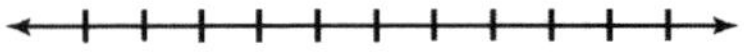

11. −5.67

12. −13.794

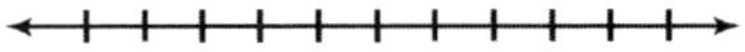

Objective 4 Use < or > to write a true statement.
For extra help, see Example 4 on page 303 of your text and the Section 6.1 lecture video.
Use < or > to write a true statement.

13. 541.11 ? 541.101

14. 0.003247 ? 0.003249

15. −2.3054 ? −2.0345

16. −56.94103 ? −56.94102

Objective 5 Round decimal numbers to a specified place.
For extra help, see Example 5 on page 304 of your text and the Section 6.1 lecture video.
Round 21.06281 to the specified place.

17. ten-thousandths

18. hundredths

19. whole number

20. tens

Name: Date:
Instructor: Section:

Chapter 6 DECIMALS

6.2 Adding and Subtracting Decimal Numbers

Learning Objectives
1 Add decimal numbers.
2 Subtract decimal numbers.
3 Add and subtract signed decimal numbers.
4 Combine like terms.
5 Solve equations using the addition principle.
6 Solve applications.

GUIDED EXAMPLES AND PRACTICE

Objective 1 Add decimal numbers.

Review this example for Objective 1:

1. Add $0.42 + 3.121$

$$\begin{array}{r} 0.42 \\ \underline{+3.121} \\ 3.541 \end{array}$$

Line up the decimal points.

Practice this exercise:

1. Add
$34.5289 + 96.2204$

Objective 2 Subtract decimal numbers.

Review this example for Objective 2:

2. Subtract $19.34 - 15.2$

$$\begin{array}{r} 19.34 \\ \underline{-15.20} \\ 4.14 \end{array}$$

Line up the decimal points.
Add 0's as placeholders.

Practice this exercise:

2. Subtract
$33.92 - 9.83$

Name: Date:
Instructor: Section:

Objective 3 Add and subtract signed decimal numbers.

Review this example for Objective 3:

3. Add or subtract $9.591+(-35.36)$

Stack the number with the greater absolute value on top and align the decimal points

$$\begin{array}{r} \overset{2}{\not 3}\,\overset{14}{\not 5}.\overset{12}{\not 3}\,\overset{15}{\not 6}\,\overset{10}{\not 0} \\ -\quad 9.5\,9\,1 \\ \hline 2\,5.7\,6\,9 \end{array}$$

Since -35.36 has the higher absolute value, the solution is -25.769.

Practice this exercise:

3. Add or subtract
$-38.7-9.806$

Objective 4 Combine like terms.

Review these examples for Objective 4:

4. Simplify $6+9.8b+2.7-6b+7.3b-3$

Combine like terms
$9.8b-6b+7.3b+6+2.7-3$
$11.1b \quad + \quad 5.7$

Practice these exercises:

4. Simplify
$3+7.7d+4.4-3d+7.7d-2$

Objective 5 Solve equations using the addition principle.

Review this example for Objective 5:

5. Solve and check $81.36+r=978.23$

$$\begin{aligned} 81.36+r &= 978.23 \\ -81.36 \qquad & -81.36 \quad \text{Subtract } 81.36. \\ r &= 896.87 \end{aligned}$$

Check:

$$81.36+896.87 \stackrel{?}{=} 978.23$$
$$978.23 = 978.23 \quad \text{True, so 896,87 is the solution.}$$

Practice this exercise:

5. $x-31.91=-47.6$

Objective 6 Solve applications.

Review this example for Objective 6:

6. A businesswoman has $1527.32 in her checking account. She writes checks of $23.03, $590.99, and $63.79 to pay some bills. She then deposits a paycheck of $524.15. How much is in her account after these changes?

Practice this exercise:

6. The perimeter of the triangle shown is 65.9 millimeters. Find the length of the missing side.

Name: Date:
Instructor: Section:

Subtract the sum of all the checks (debits) from the initial balance and add the deposit (credit).

Balance = initial balance − sum of all debts + credit

$B = 1527.32 - (23.03 + 590.99 + 63.79) + 524.15$

$B = 1527.32 - 677.81 + 524.15$

$B = 849.51 + 524.15$

$B = 1373.66$

The final balance is \$1373.66.

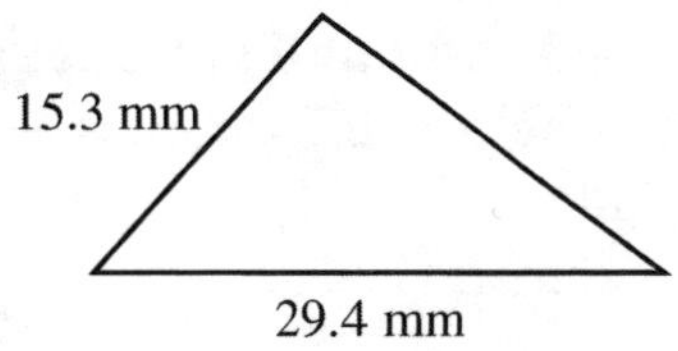

ADDITIONAL EXERCISES

Objective 1 Add decimal numbers.

For extra help, see Example 1 on pages 307–308 of your text and the Section 6.2 lecture video.

Add.

1. $65.72 + 12.098$

2. $5 + 81.12 + 34.0854$

3. $94.1903 + 73.4758$

4. $87 + 99.63 + 2.206$

Objective 2 Subtract decimal numbers.

For extra help, see Example 2 on pages 308–309 of your text and the Section 6.2 lecture video.

Subtract.

5. $79.84 - 1.63$

6. $56.1254 - 12.5891$

7. $7 - 3.529$

8. $95.6742 - 68.6783$

Name: Date:
Instructor: Section:

Objective 3 Add and subtract signed decimal numbers.
For extra help, see Examples 3–4 on pages 309–310 of your text and the Section 6.2 lecture video.
Add or subtract.

9. $-94+(-40.3)$

10. $71.2+(-21.5)$

11. $1.09-(-12.37)$

12. $-13.9-(-21.5)$

Objective 4 Combine like terms.
For extra help, see Example 5 on page 310 of your text and the Section 6.2 lecture video.
Simplify.

13. $4.8+n-2.6-5.1n+8.4n+1$

14. $8.2y-3.7y^2-2.3y+45+6.4y-3.6$

Objective 5 Solve equations using the addition principle.
For extra help, see Example 6 on page 311 of your text and the Section 6.2 lecture video.
Solve and check

15. $a+60.73=-48.432$

16. $b-(-51.9)=9.34$

Objective 6 Solve applications
For extra help, see Example 7 on pages 311-312 of your text and the Section 6.2 lecture video.

17. A business man has \$1628.67 is his checking account. He writes checks of \$623.11, \$34.56, \$63.98, and \$871.47 to pay some bills. He then deposits a paycheck of \$721.68. How much is in his account after these changes?

18. A triangle has a perimeter of 20 cm. Two of the sides are 6.3 cm and 9.1 cm. What is the length of the third side?

Name: Date:
Instructor: Section:

Chapter 6 DECIMALS

6.3 Multiplying Decimal Numbers; Exponents with Decimal Bases

Learning Objectives

1 Multiply decimal numbers.
2 Multiply signed decimal numbers.
3 Evaluate exponential forms with decimal bases.
4 Write a number in scientific notation in standard form.
5 Write standard form numbers in scientific notation.
6 Solve applications.

GUIDED EXAMPLES AND PRACTICE

Objective 1 Multiply decimal numbers.

Review this example for Objective 1:

1. Multiply $(5.4)(4.3)$

5.4	1 decimal place
$\times 4.3$	$\underline{+1}$ decimal place
162	
+216	
23.22	2 decimal places

Practice this exercise:

1. Multiply $206.57(0.01)$

Objective 2 Multiply signed decimal numbers.

Review this example for Objective 2:

2. Multiply $-2.6(-2.18)$

2.18	2 decimal places
$\times 2.6$	$\underline{+1}$ decimal place
1308	
+436	
5.668	3 decimal places

Since the two numbers have the same sign, the product is positive, so the answer is 5.668.

Practice this exercise:

2. $100(-0.16)$

Name: Date:
Instructor: Section:

Objective 3 Evaluate exponential forms with decimal bases.

Review this example for Objective 3:

3. Evaluate $(0.6)^2$.

Multiply the number of decimal places in the base by the exponent to get the number of decimal places in the product.

$$\begin{aligned}(0.6)^2 &= (0.6)(0.6)\\ &= .36\end{aligned}$$

Practice this exercise:

3. Evaluate $(-0.1)^3$.

Objective 4 Write a number in scientific notation in standard form.

Review this example for Objective 4:

4. Write 4×10^6 in standard form; then write the word name.

$$\begin{aligned}4\times10^6 &= 4\underbrace{0}_{1}\underbrace{0}_{2}\underbrace{0}_{3}\underbrace{0}_{4}\underbrace{0}_{5}\underbrace{0}_{6}\\ &= 4{,}000{,}000\end{aligned}$$

word name: 4 million

Practice this exercise:

4. Write 2.67×10^{11} in standard form; then write the word name.

Objective 5 Write standard form numbers in scientific notation.

Review this example for Objective 5:

5. Write 564,500,000 in scientific notation.

Move the decimal to express a number whose absolute value is greater than or equal to 1 but less than 10.

$$564{,}500{,}000 = 5.645\times10^8$$

There are 8 places between the new decimal position and the original position.

Practice this exercise:

5. Write 7,895,450,000 in scientific notation.

Name: Date:
Instructor: Section:

Objective 6 Solve applications.

Review this example for Objective 6:

6. Reginald is a real estate broker who helps people sell their homes. His fee is 0.09 times the price of the home. What is his fee for selling a $120,700 home?

Multiply the fee by the price of the home.

$$\text{Total fee} = 0.09 \times 120,700$$
$$= 10,863$$

Thus his fee is $10,683.

Practice this exercise:

6. What is the cost, in dollars, of 13.6 gallons of lead-free gasoline at 312.9 cents per gallon?

ADDITIONAL EXERCISES

Objective 1 Multiply decimal numbers.

For extra help, see Example 1 on pages 317–318 of your text and the Section 6.3 lecture video.

Multiply.

1. $(54.65)(78.49)$

2. $(1000)(5.1248)$

3. (3965.34)(.001)

4. $(12.68)(53.94)$

Objective 2 Multiply signed decimal numbers.

For extra help, see Example 2 on page 319 of your text and the Section 6.3 lecture video.

Multiply.

5. $42.9(-0.01)$

6. $(-0.416)(-0.9)$

7. $-7(5.29)$

8. $(-3.1415)(-1000)$

Name: Date:
Instructor: Section:

Objective 3 Evaluate exponential forms with decimal bases.
For extra help, see Example 3 on page 319 of your text and the Section 6.3 lecture video.
Evaluate.

9. $(1.5)^3$

10. $(-5.46)^2$

11. $(0.2)^4$

12. $(9.72)^2$

Objective 4 Write a number in scientific notation in standard form.
For extra help, see Example 4 on page 320 of your text and the Section 6.3 lecture video.
Write each number in standard form; then write the word name.

13. 5.897×10^7

14. 7.3814×10^4

Objective 5 Write standard form numbers in scientific notation.
For extra help, see Example 5 on page 321 of your text and the Section 6.3 lecture video.
Write each number in scientific notation.

15. 301,571,500,000

16. 3,600,000

Objective 6 Solve applications.
For extra help, see Example 6 on page 322 of your text and the Section 6.3 lecture video.
Solve.

17. Carol wants to build a square deck in her backyard. Find the area of the deck if the length of one side is 3.8 meters.

18. Find the area.

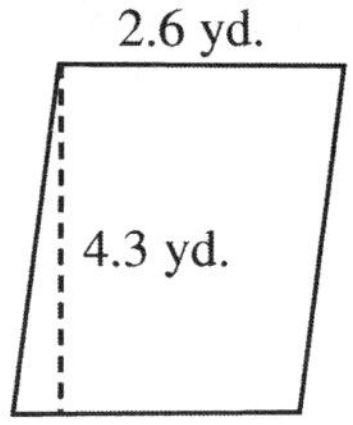

Name: Date:
Instructor: Section:

Chapter 6 DECIMALS

6.4 Dividing Decimal Numbers; Square Roots with Decimals

Learning Objectives
1 Divide decimal numbers.
2 Write fractions and mixed numbers as decimal numbers.
3 Evaluate square roots.
4 Solve equations using the multiplication principle.
5 Solve applications.

Key Terms
Use the vocabulary terms listed below to complete each statement in Exercises 1–4.

rational numbers	**real numbers**	**integers**
irrational numbers	**whole numbers**	**decimal numbers**

1. An irrational number cannot be expressed in the form $\frac{a}{b}$, where a and b are ____________________ and $b \neq 0$.

2. The ____________________ and the ____________________ together form the set of real numbers.

3. To divide a decimal number by a whole number, divide the divisor into the dividend as if both numbers were ____________________.

4. All non-terminating ____________________ with repeating digits can be expressed as fractions.

Name: Date:
Instructor: Section:

GUIDED EXAMPLES AND PRACTICE

Objective 1 Divide decimal numbers.

Review these examples for Objective 1:

1. Divide $38.4 \div 16$

$$\begin{array}{r} 2.4 \\ 16\overline{)38.4} \\ -32 \\ \hline 64 \\ -64 \\ \hline 0 \end{array}$$

2. Divide $-18.6 \div -0.6$

$0.6\overline{)18.6} = 6\overline{)186}$ Move the decimal 1 place to the right in both numbers.

$$\begin{array}{r} 31 \\ 6\overline{)186} \\ -18 \\ \hline 06 \\ -6 \\ \hline 0 \end{array}$$

Since both numbers share the same sign, the quotient is positive.

Practice these exercises:

1. Divide $0.45 \div 30$

2. Divide $0.01 \div 0.0005$

Name: Date:
Instructor: Section:

Objective 2 Write fractions and mixed numbers as decimal numbers.

Review this example for Objective 2:

3. Write $7\frac{3}{4}$ as a decimal number.

7 is written to the left of the decimal point. The decimal equivalent of $\frac{3}{4}$ is written to the right of the decimal.

$$\begin{array}{r} .75 \\ 4\overline{)3.00} \\ \underline{-28} \\ 20 \\ \underline{-20} \\ 0 \end{array}$$

So the answer is 7.75.

Practice this exercise:

3. Write $\frac{5}{6}$ as a decimal number.

Objective 3 Evaluate square roots.

Review this example for Objective 3:

4. Evaluate the square root $\sqrt{0.16}$. If the root is irrational, approximate the square root to the nearest hundredth.

16 is a perfect square, and 0.16 has an even number of decimal places; so its square root has half the number of decimal places and the same digits as the square root of 16, which is 4.

.4

Practice this exercise:

4. Evaluate the square root $\sqrt{350}$. If the root is irrational, approximate the square root to the nearest hundredth.

Name: Date:
Instructor: Section:

Objective 4 Solve equations using the multiplication principle.

Review this example for Objective 4:

5. Solve and check. $13.15x = 21.04$

$$\frac{13.15x}{13.15} = \frac{21.04}{13.15} \quad \text{Divide both sides by 13.15}$$

$$1x = 1.6$$

$$x = 1.6$$

Check

$$13.15x = 21.04$$

$$13.15(1.6) \stackrel{?}{=} 21.04 \quad \text{Replace } x \text{ with 1.6.}$$

$$21.04 = 21.04 \quad \text{True; so 1.6 is the solution.}$$

Practice this exercise:

5. Solve and check.
$-0.75b = 6.18$

Objective 5 Solve applications.

Review this example for Objective 5:

6. A car loan of $13,661.28 is to be paid off in 48 monthly payments. How much is each payment?

To calculate the amount of each payment (P), divide $13,661.28 by 48

$$P = 13,661.28 \div 48$$

$$P = 284.61$$

Each payment is $284.61

Practice this exercise:

6. To protect an elm tree in your backyard, you need gypsy moth caterpillar tape around the trunk. The tree has a 1.1-foot diameter. What length of tape is needed? Use 3.14 to approximate π.

ADDITIONAL EXERCISES

Objective 1 Divide decimal numbers.

For extra help, see Examples 1–3 on pages 327–328 of your text and the Section 6.4 lecture video.

Divide.

1. $6.936 \div 17$

2. $7560 \div 3.6$

3. $18,468 \div 5.7$

4. $-45.05 \div -0.05$

5. $1 \div 0.0004$

6. $125.5228 \div (-0.202)$

Objective 2 Write fractions and mixed numbers as decimal numbers.
For extra help, see Examples 4–5 on page 329 of your text and the Section 6.4 lecture video.
Write as a decimal number.

7. $14\frac{1}{2}$

8. $\frac{2}{3}$

9. $-5\frac{2}{5}$

10. $-4\frac{5}{6}$

Objective 3 Evaluate square roots.
For extra help, see Examples 6–7 on pages 330–331 of your text and the Section 6.4 lecture video.
Evaluate the square root. If the root is irrational, approximate the square root to the nearest hundredth.

11. $\sqrt{0.0625}$

12. $\sqrt{51}$

Objective 4 Solve equations using the multiplication principle.
For extra help, see Example 8 on page 332 of your text and the Section 6.4 lecture video.
Solve and check.

13. $0.9x = 6.3$

14. $6 = 7.5y$

15. $8.7z = -59.595$

16. $-65.038 = -12.4t$

Name: Date:
Instructor: Section:

Objective 5 Solve applications.

For extra help, see Examples 9–10 on pages 332–333 of your text and the Section 6.4 lecture video.

Solve.

17. A car loan of \$16,257.96 is to be paid off in 36 monthly payments. How much is each payment?

18. A tree has a circumference of 46.5 inches. Find the diameter of the tree.

Name: Date:
Instructor: Section:

Chapter 6 DECIMALS

6.5 Order of Operations and Applications in Geometry

Learning Objectives
1 Simplify numerical expressions using the order of operations agreement.
2 Simplify expressions containing fractions and decimals.
3 Solve application problems requiring more than one operation.
4 Find weighted means.
5 Evaluate expressions.
6 Find the area of a triangle, trapezoid, and circle.
7 Find the volume of a cylinder, pyramid, cone, and sphere.
8 Find the area and volume of a composite form.

GUIDED EXAMPLES AND PRACTICE

Objective 1 Simplify numerical expressions using the order of operations agreement.

Review this example for Objective 1:

1. Simplify $18\times(47.3+92.2)$

$18\times(47.3+92.2)=18\times(139.5)$ Add in the parentheses.

$=2511$ Multiply.

Practice this exercise:

1. Simplify $0.25+5.8(0.3)$

Objective 2 Simplify expressions containing fractions and decimals.

Review this example for Objective 2:

2. Simplify $\frac{1}{2}(-0.66)$

$\frac{1}{2}(-0.66)=\frac{1}{2}\cdot\frac{-66}{100}$ Write -0.66 as $\frac{-66}{100}$.

$=\frac{1}{\cancel{2}_1}\cdot\frac{\cancel{-66}^{-33}}{100}$ Divide out common factors.

$=\frac{-33}{100}$, or -0.33 Multiply.

Practice this exercise:

2. Simplify $\frac{1}{12}+\frac{3}{4}(0.6)$

Name: Date:
Instructor: Section:

Objective 3 Solve application problems requiring more than one operation.

Review this example for Objective 3:

3. The Perla family plans a vacation using a travel agent. The components of their trip planned by the travel agent are listed in the table. If they want to make 5 equal payments to cover the costs for these plans, how much will each payment be?

Description	**Amount**
Flights	\$1444.61
Hotel	\$1323.72
Rental car	\$347.78
Theme park tickets	\$1932.89

Add all the amounts together and divide by 5 to determine how much each payment will be.

$$P = (1444.61 + 1323.72 + 347.78 + 1932.89) \div 5$$
$$P = 5049 \div 5$$
$$P = 1009.80$$

Each payment will be \$1009.80.

Practice this exercise:

3. Jonathan is traveling for business using his own car. His employer will reimburse him \$0.41 per mile driven and cover all costs for meals and hotel. If he drives 372 miles and presents receipts for \$72.19 for meals and \$355.62 for hotel charges, how much will his employer reimburse him?

Objective 4 Find weighted means.

Review this example for Objective 4:

4. Calculate the GPA for the student whose grade reports are given.

Course	**Credits**	**Grade**
MATH 203	4.0	A
ENG 211	4.0	B
ECO 214	3.0	B+
MUS 216	2.0	B+

The following list shows commonly used grade point equivalents for each letter grade.

Practice this exercise:

4. Calculate the GPA for the student whose grade reports are given.

Course	**Credits**	**Grade**
FRE 300	4.0	C
MATH 305	4.0	B
PHY 250	4.0	B
ENG 300	3.0	B

$$
\begin{aligned}
A &= 4 \\
B+ &= 3.5 \\
B &= 3 \\
C+ &= 2.5 \\
C &= 2 \\
D+ &= 1.5 \\
D &= 1 \\
F &= 0
\end{aligned}
$$

MATH 203 $4(4.0) = 16$
ENG 211 $4(3.0) = 12$
ECO 214 $3(3.5) = 10.5$
MUS 216 $2(3.5) = 7$

$4 + 4 + 3 + 2 = 13$

$$
\begin{aligned}
\text{GPA} &= \frac{\text{total grade points}}{\text{total credits}} \\
&= \frac{16 + 12 + 10.5 + 7}{13} \\
&= \frac{45.5}{13} \\
&= 3.5
\end{aligned}
$$

Name: Date:
Instructor: Section:

Objective 5 Evaluate expressions.

Review this example for Objective 5:

5. Evaluate the expression using the given values. $\frac{1}{2}mv^2$; $m = 4.25$, $v = -12$

$\frac{1}{2}(4.25)(-12)^2$ Replace m with 4.25, v with -12.

$0.5(4.25)(-12)^2$ Write $\frac{1}{2}$ as 0.5.

0.5(4.25)(144) Evalute exponential.

306 Multiply each number together.

Practice this exercise:

5. Evaluate the expression using the given values.
mc^2; $m = 4.7\times10^4$, $c = 3\times10^8$

Objective 6 Find the area of a triangle, trapezoid, and circle.

Review these examples for Objective 6:

6. Find the area of the triangle shown.

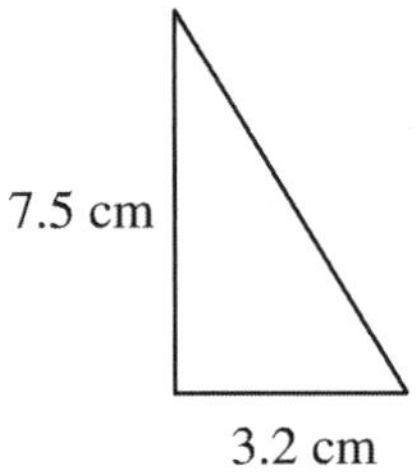

$A = \frac{1}{2}bh$

$A = \frac{1}{2}(3.2)(7.5)$ Replace b with 3.2 and h with 7.5

$A = \frac{1}{2}(24)$ Multiply 3.2 and 7.5

$A = \frac{24}{2}$ Multiply.

$A = 12$ Divide 24 by 2.

Therefore, the area is 12 cm.

Practice these exercises:

6. Find the area of the triangle shown.

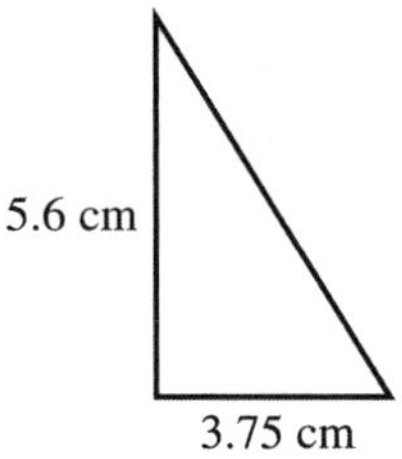

Name: Date:
Instructor: Section:

7. Find the area of the trapezoid shown.

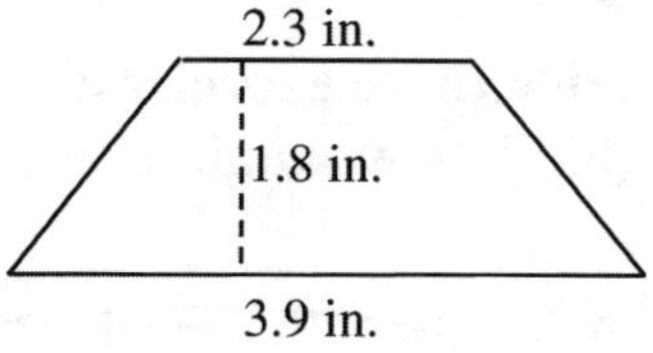

$A = \frac{1}{2}h(a+b)$

$A = \frac{1}{2}(1.8)(2.3+3.9)$ Replace h, a, and b.

$A = \frac{1}{2}(1.8)(6.2)$ Add 2.3 and 3.9.

$A = \frac{1}{2}(11.16)$ Multiply 1.8 and 6.2.

$A = \frac{11.16}{2}$ Multiply.

$A = 5.58$ Divide 11.16 by 2.

Therefore, the area is 5.58 in.

8. Find the area of the circle shown. Use 3.14 for π.

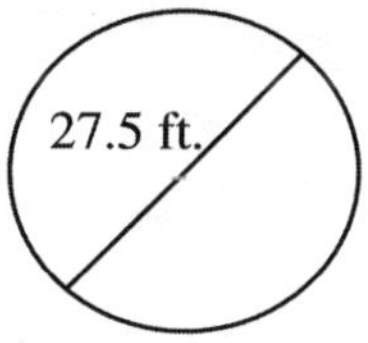

$r = \frac{27.5}{2}$

$r = 13.75$

$A = \pi r^2$

$A \approx (3.14)(13.75)^2$ Replace r with 13.75.

$A \approx (3.14)(189.0625)$ Square 13.75.

$A \approx (593.65625)$ Multiply 13.75 and 3.14.

Therefore, the area is approximately 593. 66 ft^2.

7. Find the area of the trapezoid shown.

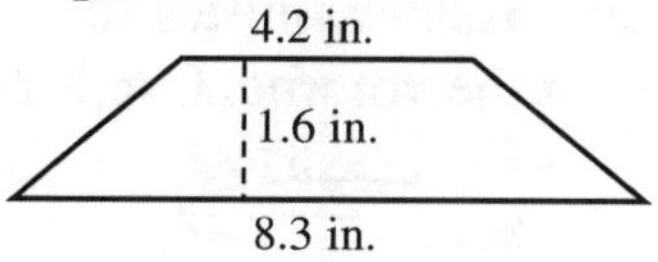

8. Find the area of the circle shown. Use 3.14 for π

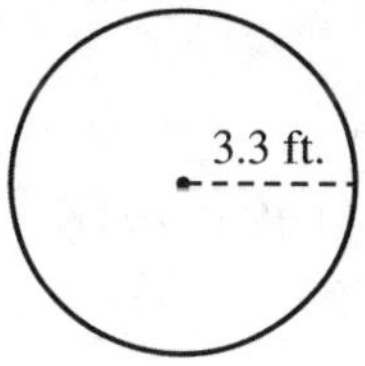

Name: Date:
Instructor: Section:

Objective 7 Find the volume of a cylinder, pyramid, cone, and sphere.

Review these examples for Objective 7:

9. Find the volume. Use 3.14 for π.

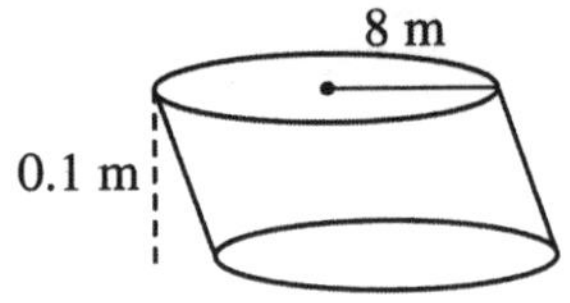

$V = \pi r^2 h$

$V \approx (3.14)(8)^2(0.1)$ Replace the values known.

$V \approx (3.14)(64)(0.1)$ Square 8.

$V \approx (3.14)(6.4)$ Multiply 64 and 0.1.

$V \approx 20.096$ Multiply 3.14 and 6.4.

Therefore, the volume is 20.096 m^3.

10. One of the ancient stone pyramids in Egypt has a square base that measures 149 meters on each side. The height is 99 meters. What is the volume of the pyramid?

$V = \frac{1}{3}lwh$

$V = \frac{1}{3}(149)(149)(99)$ Replace the values known.

$V = \frac{1}{3}(22,201)(99)$ Multiply 149 and 149.

$V = \frac{1}{3}(2,197,899)$ Multiply 22,201 and 99.

$V = \frac{2,197,899}{3}$ Divide by 3.

$V = 732,633$

Therefore, the volume is 732,633 m^3.

Practice these exercises:

9. Find the volume. Use 3.14 for π.

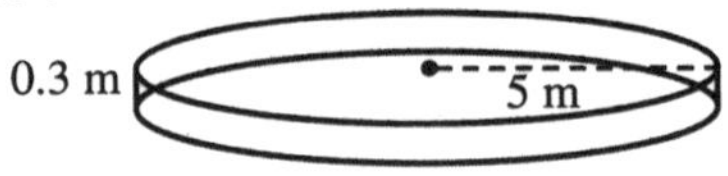

10. Find the volume.

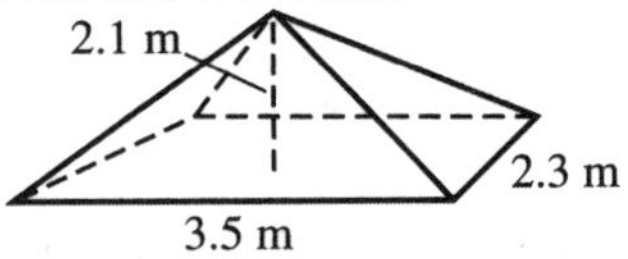

Name: Date:
Instructor: Section:

11. Find the volume. Use 3.14 for π.

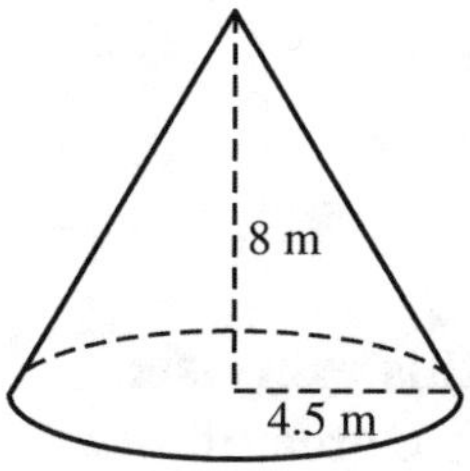

$$V = \frac{1}{3}\pi r^2 h$$

$V \approx \frac{1}{3}(3.14)(4.5)^2(8)$ Replace with values known.

$V \approx \frac{1}{3}(3.14)(20.25)(8)$ Square 4.5.

$V \approx \frac{1}{3}(3.14)(162)$ Multiply 20.25 and 8.

$V \approx (3.14)\left(\frac{162}{3}\right)$ Divide 162 and 3.

$V \approx (3.14)(54)$ Multiply 3.14 and 54.

$V \approx 169.56$

Therefore, the volume is approximately 169.56 m^3.

12. Find the volume. Use 3.14 for π.

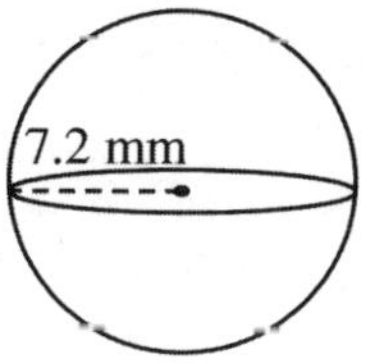

$$V = \frac{4}{3}\pi r^3$$

$V \approx \frac{4}{3}(3.14)(7.2)^3$ Replace the values known.

$V \approx \frac{4}{3}(3.14)(373.248)$ Cube 7.2.

$V \approx (3.14)\left(\frac{1492.992}{3}\right)$ Multiply 373.248 and 4.

$V \approx (3.14)(497.664)$ Divide 1492.992 by 3.

$V \approx 1562.66$ Multiply 3.14 and 497.664.

11. Find the volume. Use 3.14 for π.

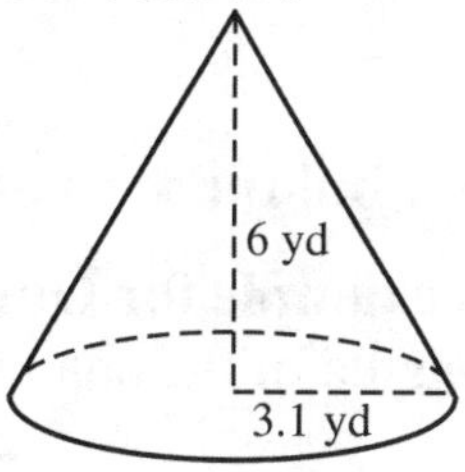

12. Find the volume. Use 3.14 for π.

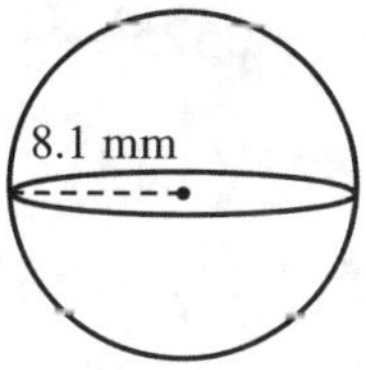

Name: Date:
Instructor: Section:

Therefore, the volume is approximately 1562.66 mm^3.

Objective 8 Find the area and volume of a composite form.

Review this example for Objective 8:

13. Find the area of the shaded region. Use 3.14 for π.

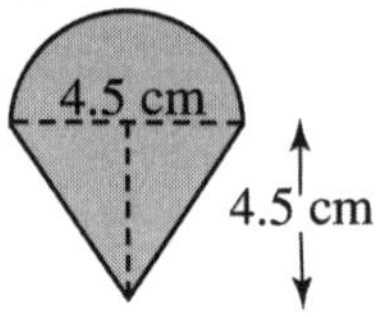

The figure is a half circle and a triangle, so add their areas together. Note that the radius is half of 4.5, which is 2.25. Create the area formula, replace with the values known, then simplify.

$$A = \text{area of half circle} + \text{area of triangle}.$$

$$A = \frac{1}{2}\pi r^2 + \frac{1}{2}bh$$

$$A \approx \frac{1}{2}(3.14)(2.25)^2 + \frac{1}{2}(4.5)(4.5)$$

$$A \approx \frac{1}{2}(3.14)(5.0625) + \frac{1}{2}(4.5)(4.5)$$

$$A \approx 0.5(5.0625)(3.14) + 0.5(4.5)(4.5)$$

$$A \approx 2.53125(3.14) + 10.125$$

$$A \approx 7.948125 + 10.125$$

$$A \approx 18.073125$$

The area is about 18.07 cm^2.

Practice this exercise:

13. Find the area of the shaded region. Use 3.14 for π.

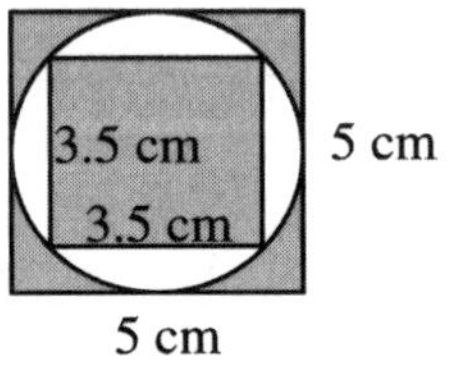

ADDITIONAL EXERCISES

Objective 1 Simplify numerical expressions using the order of operations agreement.

For extra help, see Example 1 on page 337 of your text and the Section 6.5 lecture video.

Simplify.

1. $(0.5)^2 - 5.4 \div 0.1(6.3)$

2. $\sqrt{0.0009} + 116.4 \div 4(2.5 - 5.8)$

Name: Date:
Instructor: Section:

3. $8.1+3.3\sqrt{0.16}-41.2\div 2$

4. $\left[-6.27\div(2.93+.07)\right]+0.4\sqrt{1.44}$

Objective 2 Simplify expressions containing fractions and decimals.
For extra help, see Example 2 on page 338 of your text and the Section 6.5 lecture video.
Simplify.

5. $1\frac{1}{4}(2.5)-\frac{1}{8}$

6. $\frac{1}{6}\div(-0.5)+\left(\frac{1}{5}\right)^2$

7. $\left(2\frac{3}{4}\right)^2(4.2)-5.5625$

8. $-0.7\div\frac{2}{5}-4\frac{1}{5}$

Objective 3 Solve application problems requiring more than one operation.
For extra help, see Example 3 on pages 339–340 of your text and the Section 6.5 lecture video.
Solve.

9. A student has gotten the following scores on the past 5 tests, 90, 85, 95, 73, 86. What is the mean of these scores?

10. An employee is traveling for business using his own car. His employer will reimburse him \$0.48 per mile driven and cover all costs for meals and hotel. If he drives 462 miles and presents receipts for \$82.90 for meals and \$471.29 for hotel charges, how much will his employer reimburse him?

Objective 4 Find weighted means.
For extra help, see Example 4 on pages 340–341 of your text and the Section 6.5 lecture video.
Calculate the GPA for the students whose grade reports are given.

11.

Course	Credits	Grade
Math 100	4	C+
ENG 101	4	A
PSY 101	3	B
BIO 101	3	B+

12.

Course	Credits	Grade
Math 400	5	B+
SPA 401	4	A
PSY 405	3	C
MUS 425	3	C+

Name: Date:
Instructor: Section:

Objective 5 Evaluate expressions.

For extra help, see Example 5 on page 341 of your text and the Section 6.5 lecture video.

Evaluate each expression using the given values.

13. $\frac{1}{2}mv^2$; $m = 3.75$, $v = -16$

14. mc^2; $m = 2.4\times10^5$, $c = 3\times10^6$

Objective 6 Find the area of a triangle, trapezoid, and circle.

For extra help, see Example 6 on pages 342–343 of your text and the Section 6.5 lecture video.

Find each area. Use 3.14 for π.

15.

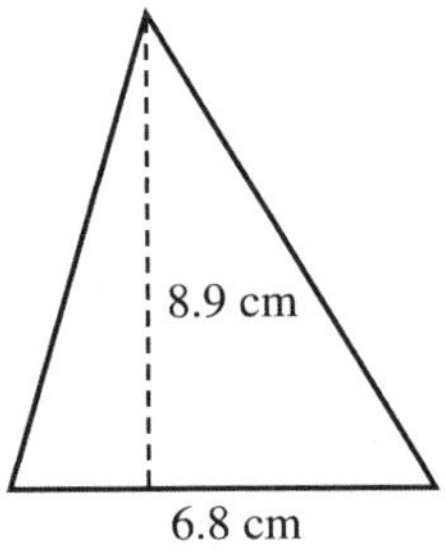

16.

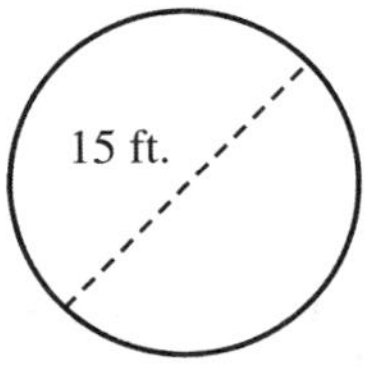

17.

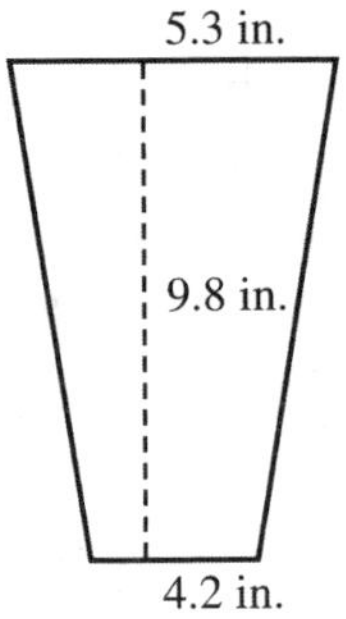

18.

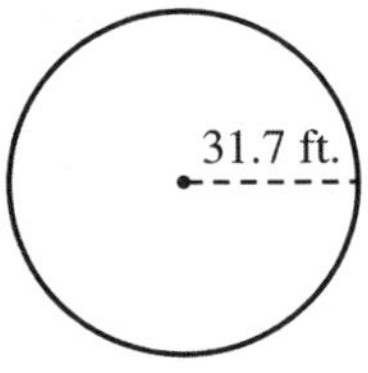

Name: Date:
Instructor: Section:

Objective 7 Find the volume of a cylinder, pyramid, cone, and sphere.
For extra help, see Examples 7–10 on pages 343–345 of your text and the Section 6.5 lecture video.
Find each volume. Use 3.14 for π.

19.

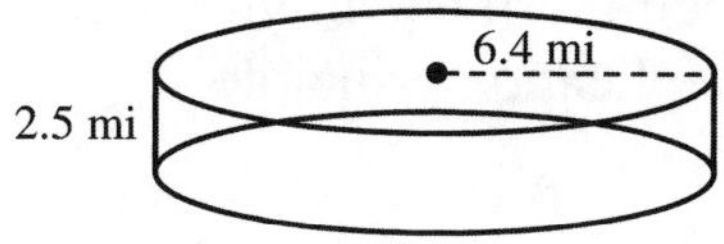

20.

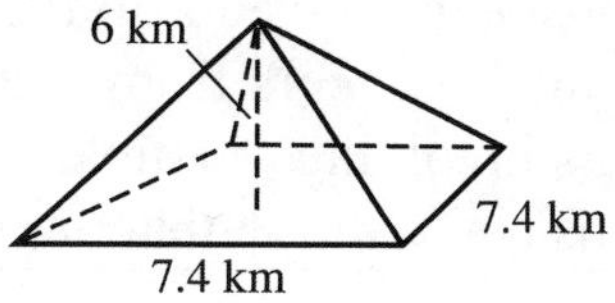

21.

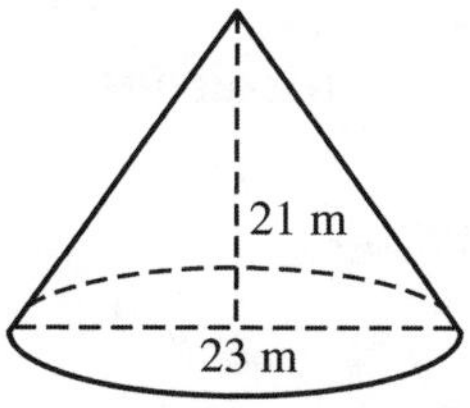

22.

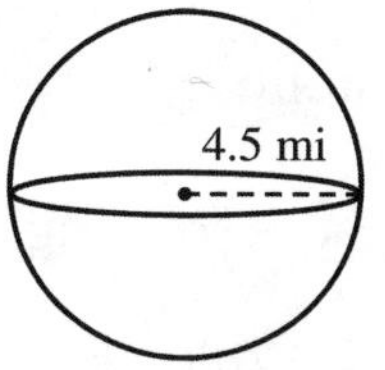

Objective 8 Find the area and volume of a composite form.
For extra help, see Examples 11–12 on pages 346–347 of your text and the Section 6.5 lecture video.
Find the area or volume of the region. Use 3.14 for π.

23.

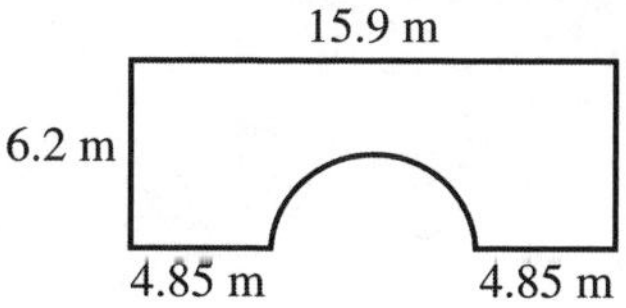

24.

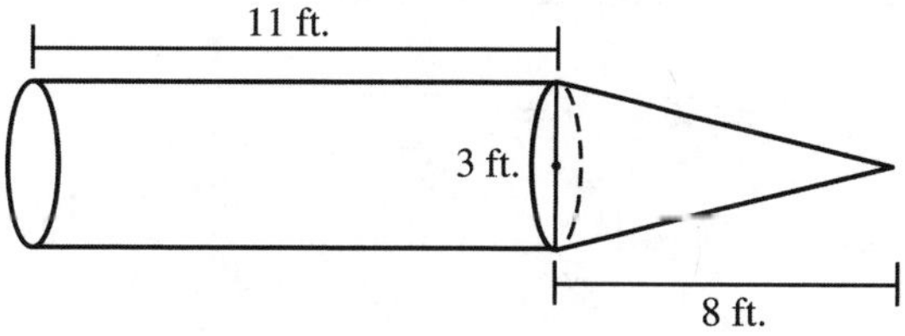

Name: Date:
Instructor: Section:

Chapter 6 DECIMALS

6.6 Solving Equations and Problem Solving

Learning Objectives

1. Solve equations using the addition and multiplication principles of equality.
2. Eliminate decimal numbers from equations using the multiplication principle.
3. Solve problems involving one unknown.
4. Solve problems using the Pythagorean theorem.

Key Terms

Use the terms listed below to complete each statement in Exercises 1–2.

right **hypotenuse** **sum** **leg** **difference**

1. The __________________ of a right triangle is opposite the 90° angle.

2. The Pythagorean theorem states that given a right triangle, the __________________ of the squares of the lengths of the legs equals the square of the hypotenuse.

GUIDED EXAMPLES AND PRACTICE

Objective 1 Solve equations using the addition and multiplication principles of equality.

Review this example for Objective 1:

1. Solve and check. $4.6a + 4.17 = -27.57$

$4.6a + 4.17 = -27.57$

$\underline{-4.17} \quad \underline{-4.17}$ Subtract 4.17 from both sides.

$\dfrac{4.6a + 0}{4.6} = \dfrac{-31.74}{4.6}$ Divide by 4.6

$a = -6.9$

Check

$4.6(-6.9) + 4.17 \stackrel{?}{=} -27.57$

$-31.74 + 4.17 \stackrel{?}{=} -27.57$

$-27.57 = -27.57$

True, so -6.9 is the solution.

Practice this exercise:

1. Solve and check.
$2.9y - 34 = 3.1y - 34.2$

Name: Date:
Instructor: Section:

Objective 2 Eliminate decimal numbers from equations using the multiplication principle.

Review this example for Objective 2:

2. Solve and check. $2.5(x+8)=14.4+1.8x$

$$2.5(x+8)=14.4+1.8x$$

$2.5x+20=14.4+1.8x$ Distribute.

$10\cdot 2.5x+10\cdot 20=10\cdot 14.4+10\cdot 1.8x$ Multiply by 10.

$25x+200=144+18x$ Combine like terms.

$\underline{-18x} \qquad \underline{-18x}$ Subtract $18x$.

$7x+200=144+0$

$\underline{-200} \quad \underline{-200}$ Subtract 200.

$\dfrac{7x+0}{7}=\dfrac{-56}{7}$ Divide by 7.

$x=-8$

Check

$2.5(-8+8)\overset{?}{=}14.4+1.8(-8)$

$2.5\cdot 0\overset{?}{=}14.4-14.4$

$0=0$

True, so -8 is the solution.

Practice this exercise:

2. Solve and check.

$1.5(x+4)=5.7+0.6x$

Objective 3 Solve problems involving one unknown.

Review this example for Objective 3:

3. Solve. 1.68 plus 7.7 times p is equal to 8.9 times p.

Practice this exercise:

3. Solve. 4 times h decreased by 8.125 is the product of h and -2.5

Name: Date:
Instructor: Section:

$\underline{\text{1.68 plus 7.7 times } p}$ $\underline{\text{is equal to}}$ $\underline{\text{8.9 times } p}$

$$\overbrace{1.68+7.7p} \quad \overbrace{=} \quad \overbrace{8.9p}$$

$100\cdot1.68+100\cdot7.7p=100\cdot8.9p$ Multiply by 100.

$$\begin{aligned} 168+\ 770p &= 890p \\ -770p &\quad -770p \end{aligned}$$ Subtract 770*p*.

$$\frac{168+0}{120}=\frac{120p}{120}$$ Divide by 120.

$$1.4=p$$

Objective 4 Solve problems using the Pythagorean theorem.

Review this example for Objective 4:

4. Find the length of the missing side.

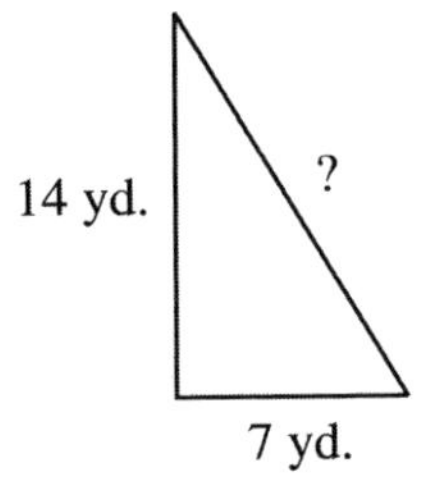

$a^2+b^2=c^2$

$(7)^2+(14)^2=c^2$ Replace with values known.

$49+196=c^2$ Square 7 and 14.

$245=c^2$ Add 49 and 196.

$\sqrt{245}=c$ Take the square root of both sides.

$15.7\approx c$

The length of the missing side is about 15.7 yd.

Practice this exercise:

4. A slow-pitch softball diamond is actually a square 66 feet on a side. How far is it from home to second base?

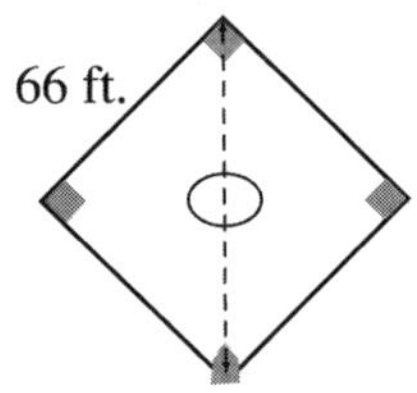

Name: Date:
Instructor: Section:

ADDITIONAL EXERCISES

Objective 1 Solve equations using the addition and multiplication principles of equality.

For extra help, see Example 1 on page 355 of your text and the Section 6.6 lecture video.
Solve and check.

1. $6.1y-8=-5.2+8.9y$

2. $4.2x-0.44=0.84-2.2x$

3. $5.3x-10.5=7.95-6.7x$

4. $10.5y-94.645=-64.825+3.4y$

Objective 2 Eliminate decimal numbers from equations using the multiplication principle.

For extra help, see Examples 2–3 on pages 356–357 of your text and the Section 6.6 lecture video.
Solve and check.

5. $3m-1.49=1.4(m-1)$

6. $7(2.75-x)-5.4x=8.9-(4+6.1x)$

Objective 3 Solve problems involving one unknown.

For extra help, see Examples 4–5 on pages 357–358 of your text and the Section 6.6 lecture video.
Solve.

7. 64.6 less than 8 times n is equal to 200 plus 1.7 times n.

8. Tommy uses a cell phone service that charges \$59.99 per month for 550 minutes. After 550 minutes, it costs \$0.55 for each additional minute. If Tommy's bill is \$117.74, how many total minutes did he use?

Name: Date:
Instructor: Section:

Objective 4 Solve problems using the Pythagorean theorem.
For extra help, see Example 6 on pages 360–361 of your text and the Section 6.6 lecture video.
Solve.

9. Holly's new cordless telephone has clear reception up to 190 feet from its base. Her phone base is located near a window in her apartment, 114 feet above street level. How far into the backyard can Holly use her phone?

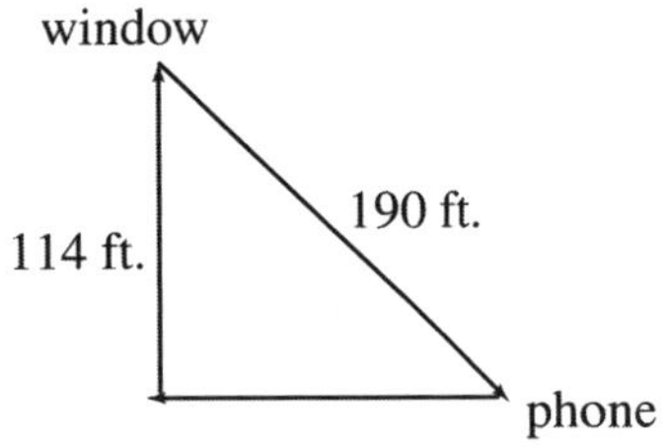

10. A baseball diamond is actually a square 27.4 meters on a side. How far is it from home to second base?

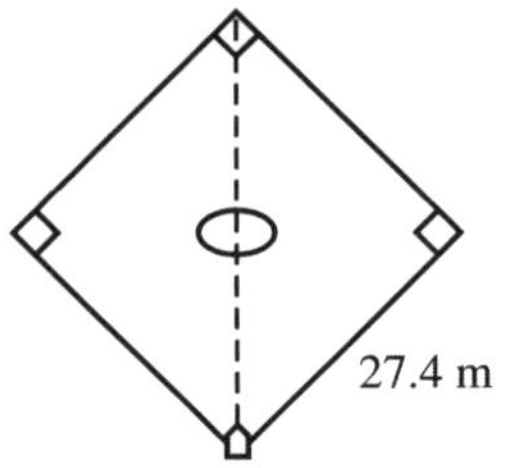

Name: Date:
Instructor: Section:

Chapter 6 DECIMALS

6.7 Points, Lines, and Angles

Learning Objectives
1 Identify points, lines, line segments, rays, planes, and parallel lines.
2 Identify angles.
3 Solve problems involving angles formed by intersecting lines.
4 Solve problems involving angles in a triangle.
5 Solve problems involving congruent triangles.

Key Terms
Use the terms listed below to complete each statement in Exercises 1–8.

supplementary	**parallel**	**segment**	**point**	**exterior**	**plane**
perpendicular	**acute**	**right**	**obtuse**	**line**	**ray**
complementary	**vertical**	**angle**	**intersect**	**congruent**	
corresponding	**alternate**	**interior**	**straight**	**consecutive**	

1. Lines that intersect to form a 90° angle are ____________________ lines.

2. A(n) ____________________ triangle has three angles that measure less than 90°.

3. A straight, one-dimensional figure extending forever in one direction from a single point is called a(n) ____________________.

4. ____________________ lines are in the same plane and never intersect.

5. A(n) ____________________ angle has measure 180°.

6. The sum of the measures of two ____________________ angles is 90°.

7. Vertical angles are ____________________ angles formed by the intersection of two lines.

8. In a figure formed by a transversal intersection two lines, a pair of angles between the lines on the same side of the transversal are called consecutive ____________________ angles.

Name: Date:
Instructor: Section:

GUIDED EXAMPLES AND PRACTICE

Objective 1 Identify points, lines, line segments, rays, planes, and parallel lines.

Review this example for Objective 1:

1. Identify the figure.

The two points, M and N, fix the line in place and are used to name the line.

line $\overleftrightarrow{MN}$

Practice this exercise:

1. Identify the figure.

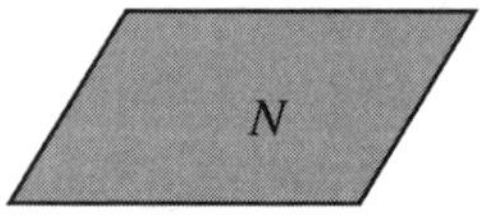

Objective 2 Identify angles.

Review this example for Objective 2:

2. State whether each angle is acute, right, obtuse, or straight.

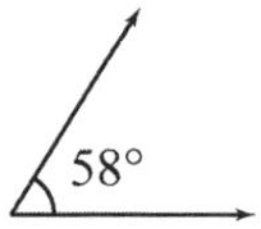

The angle is acute because its measure is between 0° and 90°.

Practice this exercise:

2. State whether each angle is acute, right, obtuse, or straight.

Objective 3 Solve problems involving angles formed by intersecting lines.

For Exercise 3, use the following figure. Lines l and m are parallel.

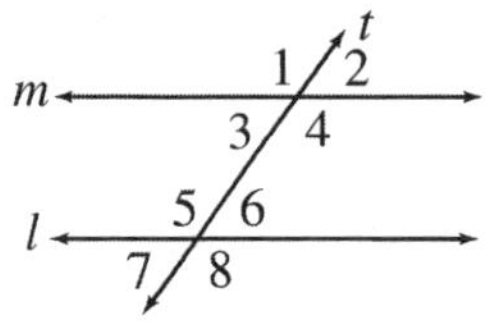

Review these examples for Objective 3:

3. Are $\angle 1$ and $\angle 3$ congruent or supplementary? Explain why.

Supplementary; because they form a straight angle.

Practice these exercises:

3. Are $\angle 2$ and $\angle 3$ congruent or supplementary? Explain why.

Name: Date:
Instructor: Section:

4. Four angles are formed by the intersecting lines. Find the measure of each angle.

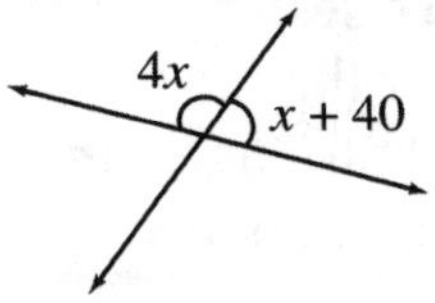

Because the two angles with variable expressions form a straight line, they are supplementary; so the sum of the measure is 180°.

$$4x+(x+40)=180 \quad \text{Translate to an equation.}$$
$$5x+40=180 \quad \text{Combine like terms.}$$
$$\underline{-40} \quad \underline{-40} \quad \text{Subtract 40.}$$
$$5x+0=140$$
$$5x=140$$
$$\frac{5x}{5}=\frac{140}{5} \quad \text{Divide by 5.}$$
$$x=28$$

The first angle measures $4x=4\cdot 28=112^\circ$. The second angle measures $x+40=28+40=68^\circ$. Because the two intersecting lines form two pairs of vertical angles, the measures of the other angles are also 112° and 68°.

Check.
The sum of the measures of supplementary angles is $112^\circ+68^\circ=180^\circ$

4. Four angles are formed by the intersecting lines. Find the measure of each angle.

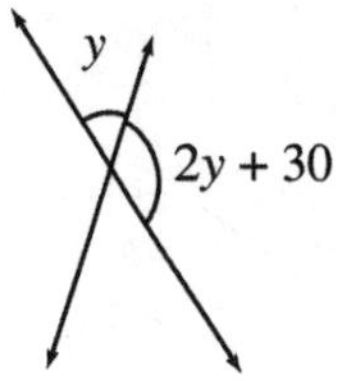

Name: Date:
Instructor: Section:

Objective 4 Solve problems involving angles in a triangle.

Review these examples for Objective 4:

5. Indicate whether the triangle is a right triangle, an acute triangle, or an obtuse triangle.

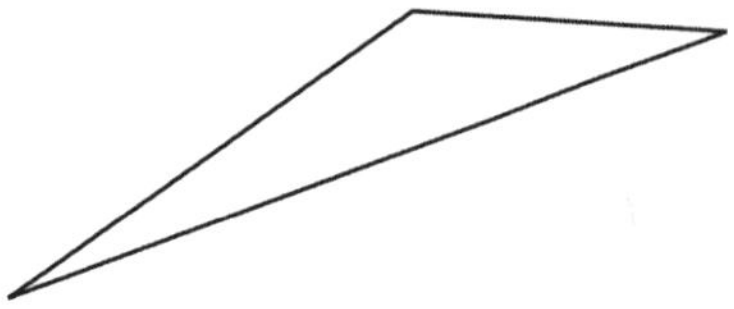

Obtuse triangle; the measure of one angle is greater than 90°.

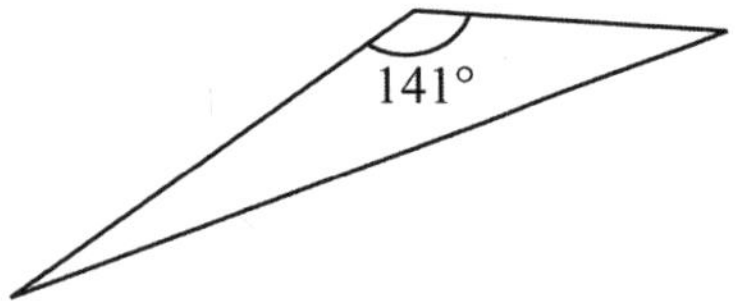

6. Find the measure of each angle in the following triangle.

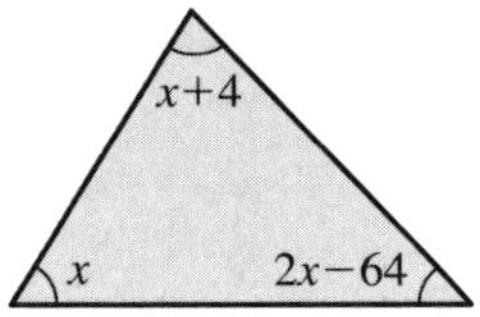

The angles of a triangle add up to 180°.

$$x+(x+4)+(2x-64)=180$$
$$4x-60=180$$
$$4x=240$$
$$x=60^\circ$$

Substitute $x=60$ to find the other two angles.

$x+4$, $60+4$, 64° and $2x-64$, $2(60)-64$, $120-64$, 56°

The three angles are 60°, 64°, and 56°

Practice these exercises:

5. Indicate whether the triangle is a right triangle, an acute triangle, or an obtuse triangle.

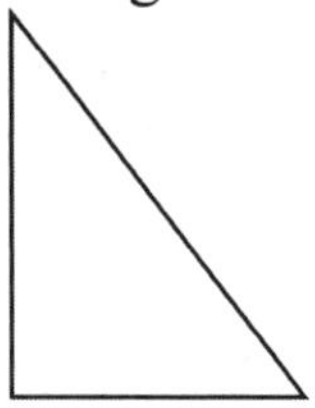

6. Find the measure of each angle in the following triangle.

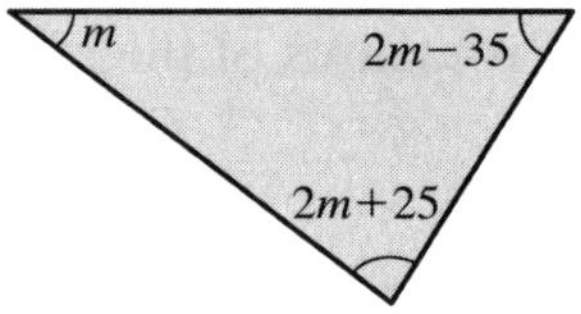

Name: Date:
Instructor: Section:

Objective 5 Solve problems involving congruent triangles.

Review these examples for Objective 5:

7. Determine whether the triangles are congruent. If they are congruent, state the rule that explains why.

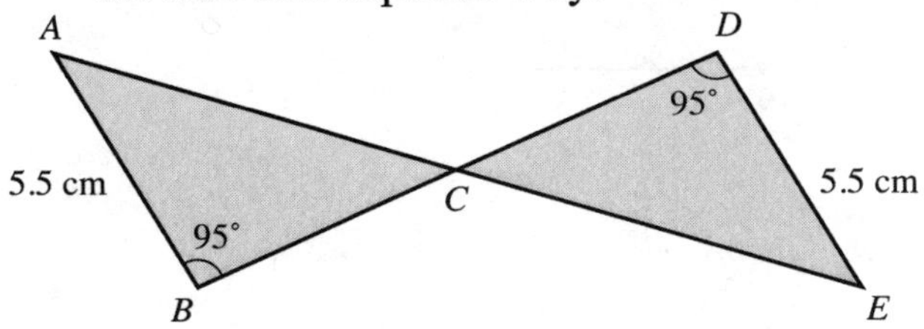

Sides $\overline{AB}$ and $\overline{DE}$ are congruent between the two triangles.
$\angle B$ and $\angle D$ are congruent.
The vertical angles at C are congruent,
The triangles are congruent by the AAS rule.

8. The following triangles are congruent. Find the measure of the angles.

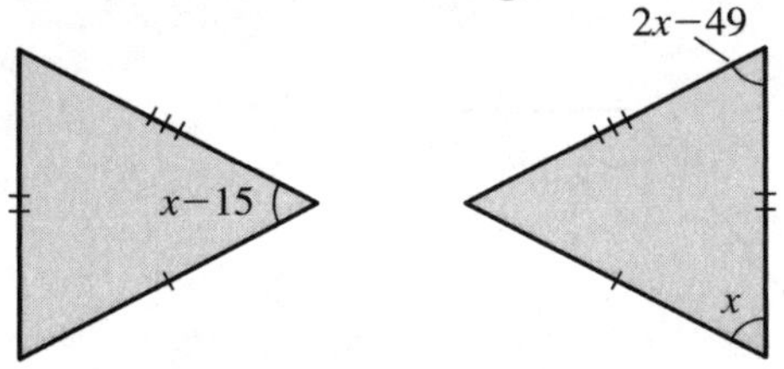

The given angle in the left triangle corresponds to the missing angle in the right triangle. The missing angle in the right triangle is $x-15$.

The angles of a triangle add up to 180°.

$$x+(x\quad 15)+(2x-49)=180$$
$$4x-64=180$$
$$4x=244$$
$$x=61^\circ$$

Substitute $x=61$ to find the other two angles.

$x-15$
$61-15$ and
46°

$2x-49$
$2(61)-49$
$122-49$
73°

The three angles are 61°, 46°, and 73°.

Practice these exercises:

7. Determine whether the triangles are congruent. If they are congruent, state the rule that explains why.

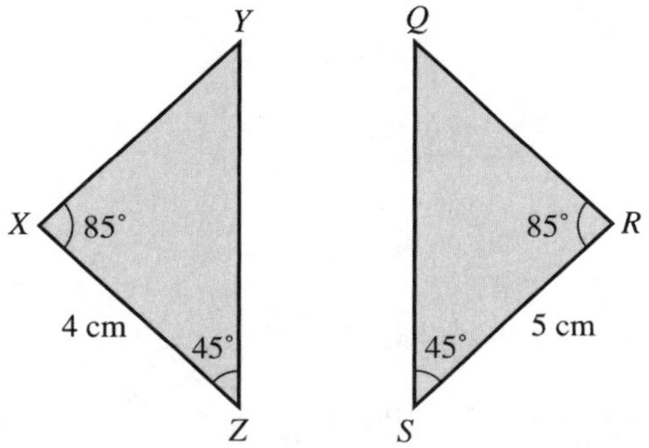

8. The following triangles are congruent. Find the measure of the angles.

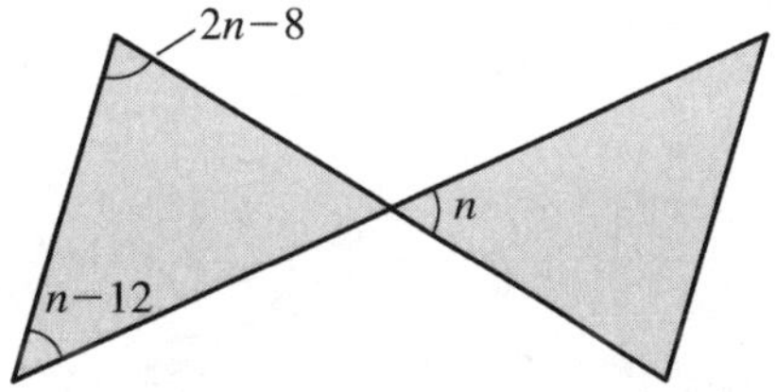

Name: Date:
Instructor: Section:

ADDITIONAL EXERCISES

Objective 1 Identify points, lines, line segments, rays, planes, and parallel lines.

For extra help, see Example 1 on page 365 of your text and the Section 6.7 lecture video.

Identify each figure.

1.

2.

3.

4.

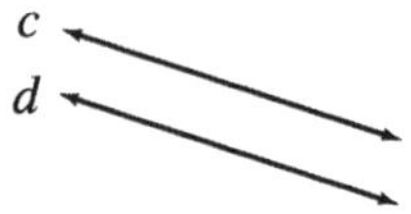

Objective 2 Identify angles.

For extra help, see Examples 2-4 on pages 366–370 of your text and the Section 6.7 lecture video.

State whether each angle is acute, right, obtuse, or straight.

5.

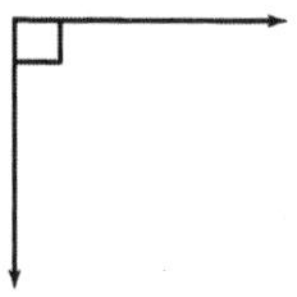

6.

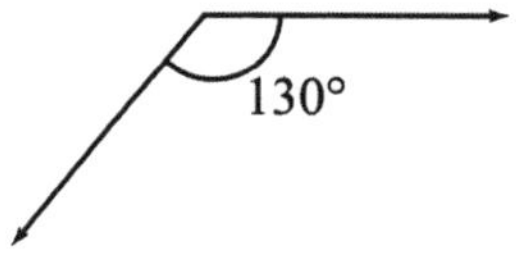

Objective 3 Solve problems involving angles formed by intersecting lines.

For extra help, see Examples 5–6 on pages 370–372 of your text and the Section 6.7 lecture video.

For exercises 7–10, use the following figure. Lines l and m are parallel.

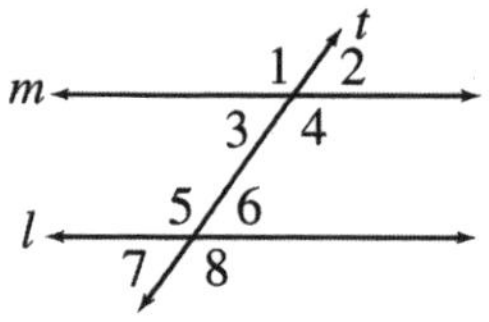

7. Are $\angle 4$ and $\angle 5$ congruent or supplementary? Explain why.

8. Are $\angle 4$ and $\angle 8$ congruent or supplementary? Explain why.

Name: Date:
Instructor: Section:

9. If $\angle 1 = 116^\circ$, what is the measure of $\angle 8$?

10. If $\angle 6 = 52^\circ$, what is the measure of $\angle 2$?

11. If $\angle 3 = 55^\circ$, what is the measure of $\angle 6$?

12. If $\angle 5 = 121^\circ$, what is the measure of $\angle 1$?

13. Lines h and k are parallel and the transversal t forms eight angles. Find the measure of each angle.

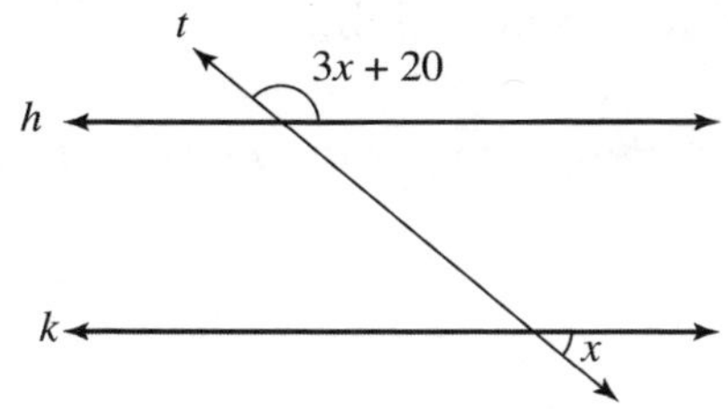

14. Lines l and m are parallel and the transversal t forms eight angles. Find the measure of each angle.

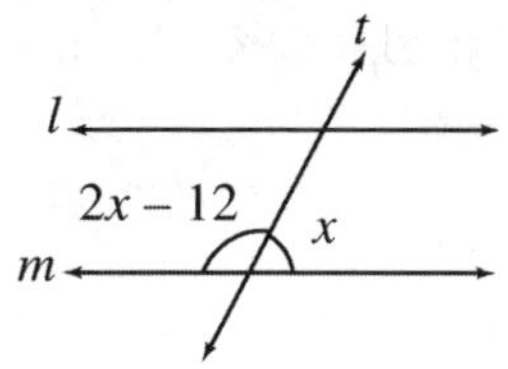

Objective 4 Solve problems involving angles in a triangle.
For extra help, see Examples 7–8 on pages 372–373 of your text and the Section 6.7 lecture video.
Indicate whether each triangle is a right triangle, an acute triangle, or an obtuse triangle. Explain why.

15.

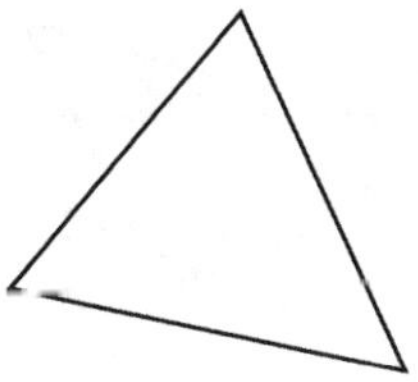

16.

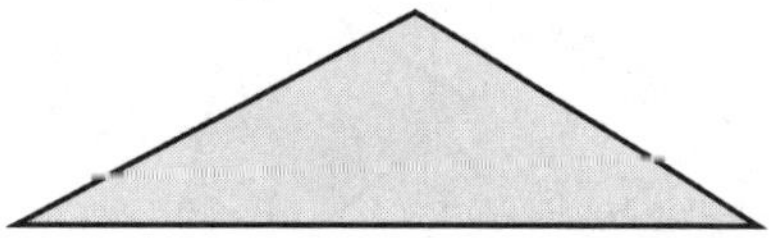

Name: Date:
Instructor: Section:

Find the measure of each angle in the following triangle.

17.

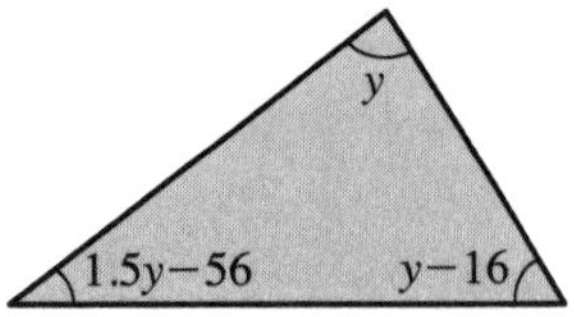

18.

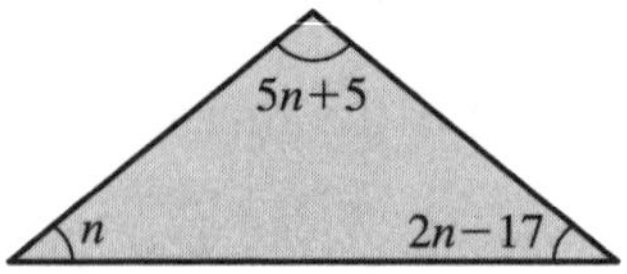

Objective 5 Solve problems involving congruent triangles.

For extra help, see Examples 9–10 on pages 375–376 of your text and the Section 6.7 lecture video.

Determine whether the triangles are congruent. If they are congruent, state the rule that explains why.

19.

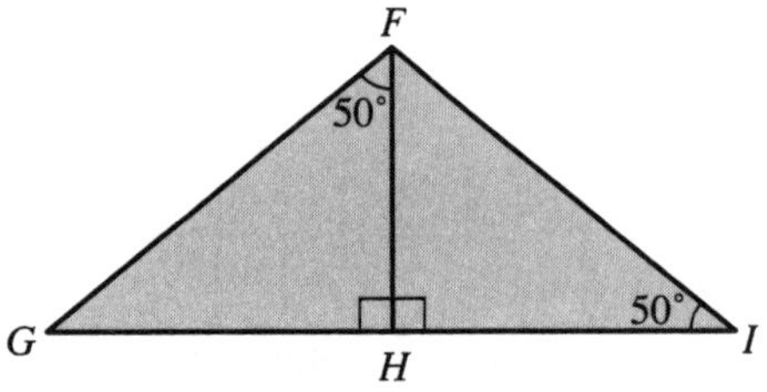

20.

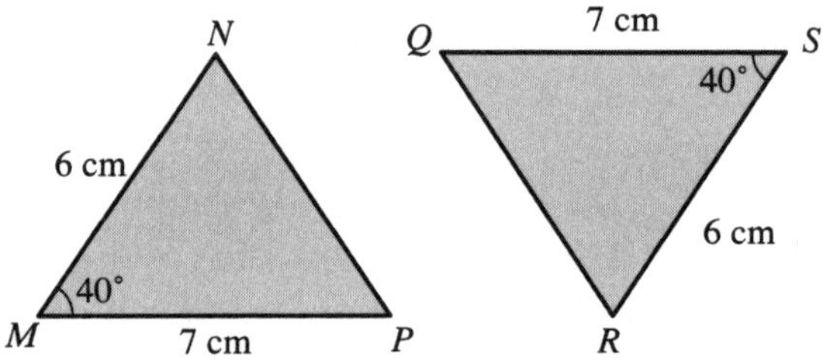

The following triangles are congruent. Find the measure of the angles.

21.

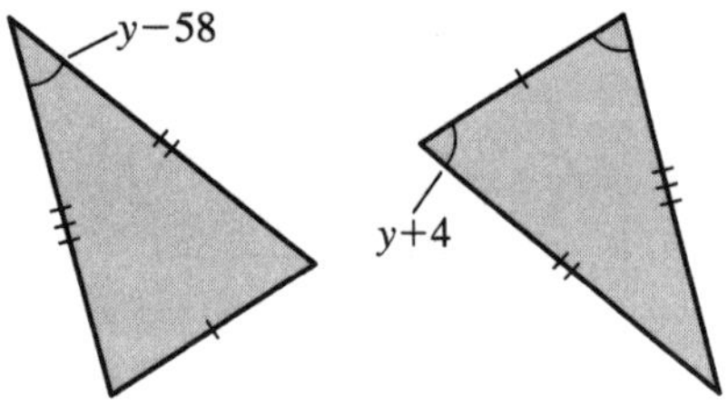

22.

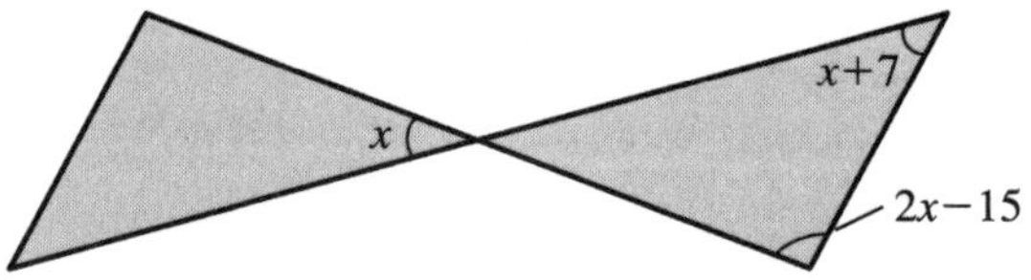

Name: Date:
Instructor: Section:

Chapter 7 RATIOS, PROPORTIONS, AND MEASUREMENT

7.1 Ratios, Probability, and Rates

Learning Objectives
1 Write ratios in simplest form.
2 Determine probabilities.
3 Calculate unit ratios.
4 Write rates in simplest form.
5 Use unit price to determine the better buy.

Key Terms
Use the terms listed below to complete each statement in Exercises 1–4.

possible **rate** **ratio** **product** **quotient** **favorable** **unit**

1. A ratio comparing two different measurements is called a(n) __________________.

2. When the denominator is 1, the ratio is called a(n) __________________ ratio.

3. The theoretical probability of equally likely outcomes is the ratio of the number of __________________ outcomes to the number of __________________ outcomes.

4. A ratio is a comparison of two quantities using a(n) __________________.

GUIDED EXAMPLES AND PRACTICE

Objective 1 Write ratios in simplest form.

Review this example for Objective 1:

1. Write each ratio in simplest form. There are 36 males and 39 females in a school marching band.

a. What is the ratio of female band members to total band members?

b. What is the ratio of male band members to female band members?

Practice this exercise:

1. Write each ratio in simplest form. A theater sells tickets at different price levels depending on the location of the seats. The following table shows the number of seats sold at each price level for a particular show.

Price	Number of Seats
$80	875
$65	180
$45	375

Name: Date:
Instructor: Section:

$$\frac{\text{female band members}}{\text{total band members}}$$

$\frac{39}{39+36}$ Replace the values known.

$\frac{39}{75}$ Add 39 and 36.

$\frac{13}{25}$ Divide the numerator and denominator by 3.

b. The ratio of male band members to female band member is the following:

$$\frac{\text{male band members}}{\text{female band members}}$$

$\frac{36}{39}$ Replace the values known.

$\frac{12}{13}$ Divide the numerator and denominator by 3.

a. What is the ratio of the number of \$80 seats to the number of \$45 seats?

b. What is the ratio of the number of \$65 seats to the total number of seats?

Objective 2 Determine probabilities.

Review this example for Objective 2:

2. What is the probability of selecting a 10 in a standard deck of cards? Write the result in simplest form.

$$\frac{\text{Number of tens in deck}}{\text{Total cards in deck}}$$

$\frac{4}{52}$ There are 4 tens, one of each suit.

$\frac{1}{13}$ Divide the numerator and denominator by 4.

Practice this exercise:

2. What is the probability of selecting a spade in a standard deck of cards? Write the result in simplest form.

Name: Date:
Instructor: Section:

Objective 3 Calculate unit ratios.

Review this example for Objective 3:

3. A family has a total debt of \$5923 and a gross income of \$25,525. What is their debt-to-income ratio? Interpret the result.

$$\text{Debt-to-income ratio} = \frac{5923}{25,525} \approx 0.23$$

The family's debt to income ratio is approximately 0.23, which means it owes about \$0.23 for every \$1 of income.

Practice this exercise:

3. The stock of Pest B Gone has a current share price of $67\frac{9}{16}$ and annual net earnings per share of \$3.89. What is the ratio of selling price to annual earning per share? Interpret the result.

Objective 4 Write rates in simplest form.

Review this example for Objective 4:

4. A long-distance phone call between two cities costs \$2.21 for 17 minutes. What is the unit price in dollars per minute?

Write the number of dollars in the numerator and the number of minutes in the denominator.

$$\frac{\$2.21}{17 \text{ minutes}}$$

$$\frac{\$0.13}{1 \text{ minute}} \quad \text{Divide 2.21 and 17.}$$

\$0.13 / minute

Practice this exercise:

4. A $\frac{1}{3}$-pound package of cheese costs \$0.59. What is the unit price in dollars per pound?

Objective 5 Use unit price to determine the better buy.

Review this example for Objective 5:

5. Determine the better buy: a 3.9-ounce package of instant pudding for \$0.64 or a 5.9-ounce package of instant pudding for \$0.82

Practice this exercise:

5. Determine the better buy: 3 bars of brand A soap for \$1.11 or 4 bars of brand B soap for \$1.75

Name: Date:
Instructor: Section:

$$\frac{\$0.64}{3.9 \text{ oz.}}$$
\$0.1641 / oz

versus

$$\frac{\$0.82}{5.9 \text{ oz.}}$$
\$0.1390 / oz.

Because \$0.14 is the smaller unit price, buying the 5.9-ounce package is the better buy.

ADDITIONAL EXERCISES

Objective 1 Write ratios in simplest form.

For extra help, see Examples 1–3 on pages 401–403 of your text and the Section 7.1 lecture video.

Write each ratio in simplest form.

1. The back wheel of a bicycle rotates $1\frac{3}{7}$ times with $2\frac{2}{5}$ rotations of the pedals. Write the ratio of back wheel rotations to pedal rotations in simplest form.

2. The following table shows the number of each type of song stored on an MP3 player.

Genre	**Number of Songs**
Rock	1650
Rap	54
Pop	1320
Country	16

a. What is the ratio of rock songs to rap songs?

b. What is the ratio of pop songs to rock songs?

c. What is the ratio of country songs to total songs?

Name: Date:
Instructor: Section:

Objective 2 Determine probabilities.

For extra help, see Example 4 on page 404 of your text and the Section 7.1 lecture video.

Determine each probability. Write the result in simplest form.

3. What is the probability of selecting a 7 or a Jack in a standard deck of cards?

4. What is the probability of selecting a black queen in a standard deck of cards?

5. What is the probability of tossing an odd number on a six-sided game die?

6. A box of gumdrops was found to contain the number of gumdrops listed in the table.

Type	Number
Lemon	6
Orange	2
Cherry	6
Grape	6

a. If one gumdrop is drawn from the box, what is the probability of drawing an orange?

b. What is the probability of drawing a blueberry?

Objective 3 Calculate unit ratios.

For extra help, see Example 5 on page 405 of your text and the Section 7.1 lecture video.

Calculate each ratio as a unit ratio and interpret the result.

7. A family has a total debt of $5473 and a gross income of $13,031. What is their debt-to-income ratio? Interpret the result.

8. A community college has 12,315 students and 730 faculty. What is the student-to-faculty ratio? Interpret the result.

Name: Date:
Instructor: Section:

Objective 4 Write rates in simplest form.
For extra help, see Examples 6–7 on pages 406–407 of your text and the Section 7.1 lecture video.
Solve.

9. A 20-ounce box of cereal costs \$5.05. What is the unit price in cents per ounce?

10. Anna drove 246.5 miles in $4\frac{1}{4}$ hours. What was Anna's average rate in miles per hour?

Objective 5 Use unit price to determine the better buy.
For extra help, see Example 8 on page 407 of your text and the Section 7.1 lecture video.
Determine the better buy.

11. Half a pound of brand A deli meat for \$1.79 or a quarter pound of brand B deli meat for \$0.90

12. Three 15-ounce cans of peas for \$1.26 or two 20-ounce cans of peas for \$1.20

Name: Date:
Instructor: Section:

Chapter 7 RATIOS, PROPORTIONS, AND MEASUREMENT

7.2 Proportions

Learning Objectives
1 Determine whether two ratios are proportional.
2 Solve for an unknown value in a proportion.
3 Solve proportion problems.
4 Use proportions to solve for unknown lengths in similar figures.

Key Terms
Use the terms listed below to complete each statement in Exercises 1–4.

proportional **cross products** **quotients** **similar** **congruent**

1. Two figures are similar if they have __________________ angles and proportional side lengths.

2. If two ratios are __________________, their cross products are equal.

3. To solve a proportion, calculate the __________________ and set them equal to one another.

4. Two squares are always __________________.

GUIDED EXAMPLES AND PRACTICE

Objective 1 Determine whether two ratios are proportional.

Review this example for Objective 1:

1. Determine whether the ratios are proportional.

$$\frac{10.5}{2.5}=\frac{21}{5}$$

Take the cross product of the ratios.

$$\frac{10.5}{2.5}=\frac{21}{5}$$

$$(5)(10.5)\stackrel{?}{=}(2.5)(21)$$

$$52.5=52.5$$

The ratios are proportional.

Practice this exercise:

1. Determine whether the ratios are proportional.

$$\frac{7.25}{1.75}=\frac{29}{7}$$

Name: Date:
Instructor: Section:

Objective 2 Solve for an unknown value in a proportion.

Review this example for Objective 2:

2. Solve. $\frac{5}{7}=\frac{h}{42}$

Calculate the cross products and set them equal to each other.

$$\frac{5}{7}=\frac{h}{42}$$

$$(42)(5)=(7)(h)$$

$210=7h$ Multiply both sides.

$30=h$ Divide 210 and 7.

Practice this exercise:

2. Solve. $\frac{7}{8}=\frac{q}{40}$

Objective 3 Solve proportion problems.

Review this example for Objective 3:

3. If a car travels 110 miles in 3 hours, how far will it go in 7 hours?

$\frac{110\text{ miles}}{3\text{ hours}}=\frac{x\text{ miles}}{7\text{ miles}}$ Set up the proportion.

$(7)(110)=(3)(x)$ Cross multiply.

$770=3x$ Multiply.

$256.\overline{6}=x$ Divide by 3.

In 7 hours the car will travel approximately 256.7 miles.

Practice this exercise:

3. If it costs \$2 to buy 15 apples, how many apples will \$7 buy?

Objective 4 Use proportions to solve for unknown lengths in similar figures.

Review this example for Objective 4:

4. Find the unknown length in the similar figures.

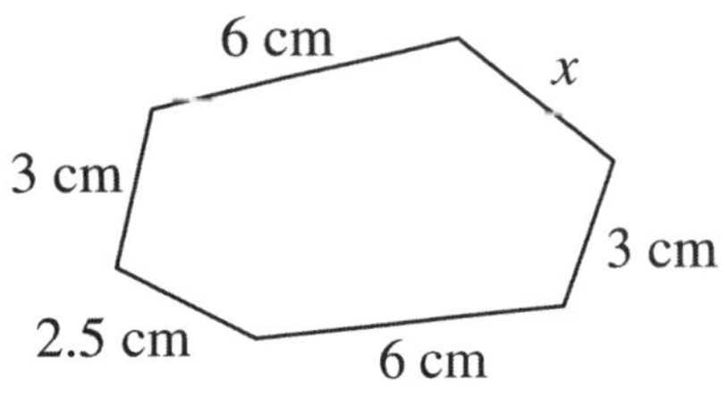

Practice this exercise:

4. The Eiffel Tower is found to have a shadow measuring 200 meters in length. If Hugh is standing right next to the Eiffel Tower, his height is 1.2 meters and his shadow is 1.6 meters long, how tall is the Eiffel Tower?

Name: Date:
Instructor: Section:

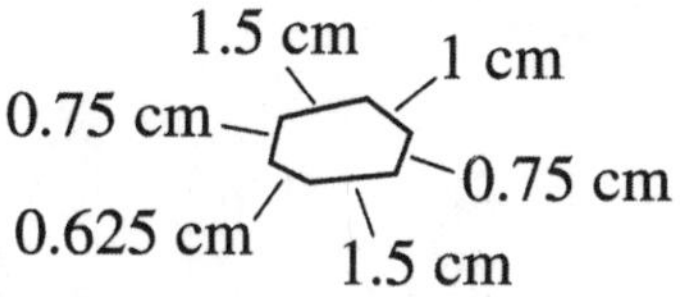

Choose any side to set up the proportion to find the unknown value.

$$\frac{6}{1.5} = \frac{x}{1}$$
$$(6)(1) = (1.5)(x)$$
$$6 = 1.5x$$
$$4 = x$$

The unknown length is 4 cm.

ADDITIONAL EXERCISES

Objective 1 Determine whether two ratios are proportional.

For extra help, see Example 1 on pages 413–414 of your text and the Section 7.2 lecture video.

Determine whether the ratios are proportional.

1. $\frac{7}{12.2} = \frac{28.7}{49.8}$

2. $\frac{5.7}{16} = \frac{17.1}{49}$

3. $\frac{8.5}{4.02} = \frac{17}{8.04}$

4. $\frac{2\frac{1}{4}}{8\frac{1}{8}} = \frac{5\frac{5}{8}}{18\frac{1}{4}}$

Objective 2 Solve for an unknown value in a proportion.

For extra help, see Examples 2–3 on pages 414–415 of your text and the Section 7.2 lecture video.

Solve.

5. $\frac{8}{n} = \frac{3}{7}$

6. $\frac{9}{2} = \frac{63}{c}$

Name: Date:
Instructor: Section:

7. $\frac{5}{13}=\frac{x}{65}$

8. $\frac{8.5}{y}=\frac{17}{2}$

Objective 3 Solve proportion problems.

For extra help, see Example 4 on page 416 of your text and the Section 7.2 lecture video.

Solve.

9. Joan's family drinks 2 gallons of milk in 5 days. How many gallons does her family drink in 3 days?

10. Clarise pays \$2014 in taxes on her house, which is valued at \$240,000. She is planning to buy a house valued at \$300,000. At the same rate, how much will her taxes be on the new house?

11. If 20 pounds of dog food costs \$16.17, how much would a 30-pound bag cost?

12. On a map, 1.5 inches represents 10 miles. How many miles does 7 inches represent?

Objective 4 Use proportions to solve for unknown lengths in similar figures.

For extra help, see Examples 5–6 on pages 417–418 of your text and the Section 7.2 lecture video.

Find the unknown lengths in the similar figures.

13.

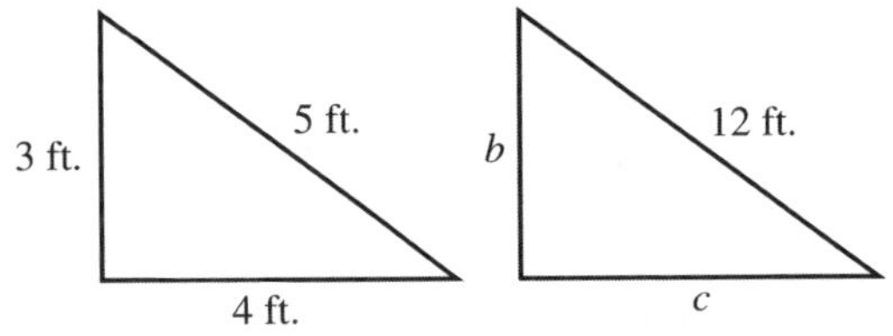

14. A skyscraper has a shadow measuring 255 meters in length. If a man is standing right next to the skyscraper, his height is 1.3 meters and his shadow is 1.7 meters long, how tall is the skyscraper?

Name: Date:
Instructor: Section:

Chapter 7 RATIOS, PROPORTIONS, AND MEASUREMENT

7.3 American Measurement; Time

Learning Objectives
1 Convert American units of length.
2 Convert American units of area.
3 Convert American units of capacity.
4 Convert American units of weight.
5 Convert units of time.
6 Convert American units of speed.

GUIDED EXAMPLES AND PRACTICE

Objective 1 Convert American units of length.

Review this example for Objective 1:

1. Convert 24 yards to feet using dimensional analysis.

$\frac{24 \text{ yd.}}{1}$

$\frac{24 \cancel{\text{yd.}}}{1} \cdot \frac{3 \text{ ft.}}{1 \cancel{\text{yd.}}}$ Multiply by the conversion factor.

72 ft.

Practice this exercise:

1. Convert 4 feet to inches using dimensional analysis.

Objective 2 Convert American units of area.

Review this example for Objective 2:

2. Convert 24 square yards to square feet using dimensional analysis.

$\frac{24 \text{ yd.}^2}{1}$

$\frac{24 \cancel{\text{yd.}^2}}{1} \cdot \frac{9 \text{ ft.}^2}{1 \cancel{\text{yd.}^2}}$ Multiply by the conversion factor.

216 ft.^{2}

Practice this exercise:

2. Convert 73 square yards to square feet using dimensional analysis.

Name: Date:
Instructor: Section:

Objective 3 Convert American units of capacity.

Review this example for Objective 3:

3. Convert 8 pints to cups using dimensional analysis.

$$\frac{8 \text{ pints}}{1}$$

$$\frac{8 \cancel{\text{pints}}}{1} \cdot \frac{2 \text{ cups}}{1 \cancel{\text{pint}}}$$ Multiply by the conversion factor.

16 cups

Practice this exercise:

3. Convert 12 quarts to pints using dimensional analysis.

Objective 4 Convert American units of weight.

Review this example for Objective 4:

4. Convert 8.3 tons to pounds using dimensional analysis.

$$\frac{8.3 \text{ tons}}{1}$$

$$\frac{8.3 \cancel{\text{tons}}}{1} \cdot \frac{2000 \text{ pounds}}{1 \cancel{\text{ton}}}$$ Multiply by the conversion factor.

16,600 pounds

Practice this exercise:

4. Convert 4 pounds to ounces using dimensional analysis.

Objective 5 Convert units of time.

5. Convert 16.3 minutes to seconds using dimensional analysis.

$$\frac{16.3 \text{ minutes}}{1}$$

$$\frac{16.3 \cancel{\text{minutes}}}{1} \cdot \frac{60 \text{ seconds}}{1 \cancel{\text{minutes}}}$$ Multiply by the conversion factor.

978 seconds

Practice this exercise:

5. Convert 300 minutes to hours using dimensional analysis.

Name: Date:
Instructor: Section:

Objective 6 Convert American units of speed.

6. The speed of sound in dry air at 5° C is approximately 1098 feet per second. Calculate the speed of sound at 5° C in miles per minute.

$$\frac{1098 \text{ feet}}{1 \text{ second}}$$

$$\frac{1098 \text{ feet}}{1 \cancel{\text{second}}} \cdot \frac{60 \cancel{\text{seconds}}}{1 \text{ minute}} \quad \text{Convert seconds to minutes.}$$

$$\frac{65{,}880 \cancel{\text{feet}}}{1 \text{ minute}} \cdot \frac{1 \text{ mile}}{5280 \cancel{\text{feet}}} \quad \text{Convert feet to miles.}$$

$$\approx 12.477 \text{ miles / minute}$$

Practice this exercise:

6. The escape velocity for the Sun is approximately 2,025,936 feet per second. Calculate this speed in miles per hour.

ADDITIONAL EXERCISES

Objective 1 Convert American units of length.

For extra help, see Examples 1–2 on pages 423–425 of your text and the Section 7.3 lecture video.

Convert using dimensional analysis.

1. 60 inches to feet

2. 16 feet to yards

3. 7.5 miles to feet

4. 4250 feet to miles

Objective 2 Convert American units of area.

For extra help, see Examples 3–4 on pages 425–426 of your text and the Section 7.3 lecture video.

Convert using dimensional analysis.

5. 142 square inches to square yards

6. 432 square inches to square feet

Objective 3 Convert American units of capacity.

For extra help, see Example 5 on page 427 of your text and the Section 7.3 lecture video.

Convert using dimensional analysis.

7. 150 ounces to pints

8. 0.6 gallons to ounces

Name: Date:
Instructor: Section:

Objective 4 Convert American units of weight.
For extra help, see Example 6 on page 427 of your text and the Section 7.3 lecture video.
Convert using dimensional analysis.

9. $17\frac{1}{3}$ pounds to ounces

10. 6000 pounds to tons

Objective 5 Convert units of time.
For extra help, see Example 7 on page 428 of your text and the Section 7.3 lecture video.
Convert using dimensional analysis.

11. 60 days to years

12. 15 years to minutes

Objective 6 Convert American units of speed.
For extra help, see Example 8 on pages 428–429 of your text and the Section 7.3 lecture video.
Solve using dimensional analysis.

13. A car is traveling 65 miles per hour. Calculate this speed in feet per second.

14. A commercial airliner travels 807 feet per second. Calculate this speed in miles per hour.

Name: Date:
Instructor: Section:

Chapter 7 RATIOS, PROPORTIONS, AND MEASUREMENT

7.4 Metric Measurement

Learning Objectives
1 Convert units of metric length.
2 Convert units of metric capacity.
3 Convert units of metric mass.
4 Convert units of metric area.

GUIDED EXAMPLES AND PRACTICE

Objective 1 Convert units of metric length.

Review this example for Objective 1:

1. Convert 7.2 meters to centimeters.

Centimeters are the second unit to the right of meters, so move the decimal point 2 places to the right of its position at 7.2.

7.2 m = 720 cm

Practice this exercise:

1. Convert 0.73 meters to millimeters.

Objective 2 Convert units of metric capacity.

Review this example for Objective 2:

2. Convert 9.1 centiliters to liters.

Liters are the second unit to the left of centiliters, so move the decimal point 2 places to the left of its position at 9.1.

9.1 cl = 0.091 L

Practice this exercise:

2. Convert 7800 milliliters to centiliters.

Name: Date:
Instructor: Section:

Objective 3 Convert units of metric mass.

Review this example for Objective 3:

3. Convert 800 milligrams to grams.

Grams are the third unit to the left of milligrams, so move the decimal point 3 places to the left of its position at 800.

800 mg = 0.8 g

Practice this exercise:

3. Convert 0.8 kilograms to grams.

Objective 4 Convert units of metric area.

Review this example for Objective 4:

4. Convert 0.022 square kilometers to square meters.

Square meters are the third unit to the right of square kilometers, so move the decimal point $3 \cdot 2 = 6$ places to the right of its position at 0.022.

$0.022 \text{ km}^2 = 22{,}000 \text{ m}^2$

Practice this exercise:

4. Convert 38 square meters to square kilometers.

ADDITIONAL EXERCISES

Objective 1 Convert units of metric length.

For extra help, see Example 1 on pages 433–434 of your text and the Section 7.4 lecture video.

Convert.

1. 2.3 kilometers to meters

2. 125 centimeters to dekameters

3. 730 decimeters to hectometers

4. 12.5 centimeters to millimeters

5. 4 kilometers to decimeters

6. 538 centimeters to kilometers

Name: Date:
Instructor: Section:

Objective 2 Convert units of metric capacity.
For extra help, see Example 2 on page 435 of your text and the Section 7.4 lecture video.
Convert.

7. 0.01 dekaliters to kiloliters

8. 0.611 liters to milliliters

9. 6 kiloliters to liters

10. 4.19 hectoliters to liters

11. 15 cubic centimeters to milliliters

12. 7.07 cubic centimeters to liters

Objective 3 Convert units of metric mass.
For extra help, see Examples 3–4 on pages 435–436 of your text and the Section 7.4 lecture video.
Convert.

13. 0.0089 hectograms to milligrams

14. 0.052 kilograms to metric tons

15. 0.025 dekagrams to centigrams

16. 7200 grams to kilograms

17. 6 grams to decigrams

18. 3.66 metric tons to kilograms

Objective 4 Convert units of metric area.
For extra help, see Example 5 on pages 437–438 of your text and the Section 7.4 lecture video.
Convert.

19. 0.017 square decimeters to square centimeters

20. 8400 square millimeters to square meters

21. 13,000 square meters to square dekameters

22. 47,200 square hectometers to square kilometers

Name: Date:
Instructor: Section:

Chapter 7 RATIOS, PROPORTIONS, AND MEASUREMENT

7.5 Converting Between Systems; Temperature

Learning Objectives

1. Convert units of length.
2. Convert units of capacity.
3. Convert units of mass/weight.
4. Convert units of temperature.

GUIDED EXAMPLES AND PRACTICE

Objective 1 Convert units of length.

Review this example for Objective 1:

1. The distance from the 30-yard line to the 50-yard line on a football field is 20 yards. Convert to meters.

$$\frac{20 \text{ yards}}{1}$$

$$\frac{20 \not{\text{yards}}}{1} \cdot \frac{0.914 \text{ meters}}{1 \not{\text{yard}}}$$ Multiply by the conversion factor.

18.28 meters.

Practice this exercise:

1. The distance from the starting line of the Boston Marathon to the finish line is 138,443.5 feet. Convert to meters.

Objective 2 Convert units of capacity.

Review this example for Objective 2:

2. How many liters are in 169.6 ounces of soda?

$$\frac{169.6 \text{ ounces}}{1}$$

$$\frac{169.6 \not{\text{ounces}}}{1} \cdot \frac{1 \text{ cup}}{8 \not{\text{ounces}}}$$ Convert to cups.

$$\frac{21.2 \not{\text{cups}}}{1} \cdot \frac{1 \text{ pint}}{2 \not{\text{cups}}}$$ Convert to pints.

$$\frac{10.6 \not{\text{pints}}}{1} \cdot \frac{1 \text{ quart}}{2 \not{\text{pints}}}$$ Convert to quarts.

$$\frac{5.3 \not{\text{quarts}}}{1} \cdot \frac{0.946 \text{ liters}}{1 \not{\text{quart}}}$$ Convert to liters.

5.0138 liters

Practice this exercise:

2. The heavy duty wash cycle in a dishwasher uses 8.8 gallons of water. How many liters does it use?

Name: Date:
Instructor: Section:

Objective 3 Convert units of mass/weight.

Review this example for Objective 3:

3. A man weighs 262 pounds. How many kilograms is this?

$\frac{262 \text{ pounds}}{1}$

$\frac{262 \text{ pounds}}{1} \cdot \frac{1 \text{ kilogram}}{2.2 \text{ pounds}}$ Multiply by the conversion factor.

119.09 kilograms

Practice this exercise:

3. A person from Europe claims that her mass is 55 kilograms. What is her weight in pounds?

Objective 4 Convert units of temperature.

Review this example for Objective 4:

4. On a cold day in December, the temperature is reported to be 3° C. What is this in degrees Fahrenheit?

Use the formula $F = \frac{9}{5}C + 32$ and solve.

$F = \frac{9}{5}C + 32$

$F = \frac{9}{5}(3) + 32$ Replace C with 3.

$F = 5.4 + 32$ Multiply 3 and $\frac{9}{5}$.

$F = 37.4$ Add 5.4 and 32.

Practice this exercise:

4. A city recorded a temperature of 101° F. Convert the temperature to degrees Celsius.

ADDITIONAL EXERCISES

Objective 1 Convert units of length.

For extra help, see Examples 1–2 on pages 440–441 of your text and the Section 7.5 lecture video.

Convert.

1. The width of a piece of paper is $8\frac{1}{2}$ inches. How many millimeters is this?

2. The height of a skyscraper is 415 meters. How many yards is this?

Name: Date:
Instructor: Section:

3. How many miles are equal to 1256.345 kilometers?

4. How many feet are equal to 1,445,674 meters?

Objective 2 Convert units of capacity.
For extra help, see Example 3 on pages 441–442 of your text and the Section 7.5 lecture video.
Convert.

5. A recipe calls for $\frac{3}{4}$ cup water. How many milliliters is this?

6. How many pints are there in 0.5 liters of milk?

7. A fuel tank holds 18.6 gallons of gasoline. How many liters is this?

8. A cough syrup says to take 15 milliliters. How many ounces is this?

Objective 3 Convert units of mass/weight.
For extra help, see Example 4 on page 442 of your text and the Section 7.5 lecture video.
Convert.

9. A box of cereal weighs 340 grams. Convert to ounces.

10. A pain reliever contains 400 milligrams of ibuprofen per does. How many ounces is this?

11. In some states a child must sit in a booster seat in a vehicle until the child weighs 40 pounds. How many kilograms is this?

12. It costs $0.39 to mail a letter weighing up to 1 ounce. How many grams is this?

Objective 4 Convert units of temperature.
For extra help, see Examples 5–6 on pages 443–444 of your text and the Section 7.5 lecture video.
Convert.

13. During January, the temperature is reported to be −19° F. What is this temperature in degrees Celsius?

14. A child has a temperature of 101.2° F. Convert this temperature to degrees Celsius.

15. To cook lasagna in an oven, preheat the oven to 350° F. What is this temperature in degrees Celsius?

16. On a sunny day in June the temperature is reported to be 22° C. Convert this temperature to degrees Fahrenheit.

Name: Date:
Instructor: Section:

Chapter 7 RATIOS, PROPORTIONS, AND MEASUREMENT

7.6 Applications and Problem Solving

Learning Objectives
1 Use debt-to-income ratios to decide loan qualification.
2 Calculate the maximum monthly PITI payment that meets the front-end ratio qualification for a loan.
3 Calculate the maximum debt that meets the back-end ratio qualification for a loan.
4 Calculate medical dosages.

Key Terms
Use the terms listed below to complete each statement in Exercises 1–2.

front-end **back-end** **interest** **income** **debt** **total**

1. A PITI payment includes principle, __________________, taxes, and insurance.

2. A(n) __________________ ratio is the ratio of the total monthly debt payments to gross monthly income.

Use the following table for Guided Examples and Practice 1–3.

Factors for loan qualification	**Conventional**	**VA**	**FHA**
1. Front-end ratio should not exceed	0.28	N/A	0.29
2. Back-end ratio should not exceed	0.36	0.41	0.41
3. Credit score should be	650 or higher	N/A	N/A

GUIDED EXAMPLES AND PRACTICE

Objective 1 Use debt-to-income ratios to decide loan qualification.

Review this example for Objective 1:

1. The Haines family is trying to qualify for a conventional loan. Its gross monthly income is \$4050. If the family gets the loan, the monthly payment will be \$920. The family has monthly debt of \$512 and a credit score of 727.
   a. What is the front-end ratio?
   b. What is the back-end ratio?
   c. Does the family qualify for the loan?

Practice this exercise:

1. The Shaw family is trying to qualify for a VA loan. Its gross monthly income is \$4870. If the family gets the loan, the monthly payment will be \$928. The family has monthly debt of \$530 and a credit score of 703.
a. What is the front-end ratio?

a.

$$\text{Front end ratio} = \frac{\text{monthly PITI payment}}{\text{gross monthly income}}$$

$$\text{Front end ratio} = \frac{920}{4050}$$

$$\text{Front end ratio} \approx 0.227$$

Check:

$$4050 \cdot 0.227 = 919.35 \approx 920$$

b.

$$\text{Back-end ratio} = \frac{\text{total monthly debt payments}}{\text{gross monthly income}}$$

$$\text{Back-end ratio} = \frac{512 + 920}{4050}$$

$$\text{Back-end ratio} \approx 0.3536$$

Check:

$$4050 \cdot 0.3536 = 1432.08 \approx 512 + 920$$

c. Since the front-end ratio and back-end ratio are both below conventional standards and the family has a great credit score, they should qualify for the loan.

b. What is the back-end ratio?
c. Does the family qualify for the loan?

Objective 2 Calculate the maximum monthly PITI payment that meets the front-end ratio qualification for a loan.

Review this example for Objective 2:

2. Suppose a family has a gross monthly income of $4300. Find the maximum mortgage payment that would meet the front-end ratio qualification for an FHA loan.

Practice this exercise:

2. Suppose a family has a gross monthly income of $4800. Find the maximum mortgage payment that would meet the front-end ratio qualification for a conventional loan.

Name: Date:
Instructor: Section:

$$\text{Front end ratio} = \frac{\text{monthly PITI payment}}{\text{gross monthly income}}$$

$$0.29 = \frac{p}{4300}$$ Replace with known values.

$$4300 \cdot 0.29 = \frac{p}{\cancel{4300}} \cdot \cancel{4300}$$ Multiply both sides by 4300.

$$1247 = p$$

Check:

$$\frac{1247}{4300} = 0.29$$

Objective 3 Calculate the maximum debt that meets the back-end ratio qualification for a loan.

Review this example for Objective 3:

3. Suppose a family has a gross monthly income of \$4220. Find the maximum monthly debt that would meet the back-end ratio qualification for a conventional loan.

$$\text{Back-end ratio} = \frac{\text{total monthly debt payments}}{\text{gross monthly income}}$$

$$0.36 = \frac{d}{4220}$$

$$4220 \cdot 0.36 = \frac{d}{\cancel{4220}} \cdot \cancel{4220}$$

$$1519.20 = d$$

Check:

$$\frac{1519.20}{4220} = 0.36$$

Practice this exercise:

3. Suppose a family has a gross monthly income of \$4310. Find the maximum monthly debt that would meet the back-end ratio qualification for an FHA loan.

Objective 4 Calculate medical dosages.

Review this example for Objective 4:

4. A patient with a mass of 57 kilograms is to receive an antibiotic. The order is to administer 8.9 milligrams per kilogram of mass. How much of the antibiotic should be administered?

Practice this exercise:

4. A patient is to receive 300 milliliters of a 5% D/W solution IV over 10 hours.

Name: Date:
Instructor: Section:

total dose = patient's mass · dosage per unit of mass

$$\text{total dose} = \frac{57\ \cancel{\text{kg}}}{1} \cdot \frac{8.9\ \text{mg}}{1\ \cancel{\text{kg}}}$$

total dose = 507.3 mg

Check:

$$\frac{507.3\ \text{mg}}{57\ \text{kg}} = 8.9\ \text{mg / kg}$$

The label on the box of the IV indicates that 10 drops dissipate 1 milliliter of the solution. How many drops should the patient receive each minute?

Use the following table for Additional Exercises 1–6.

Factors for loan qualification	**Conventional**	**VA**	**FHA**
1. Front-end ratio should not exceed	0.28	N/A	0.29
2. Back-end ratio should not exceed	0.36	0.41	0.41
3. Credit score should be	650 or higher	N/A	N/A

ADDITIONAL EXERCISES

Objective 1 Use debt-to-income ratios to decide loan qualification.

For extra help, see Example 1 on pages 448–449 of your text and the Section 7.6 lecture video.

Solve and check.

1. The Martin family is trying to qualify for an FHA loan. Its gross monthly income is $3970. If the family gets the loan, the monthly payment will be $1012. The family has monthly debt of $726 and a credit score of 727.

a. What is the front-end ratio?

b. What is the back-end ratio?

c. Does the family qualify for the loan?

2. The Holmes family is trying to qualify for a conventional loan. Its gross monthly income is $4100. If the family gets the loan, the monthly payment will be $933. The family has monthly debt of $508 and a credit score of 698.

a. What is the front-end ratio?

b. What is the back-end ratio?

c. Does the family qualify for the loan?

Name: Date:
Instructor: Section:

Objective 2 Calculate the maximum monthly PITI payment that meets the front-end ratio qualification for a loan.

For extra help, see Example 2 on page 449 of your text and the Section 7.6 lecture video.
Solve and check.

3. Suppose a family has a gross monthly income of $4600. Find the maximum mortgage payment that would meet the front-end ratio qualification for an FHA loan.

4. Suppose a family has a gross monthly income of $4200. Find the maximum mortgage payment that would meet the front-end ratio qualification for a conventional loan.

Objective 3 Calculate the maximum debt that meets the back-end ratio qualification for a loan.

For extra help, see Example 3 on page 450 of your text and the Section 7.6 lecture video.
Solve.

5. Suppose a family has a gross monthly income of $4260. Find the maximum monthly debt that would meet the back-end ratio qualification for a conventional loan.

6. Suppose a family has a gross monthly income of $4380. Find the maximum monthly debt that would meet the back-end ratio qualification for a VA loan.

Objective 4 Calculate medical dosages.

For extra help, see Examples 4–5 on pages 450–451 of your text and the Section 7.6 lecture video.
Solve.

7. A patient weighing 176 pounds is to receive a total of 0.5 milligrams per kilogram per day of accutane, given in two divided doses. If each tablet contains 10 milligrams, how many tablets should the patient receive in each dose?

8. A patient is to receive 600 milliliters of a 5% D/W solution IV over 5 hours. The label on the box of the IV indicates that 6 drops dissipate 1 milliliter of the solution. How many drops should the patient receive each minute?

Name: Date:
Instructor: Section:

Chapter 8 PERCENTS

8.1 Introduction to Percent

Learning Objectives
1 Write a percent as a fraction or decimal number.
2 Write a fraction or a decimal number as a percent.
3 Draw circle graphs.

Key Terms
Use the vocabulary terms listed below to complete each statement in Exercises 1–2.

left **right**

1. When we convert from percent notation to decimal notation, we move the decimal point two places to the __________________ .

2. When we convert from decimal notation to percent notation, we move the decimal point two places to the __________________ .

GUIDED EXAMPLES AND PRACTICE
Objective 1 Write a percent as a fraction or decimal number.

Review these examples for Objective 1:

1. Write the percent as a fraction in lowest terms.

15%

$15\% = \frac{15}{100} = \frac{3}{20}$

2. Write the percent as a decimal.

16.2%

$16.2\% = \frac{16.2}{100} = 0.162$

Practice these exercises:

1. Write the percent as a fraction in lowest terms.

40%

2. Write the percent as a decimal.

95%

Name: Date:
Instructor: Section:

Objective 2 Write a fraction or decimal number as a percent.

Review these examples for Objective 2:

3. Write the fraction as a percent.

$\frac{1}{4}$

$$\frac{1}{4}=\frac{1}{\cancel{4}_1}\cdot\frac{^{25}\cancel{100}}{1}=25\%$$

4. Write the decimal as a percent.

0.85

$0.85=(0.85)(100\%)=85\%$

Practice these exercises:

3. Write the fraction as a percent.

$\frac{7}{9}$

4. Write the decimal as a percent.

0.004

Objective 3 Draw circle graphs.

Review these examples for Objective 3:

5. Movie theater ticket sales show that 25% of the movies in the summer are comedies, 40% are action/adventure movies, and the rest are other kinds of movies. Draw a circle graph that represents this data.

Lightly draw 10 equal-size wedges so that each wedge represents 10%.

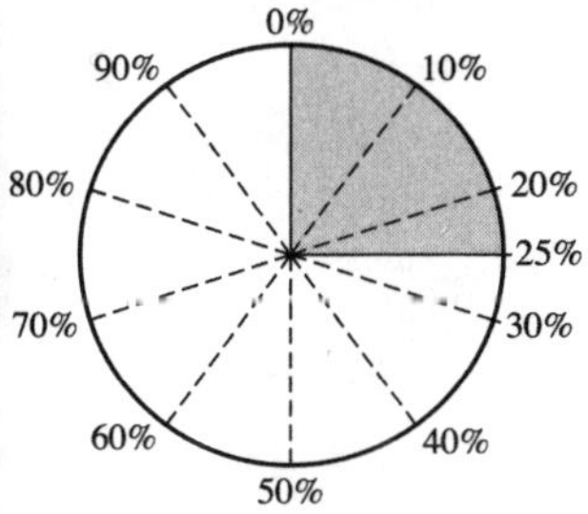

The wedge for 25% goes from 0% (straight up) to halfway between 20% and 30%.

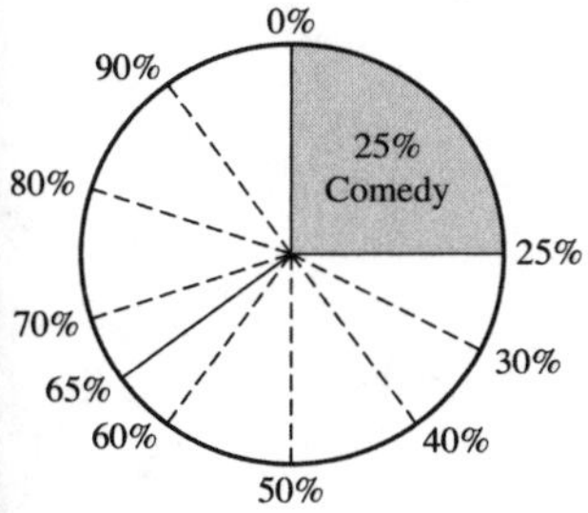

The wedge for 40% is added to the wedge for 25%. 25% +40% = 65% Draw a radius line halfway between 60% and 70%.

Practice these exercises:

5. User ratings for a television go from five stars (the best) to one star (the worst.) On a website, the ratings for a TV were 20% 5-stars, 35% 4-stars, 15% 3-stars, 20% 2-stars and 10% 1-star. Draw a circle graph that represents these ratings.

Name: Date:
Instructor: Section:

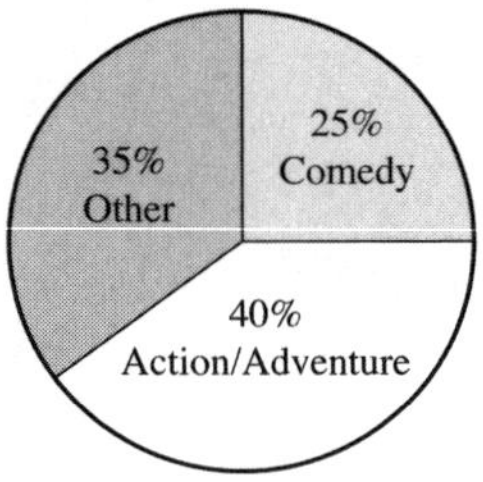

The rest of the graph represents other kinds of movies.

ADDITIONAL EXERCISES

Objective 1 Write a percent as a fraction or decimal number.

For extra help, see Examples 1–2 on pages 464–465 of your text and the Section 8.1 lecture video.

Write a percent as a fraction or decimal number.

1. 75% as a fraction.

2. 27.3% as a fraction

3. 4.8% as a decimal

4. 126% as a decimal.

Objective 2 Write a fraction or decimal number as a percent.

For extra help, see Examples 3–4 on pages 466–467 of your text and the Section 8.1 lecture video.

Write each fraction as a percent.

5. $\frac{14}{25}$

6. $\frac{7}{12}$

Write each decimal as a percent.

7. 0.83

8. 0.001

Name: Date:
Instructor: Section:

Objective 3 Draw circle graphs.
For extra help, see Example 5 on page 467 of your text and the Section 8.1 lecture video.
Draw the circle graph described.

9. The grades for a math course were 15% As, 30% Bs, 35%Cs, 10% Ds and 10% Fs

10. A company found that 64% of its income went to salaries for the workers, 26% went to facilities, and the rest went to other expenses.

Name: Date:
Instructor: Section:

Chapter 8 PERCENTS

8.2 Solving Basic Percent Sentences

Learning Objectives
1 Identify the unknown amount in a basic percent sentence.
2 Solve for the unknown part in a basic percent sentence.
3 Solve for the unknown whole in a basic percent sentence.
4 Solve for the unknown percent in a basic percent sentence.

Key Terms
Use the terms listed below to complete each statement in Exercises 1–4.

% **is** **of** **what**

1. ________________ translates to "=."

2. ________________ translates to any letter.

3. ________________ translates to "$\times\frac{1}{100}$" or "$\times 0.01$."

4. ________________ translates to "$\cdot$" or "$\times$."

GUIDED EXAMPLES AND PRACTICE

Objective 1 Identify the unknown amount in a basic percent sentence.

Review this example for Objective 1:

1. Identify the unknown amount as the percent, whole, or part.

80% of 690 is what number?
Answer: part

A percent	of	a whole	is	a part.
80%	of	690	is	what number?

Practice this exercise:

1. Identify the unknown amount as the percent, whole, or part.

56% of what number is 25?

Objective 2 Solve for the unknown part in a basic percent sentence.

Review this example for Objective 2:

2. 30% of 72 is what number?

$$\begin{array}{ccccc} 30\% & \text{of} & 72 & \text{is} & \text{what number?} \\ \downarrow & \downarrow & \downarrow & \downarrow & \downarrow \\ 0.3 & \cdot & 72 & = & n \end{array}$$

$$21.6 = n$$

Practice these exercises:

2. 90% of 72 is what number?

Name: Date:
Instructor: Section:

Objective 3 Solve for the unknown whole in a basic percent sentence.

Review this example for Objective 3:

3. 65% of what number is 130?

$$\downarrow \quad \downarrow \quad \downarrow \quad \downarrow \quad \downarrow$$

$$0.65 \cdot n = 130$$

$$0.65n = 130$$

$$\frac{0.65n}{0.65} = \frac{130}{0.65}$$

$$n = 200$$

Practice this exercise:

3. 36% of what number is 23.4?

Objective 4 Solve for the unknown percent in a basic percent sentence.

Review this example for Objective 3:

4. What percent of 350 is 70?

$$\downarrow \quad \downarrow \quad \downarrow \quad \downarrow \quad \downarrow$$

$$p \cdot 350 = 70$$

$$350p = 70$$

$$\frac{350p}{350} = \frac{70}{350}$$

$$p = 0.2$$

$$p = (0.2)(100\%) = 20\%$$

Practice this exercise:

4. What percent of 40 is 16?

ADDITIONAL EXERCISES

Objective 1 Identify the unknown amount in a basic percent sentence.

For extra help, see Example 1 on pages 470–471 of your text and the Section 8.2 lecture video.

Identify the unknown amount as the percent, whole, or part.

1. 18% of 93 is what number?

2. 5% of what number is 36?

3. What percent of 32 is 948?

4. What percent of 16 is 48?

Name: Date:
Instructor: Section:

Objective 2 Solve for the unknown part in a basic percent sentence.
For extra help, see Example 2 on pages 472–473 of your text and the Section 8.2 lecture video.
Translate to an equation and solve.

5. 45% of 120 is what number?

6. 135% of 16 is what number?

Objective 3 Solve for the unknown whole in a basic percent sentence.
For extra help, see Example 3 on page 474 of your text and the Section 8.2 lecture video.
Translate to an equation and solve.

7. 40% of what number is 62?

8. 64 is 96% of what number?

Objective 4 Solve for the unknown percent in a basic percent sentence.
For extra help, see Example 4 on pages 475–476 of your text and the Section 8.2 lecture video.
Translate to an equation and solve.

9. What percent of 81.5 is 32.6?

10. What percent of 21 is 14?

Name: Date:
Instructor: Section:

Chapter 8 PERCENTS

8.3 Solving Percent Problems (Portions)

Learning Objectives
1 Solve for the part in percent problems.
2 Solve for the whole amount in percent problems.
3 Calculate the percent.

GUIDED EXAMPLES AND PRACTICE

Objective 1 Solve for the part in percent problems.

Review this example for Objective 1:

1. Duchess earns 6% of total sales in commission. If she sells $1507 in merchandise during the week, what is her commission?

The unknown is Duchess's commission, which is the part. We are given her commission rate (6%) and her total sales, ($1507) which is the whole.

Her commission is 6% of her total sales.

$$\begin{array}{ccccc} \downarrow & \downarrow & \downarrow & \downarrow & \downarrow \\ c & = & 0.06 & \cdot & 1506 \end{array}$$

$$c = 90.36$$

Duchess's commission is $90.36.

Practice this exercise:

1. Ten percent of Americans observe the "three-second rule" when eating food that has dropped to the floor. In other words, if it is picked up within three seconds, it is okay to eat it. If 220 Americans are asked if they observe the "three-second rule" how many are expected to observe the rule?

Objective 2 Solve for the whole amount in percent problems.

Review this example for Objective 2:

2. In a basketball game, the team made 40% of the total number of attempted shots. If they made a total of 40 shots, how many shots were attempted?

Practice this exercise:

2. *Communication Arts* magazine is sold to student subscribers for 68% of the regular subscription rate. If the student subscription price is $28.56, what is the regular subscription rate?

Name: Date:
Instructor: Section:

The unknown is the number of total shots attempted. This is the whole. We are given the percentage of shots that the team made (40%) and the number they madc (40) which is the part.

40% of how many shots is 40?

$$\begin{array}{ccccc} 40\% & \text{of} & \text{how many shots} & \text{is} & 40? \\ \downarrow & \downarrow & \downarrow & \downarrow & \downarrow \\ 0.4 & \cdot & s & = & 40 \end{array}$$

$$0.4s = 40$$

$$\frac{0.4s}{0.4} = \frac{40}{0.4}$$

$$s = 100$$

The team attempted 100 shots.

Objective 3 Calculate the percent.

Review this example for Objective 3:

3. A company determines that 7 out of 125 products are defective. What percent of the products are defective?

The unknown is the percent. We are given the whole (125) and the part (7).

What percent of 125 is 7?

$$\begin{array}{cccccc} \text{What percent} & \text{of} & 125 & \text{is} & 7? \\ \downarrow & \downarrow & \downarrow & \downarrow & \downarrow \\ p & \cdot & 125 & = & 7 \end{array}$$

$$125p = 7$$

$$\frac{125p}{125} = \frac{7}{125}$$

$$p = 0.056$$

$$p = (0.056)(100\%) = 5.6\%$$

5.6% of the products are defective.

Practice this exercise:

3. Uber answers 32 out of 33 questions on his test correctly. What percent of the questions did he answer correctly?

Name: Date:
Instructor: Section:

ADDITIONAL EXERCISES

Objective 1 Solve for the part in percent problems.

For extra help, see Examples 1–2 on pages 480–481 of your text and the Section 8.3 lecture video.

Translate to an equation and solve.

1. Amarillo, Texas had a total budget of $79 million during the 2003/2004 fiscal year. If 2.8% is spent on staff services, what amount is spent on staff services?

2. According to *USA Today*, girls make up 60% of the horror movie audience. If 425 people attend the showing of *Paranormal Activity 3,* how many are expected to be girls?

Objective 2 Solve for the whole amount in percent problems.

For extra help, see Example 3 on page 481 of your text and the Section 8.3 lecture video.

Translate to an equation and solve.

3. A dishwasher salesperson's commission rate is 30%. The salesperson receives a commission of $424. How many dollars worth of dishwashers did the salesperson sell?

4. Yesterday, 10 employees in an office were absent. If these absentees constitute 25% of the employees, how many employees are there?

Objective 3 Calculate the percent.

For extra help, see Example 4 on page 482 of your text and the Section 8.3 lecture video.

Translate to an equation and solve.

5. A blouse is normally sold for $42 and is discounted $14. What is the percent of discount?

6. The National Osteoporosis Foundation reports that 2,300,000 men and 7,800,000 women suffer from osteoporosis. What percent of the total number of people who suffer from osteoporosis are women?

Name: Date:
Instructor: Section:

Chapter 8 PERCENTS

8.4 Solving Problems Involving Percent of Increase or Decrease

Learning Objectives
1 Use a percent of increase to find an increase amount and the final amount after the increase.
2 Use a percent of decrease to find a decrease amount and the final amount after the decrease.
3 Find an unknown initial amount in a percent of increase or decrease problem.
4 Find a percent of increase or decrease.

Key Terms
Use the vocabulary terms listed below to complete Exercise 1. One of the terms will not be used.

original amount **new amount** **amount of increase/decrease**

1. We find a percent of increase/decrease by dividing the ________________ by the ________________.

GUIDED EXAMPLES AND PRACTICE

Objective 1 Use a percent of increase to find an increase amount and the final amount after the increase.

Review this example for Objective 1:

1. The total dining bill is \$42.60. If you decide to leave a 15% tip, how much money should you leave to cover the bill and the tip?

First, find the amount of the tip.

What is 15% of 42.60?

$$\begin{array}{ccccc} \downarrow & \downarrow\ \downarrow & & \downarrow & \downarrow \\ x & =0.15 & & \cdot & 42.6 \end{array}$$

$$x = 6.39$$

The amount of the tip is \$6.39.

The total amount needed for the bill and the tip is $42.60 + 6.39 = \$48.99$.

Practice this exercise:

1. The sales tax rate in Massachusetts is 6.25%. If you bought a TV for \$1138, find the sales tax and the total amount of the purchase.

Name: Date:
Instructor: Section:

Objective 2 Use a percent of decrease to find a decrease amount and the final amount after the decrease.

Review this example for Objective 2:

2. A student discount of 25% is applied to a purchase at a bookstore which had an initial price of $130. Find the amount of the discount and the final price.

First, find the amount of the discount.

What is 25% of 130?

$$\begin{array}{ccccc} \downarrow & \downarrow & \downarrow & \downarrow & \downarrow \\ x & = & 0.25 & \cdot & 130 \end{array}$$
$$x = 32.5$$

The amount of the discount is $32.50.

The final price after the discount is

$130 - 32.50 = \$97.50$

Practice this exercise:

2. A sweater is on sale with a 30% discount. If the sweater has an initial price of $58.00, find the amount of the discount and the final price.

Objective 3 Find an unknown initial amount in a percent of increase or decrease problem.

Review this example for Objective 3:

3. You added a 15% tip when charging a meal to a credit card. If the total amount charged to the card was $53.13, what was the cost of the meal before the tip?

The total bill = cost of the meal + tip amount

The total bill = cost of the meal + 0.15 cost of the meal

$$53.13 = c + 0.15c$$
$$53.13 = 1.15c$$
$$\frac{53.13}{1.15} = \frac{1.15c}{1.15}$$
$$c = 46.2$$

The cost of the meal was $46.20.

Practice this exercise:

3. An employee's new salary is $16,800 after getting a 5% raise. What was the salary before the increase in pay?

Name: Date:
Instructor: Section:

Objective 4 Find a percent of increase or decrease.

Review this example for Objective 4:

4. The sales tax is $41 on the purchase of a dining room set originally priced at $820. Find the sales tax rate.

What percent of $820 is $41?

$$p \cdot 820 = 41$$

$$820p = 41$$

$$\frac{820p}{820} = \frac{41}{820}$$

$$p = 0.05$$

$$p = (0.05)(100\%) = 5\%$$

The sales tax rate is 5%.

5. A $385 suit is on sale for $231. Find the rate of discount.

$$\frac{\text{The rate of discount}}{100} = \frac{\text{discount amount}}{\text{initial amount}}$$

First, find the amount of the discount.

$$\$385 - \$231 = \$154$$

$$\frac{P}{100} = \frac{154}{385}$$

Equate the cross products to solve.

$$385P = 15400$$

$$\frac{385P}{385} = \frac{15400}{385}$$

$$p = 40$$

The rate of discount is 40%.

Practice this exercise:

4. Jackson earns $8.50 per hour. If his hourly pay is increased to $11.00, what is the percent of increase?

5. An item's original price is $700, but it is on sale for $140 less. Find the rate of discount and the sale price.

Name: Date:
Instructor: Section:

ADDITIONAL EXERCISES

Objective 1 Use a percent of increase to find an increase amount and the final amount after the increase.

For extra help, see Example 1 on pages 490–491 of your text and the Section 8.4 lecture video.

Solve.

1. The total dining bill is $23.48. If you decide to leave a 15% tip, how much money should you leave to cover the bill and the tip?

2. The sales tax rate in a city is 8.5%. Find the tax charged on a purchase of $227, and the total cost.

Objective 2 Use a percent of decrease to find a decrease amount and the final amount after the decrease.

For extra help, see Example 2 on page 492 of your text and the Section 8.4 lecture video.

Solve.

3. A $220 suit is on sale for a 10% discount. Find the amount of the discount and the sale price.

4. The state deducts 6% of the gross monthly pay for retirement. If Fitzgerald's gross monthly salary is $1586, how much is deducted for retirement? What is his pay after the deductions?

Objective 3 Find an unknown initial amount in a percent of increase or decrease problem.

For extra help, see Examples 3–4 on pages 493–494 of your text and the Section 8.4 lecture video.

Solve.

5. An item is marked down 20% from its original price. If this means that the item is selling for $21.40, what was the original price?

6. An amusement park offers a 15% military discount. If a soldier pays $55.25 for admission, what is the regular price?

Objective 4 Find a percent of increase or decrease.

For extra help, see Examples 5–6 on pages 558–560 of your text and the Section 8.4 lecture video.

Solve.

7. The amount in a savings account increased from $300 to $309. What was the percent of increase?

8. The population of a town decreased from 35,000 to 30,100. What was the percent decrease?

Name: Date:
Instructor: Section:

Chapter 8 PERCENTS

8.5 Solving Problems Involving Interest

Learning Objectives
1 Solve applied problems involving simple interest.
2 Solve applied problems involving compound interest.
3 Solve applied problems involving amortization.

Key Terms
Use the vocabulary terms listed below to complete each statement in Exercises 1–2.

compound interest **simple interest** **amortization**

1. The formula $I = P \cdot r \cdot t$ is used to find the ________________ I on principal P, invested for t years at interest rate r.

2. When interest is paid on interest we call it ________________.

3. The process of paying off a loan in installments is called ________________.

GUIDED EXAMPLES AND PRACTICE

Objective 1 Solve applied problems involving simple interest.

Review this example for Objective 1:

1. What is the interest on \$4400 invested at an interest rate of 8% for $\frac{1}{2}$ year?

Substitute \$4400 for P, 8% for r, and $\frac{1}{2}$ for t in the simple interest formula.

$$\begin{aligned} I &= P \cdot r \cdot t \\ &= \$4400 \cdot 0.08 \cdot \frac{1}{2} \\ &= \frac{\$4400 \cdot 0.08}{2} \\ &= \$176 \end{aligned}$$

The interest for $\frac{1}{2}$ year is \$176.

Practice this exercise:

1. What is the interest on \$1200 invested at 6% for $\frac{1}{4}$ year?

Name: Date:
Instructor: Section:

2. Kevin borrows \$5000 for a car loan at 4% simple interest for 3 years. Find the amount of interest due and the total amount due after 3 years.

Find the amount of interest.
Substitute \$5000 for P, 4% for r, and 3 for t in the simple interest formula.

$I = Prt$

$I = (5000)(0.04)(3)$

$I = 600$

Kevin will owe \$600 in interest.

Find the total amount owed. Add the interest to the loan amount.

$\$5000 + \$600 = \$5600$

2. Sarah borrows \$10,000 for a college loan at 6% simple interest for 5 years. Find the amount of interest due and the total amount due after 5 years.

Objective 2 Solve applied problems involving compound interest.

Review this example for Objective 2:

3. The Jensens invest \$2000 in an account paying 12%, compounded semiannually. Find the amount in the account after $1\frac{1}{2}$ years.

Substitute \$2000 for P, 12% for r, 2 for n, and $1\frac{1}{2}$ for t in the compound interest formula.

$$A = P \cdot \left(1 + \frac{r}{n}\right)^{n \cdot t}$$

$$= \$2000 \cdot \left(1 + \frac{0.12}{2}\right)^{2 \cdot \frac{3}{2}}$$

$$= \$2000 \cdot (1.06)^3$$

$$\approx \$2382.03$$

The amount in the account after $1\frac{1}{2}$ years is \$2382.03.

Practice this exercise:

3. The Shaws invest \$3600 in an account paying 8%, compounded quarterly. Find the amount in the account after 1 year.

Name: Date:
Instructor: Section:

Objective 3 Solve applied problems involving amortization.
Use the amortization table, Table 8.1 on page 508 of the text.

Review this examples for Objective 3:

4. Use Table 8.1 to find the monthly payment for \$45,600 at 9% for 30 years.

\$45,600 does not appear on the table, so write it as a sum of values that are found in the table.

$45,000 + 500 + 100$

Now use the 30-year column to find the monthly payment for each of these principals. Add the monthly payments.

$$\begin{array}{ccccc} 45,000 & + & 500 & + & 100 \\ \downarrow & & \downarrow & & \downarrow \\ 362.08 & + & 4.02 & + & 0.80 = 366.90 \end{array}$$

The monthly payment is \$366.90.

Practice this exercise:

4. Use Table 8.1 to find the monthly payment on \$375,700 at 9% for 15 years.

ADDITIONAL EXERCISES

Objective 1 Solve applied problems involving simple interest.
For extra help, see Examples 1–3 on pages 502–503 of your text and the Section 8.5 lecture video.
Find the simple interest.

	Principal	Rate of interest	Time	Simple interest
1.	\$1200	3%	1 year	
2.	\$500	4.8%	$\frac{1}{2}$ year	
3.	\$25,000	$5\frac{3}{8}\%$	$\frac{1}{4}$ year	

Name: Date:
Instructor: Section:

Sheila borrows $2400 at 6% for 90 days.

4. Find the amount of interest due.

5. Find the total amount that must be paid after 90 days.

Objective 2 Solve applied problems involving compound interest.

For extra help, see Examples 4–5 on pages 505–506 of your text and the Section 8.5 lecture video.

Solve.

6. The Hanson's deposit $1000 in an account paying 5% compounded monthly. How much is in the account after 6 months?

7. Corey invests $3000 in an account paying 8%, compounded quarterly. How much is in the account after 2 years?

8. Lucas invests $10,000 in an account paying 6%, compounded quarterly. How much is in the account after 18 months?

9. Skylar deposits $5000 in an account paying 4% compounded daily. How much is in the account after 30 days?

Objective 3 Solve applied problems involving amortization.

Use the amortization table, Table 8.1 on page 572 of the text

For extra help, see Example 6 on pages 507 of your text, and Table 8.1 on page 508 of your text and the Section 8.5 lecture video.

10. Use Table 8.1 to find the monthly payment for $82,300 at 9% for 15 years.

11. Use Table 8.1 to find the monthly payment for $50,200at 9% for 5 years.

12. Use Table 8.1 to find the monthly payment for $475,500 at 9% for 15 years.

13. Use Table 8.1 to find the monthly payment for $107,900 at 9% for 3 years.

Name: Date:
Instructor: Section:

Chapter 9 LINEAR EQUATIONS AND INEQUALITIES IN ONE VARIABLE

9.1 Linear Equations and Formulas

Learning Objectives
1 Solve linear equations in one variable.
2 Identify identities and contradictions.
3 Solve formulas for specified variables.

Key Terms
Use the vocabulary terms listed below to complete each statement in Exercises 1–6.

solution	**solution set**
linear equation in one variable	**identity**
conditional linear equation in one variable	**contradiction**

1. A linear equation with exactly one solution is called a(n) ____________________.

2. A(n) ____________________ is an equation that has no solution.

3. If an equation contains a variable for a missing number, our goal is to find its ____________________, or a number that makes an equation true when it replaces the variable in the equation.

4. A set containing all of the solutions for a given equation is called a(n) ____________________.

5. A(n) ____________________ is an equation in which every number (for which the equation is defined) is a solution.

6. If an equation can be written in the form $ax + b = c$, where a, b, and c are real numbers and $a \neq 0$, then the equation is a(n) ____________________.

GUIDED EXAMPLES AND PRACTICE

Objective 1 Solve linear equations in one variable.

Review this example for Objective 1:

1. Solve and check.

$3(m-10)+4(m-18)=10$

Distribute to clear parentheses.

$3m-30+4m-72=10$

Combine like terms.

$7m-102=10$

Use the addition principle so that all variable terms are on one side of the equation and all constants are on the other side.

$7m=112$

Use the multiplication principle to clear any remaining coefficient.

$m=16$

The solution is $m=16$.

Practice this exercise:

1. Solve and check.

$2(y-2)-1=3(y-3)$

Name: Date:
Instructor: Section:

Replace m in the original equation with the solution and verify that the equation is true.

$$3(m-10)+4(m-18)=10$$
$$3(16-10)+4(16-18)=10$$
$$3(6)+4(-2)=10$$
$$18-8=10$$
$$10=10$$

The equation is true.

Objective 2 Identify identities and contradictions.

Review this example for Objective 2:

2. Solve and check. Note that some equations may be identities or contradictions.

$$3(2x-4)-x=5x+9$$
$$6x-12-x=5x+9$$
$$5x-12=5x+9$$
$$-12=9$$

This false equation with no variable terms affirms that the equation is a contradiction.

Practice this exercise:

2. Solve and check. Note that some equations may be identities or contradictions.

$2(y-4)-3=2(y-7)$

Objective 3 Solve formulas for specified variables.

Review this example for Objective 3:

3. Solve the formula for the indicated variable.

$\frac{4}{5}+9g=\frac{d}{h}; d$

To isolate d, we must clear h. Because h is dividing d, we multiply both sides by h.

$$h\left(\frac{4}{5}+9g\right)=h\left(\frac{d}{h}\right)$$

Simplify.

$$\frac{4}{5}h+9gh=d$$

Practice this exercise:

3. Solve the formula for the indicated variable.

$\frac{Q}{t}=\pi; t$

ADDITIONAL EXERCISES

Objective 1 Solve linear equations in one variable.

For extra help, see Examples 1–5 on pages 520–523 of your text and the Section 9.1 lecture video.

Solve and check.

1. $-5x+6=41$

2. $8x-(3x+7)=28$

Name: Date:
Instructor: Section:

3. $\frac{8}{5}(m+1)=\frac{3}{5}-m$

4. $0.8y-0.3y=7.5$

Objective 2 Identify identities and contradictions.

For extra help, see Examples 6–7 on pages 523–524 of your text and the Section 9.1 lecture video.

Solve and check. Note that some equations may be identities or contradictions.

5. $2(x-6)+2=4(x+2)-10$

6. $6.5x-2.58=3.9-4.3x$

7. $\frac{3}{4}b-3=-\frac{7}{8}b+2$

8. $-2(x-5)-4x=8-6x+2$

Objective 3 Solve formulas for specified variables.

For extra help, see Example 8 on pages 524–525 of your text and the Section 9.1 lecture video.

Solve for the indicated variable.

9. $10x+9y=3; y$

10. $\frac{rs}{19}-C=t; C$

11. $9(s+5z)=g-jh; s$

12. $\frac{c}{9}+\frac{d}{7}=5; d$

13. $b=\frac{S-Q}{ny}; n$

Name: Date:
Instructor: Section:

Chapter 9 LINEAR EQUATIONS AND INEQUALITIES IN ONE VARIABLE

9.2 Solving Problems

Learning Objectives
1 Use formulas to solve problems.
2 Translate words to equations.

GUIDED EXAMPLES AND PRACTICE

Objective 1 Use formulas to solve problems.

Review this example for Objective 1:

1. Solve.
A circular tunnel is used to accelerate elementary particles. The radius of the tunnel is 7 kilometers. If a particle travels one complete revolution around the circumference of the tunnel, what distance does it travel? Use 3.14 for π. Round to the nearest tenth of a kilometer if needed.

$C = 2\pi r$
$C = 14\pi$
$C = 14(3.14)$
$C \approx 44$ kilometers

Practice this exercise:

1. Solve.
One of the large rectangular rooms in a hotel has floor dimensions of 52 feet by 119 feet. What is the total square footage of the floor of the room?

Objective 2 Translate words to equations.

Review this example for Objective 2:

2. Solve.
One-fourth of the sum of a number and two is seven. What is the number?

The phrase *one-fourth of the sum* indicates to multiply the sum of the number and two by $\frac{1}{4}$. Because the sum is being multiplied, we write it in parentheses; *is* represents an equal sign. So, the words represent the equation $\frac{1}{4}(x+2) = 7$.

Solve:

$$4 \cdot \frac{1}{4}(x+2) = 4 \cdot 7$$
$$x + 2 = 28$$
$$x = 26$$

Practice this exercise:

2. Solve.
Fifteen less than nine times a number is equal to that number added to one. What is the number?

Name: Date:
Instructor: Section:

ADDITIONAL EXERCISES

Objective 1 Use formulas to solve problems.

For extra help, see Examples 1–2 on pages 529–530 of your text and the Section 9.2 lecture video.

Solve.

1. A flight departs from city A at 7:00 A.M. EST to arrive in city B at 1:30 P.M. PST. If the plane flies an average rate of $330\frac{1}{4}$ mph, what distance does it travel, rounded to the nearest tenth of a mile? Use the formula relating distance, rate, and time: $d = rt$. (Hint: There is a three-hour time difference between EST and PST.)

2. A truck enters a highway driving 60 miles per hour. A car enters the highway at the same place 8 minutes later and drives 68 miles per hour in the same direction. From the time the car enters the highway, how many minutes will it take the car to pass the truck?

3. How many gallons of a 70% antifreeze solution must be mixed with 70 gallons of 10% antifreeze to get a mixture that is 60% antifreeze?

Objective 2 Translate words to equations.

For extra help, see Examples 3–5 on pages 531–533 of your text and the Section 9.2 lecture video.

Solve.

4. Six is the result when eight is subtracted from the ratio of a number to ten. What is the number?

5. The difference of a number and fifteen subtracted from the difference of twice the number and nine is three. What is the number?

Name: Date:
Instructor: Section:

Chapter 9 LINEAR EQUATIONS AND INEQUALITIES IN ONE VARIABLE

9.3 Solving Linear Inequalities

Learning Objectives

1 Represent solutions to linear inequalities in one variable using set-builder notation, interval notation, and graphs.
2 Solve linear inequalities in one variable.
3 Solve problems involving linear inequalities.

Key Terms

Use the vocabulary terms listed below to complete each statement in Exercise 1.

Linear inequality in one variable

1. $2x+4\leq 6$ is an example of a(n) __________________ that is written in the form $ax+b\leq c$.

GUIDED EXAMPLES AND PRACTICE

Objective 1 Represent solutions to linear inequalities in one variable using set-builder notation, interval notation, and graphs.

Review this example for Objective 1:

1. $x\geq -3$

a) Write the solution set using set-builder notation.
b) Write the solution set using interval notation.
c) Graph the solution set.

Set-builder notation: $\{x|x\geq -3\}$

Interval notation: $[-3,\infty)$

Graph:

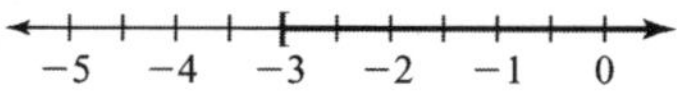

Practice this exercise:

1. $y<0$

a) Write the solution set using set-builder notation.
b) Write the solution set using interval notation.
c) Graph the solution set.

Name: Date:
Instructor: Section:

Objective 2 Solve linear inequalities in one variable.

Review this example for Objective 2:

2. $4+5x<-7x-12$

a) Write the solution set using set-builder notation.

b) Write the solution set using interval notation.

c) Graph the solution set.

Add $7x$ to both sides.

$4+12x<-12$

Subtract 4 from both sides.

$12x<-16$

Divide both sides by 12.

$x<-\frac{4}{3}$

Set-builder notation: $\left\{x \middle| x<-\frac{4}{3}\right\}$

Interval notation: $\left(-\infty, -\frac{4}{3}\right)$

Graph:

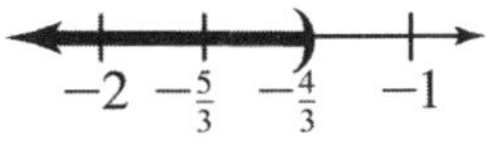

Practice this exercise:

2. $7-w<16+w$

a) Write the solution set using set-builder notation.

b) Write the solution set using interval notation.

c) Graph the solution set.

Objective 3 Solve problems involving linear inequalities.

Review this example for Objective 3:

3. Solve.
Jacqui has grades of 90 and 80 on her first two algebra tests. If she wants an average of at least 71, what possible scores can she make on her third test?

Let x be the score on her third test. Then the average of the three scores is $\frac{80+90+x}{3}$.

The inequality that portraits the problem is $\frac{80+90+x}{3}\geq 71$.

Multiply both sides by 3.

$80+90+x\geq 213$

Subtract 80 from both sides.

$90+x\geq 133$

Subtract 90 from both sides.

$x\geq 43$

Practice this exercise:

3. Solve.
The width of a rectangle is fixed at 4 centimeters. Determine those lengths for which the area will be no more than 88 square centimeters.

Name: Date:
Instructor: Section:

Therefore, she needs to make a score of at least 43 to have the average of at least 71.

ADDITIONAL EXERCISES

Objective 1 Represent solutions to linear inequalities in one variable using set-builder notation, interval notation, and graphs.

For extra help, see Example 1 on page 539 of your text and the Section 9.3 lecture video.

a) Write the solution set using set-builder notation.
b) Write the solution set using interval notation.
c) Graph the solution set.

1. $x \le -1$

2. $y > 4$

Objective 2 Solve linear inequalities in one variable.

For extra help, see Examples 2–3 on pages 540–541 of your text and the Section 9.3 lecture video.

Solve. Then:
a) Write the solution set using set-builder notation.
b) Write the solution set using interval notation.
c) Graph the solution set.

3. $7 - 9x < 61$

4. $4(4k+2) - 8(k-2) \ge 3(2k+3) - 1$

Objective 3 Solve problems involving linear inequalities.

For extra help, see Example 4 on pages 541–542 of your text and the Section 9.3 lecture video.

Translate to an inequality and then solve.

5. Twelve more than a number is greater than thirty-eight.

6. Three times a number less twenty-six is at least fifty-five.

7. The difference of sixteen times a number and ten is less than five times the number.

8. The cost to produce x units of wire is $C = 25x + 600$, while the revenue is $R = 40x$. How many units must be produced so that wire production will at least break even?

9. A metal stays solid at Fahrenheit temperatures below 1791.5°. Determine those Celsius temperatures for which the metal stays solid

Name:
Instructor:

Date:
Section:

Chapter 9 LINEAR EQUATIONS AND INEQUALITIES IN ONE VARIABLE

9.4 Compound Inequalities

Learning Objectives
1 Solve compound inequalities involving *and*.
2 Solve compound inequalities involving *or*.

Key Terms
Use the vocabulary terms listed below to complete each statement in Exercises 1–3.

compound inequality **intersection** **union**

1. For two sets A and B, the ________________ of A and B is a set containing only elements that are both in A and B.

2. A(n) ________________ is composed of two inequalities joined by either *and* or *or*.

3. For two sets A and B, the ________________ of A and B is a set containing every element in A or in B.

GUIDED EXAMPLES AND PRACTICE

Objective 1 Solve compound inequalities involving *and*.

Review these examples for Objective 1:

1. $10 < 2x + 12 \le 20$

a) Write the solution set using set-builder notation.

b) Write the solution set using interval notation.

c) Graph the solution set.

Subtract 12 from all three parts of the compound inequality.

$-2 < 2x \le 8$

Divide all three parts of the compound inequality by 2.

$-1 < x \le 4$

a) Set-builder notation: $\{x \mid -1 < x \le 4\}$

b) Interval notation: $(-1, 4]$

c) Graph:

Practice these exercises:

1. $11 < 3x + 14 \le 23$

a) Write the solution set using set-builder notation.

b) Write the solution set using interval notation.

c) Graph the solution set.

Name: Date:
Instructor: Section:

2. $x+8\geq 6$ and $2x+7\leq 9$

a) Write the solution set using set-builder notation.

b) Write the solution set using interval notation.

c) Graph the solution set.

Isolate the variable term in each inequality. Subtract 8 from both sides of the first inequality and subtract 7 from both sides of the second inequality.

$x\geq -2$ and $2x\leq 2$

Solve for x in each inequality. Divide both sides of the second inequality by 2.

$x\geq -2$ and $x\leq 1$

a) Set-builder notation: $\{x\,|\,-2\leq x\leq 1\}$

b) Interval notation: $[-2, 1]$

c) Graph:

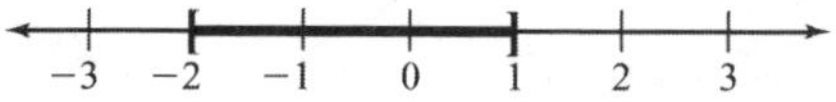

2. $0\leq x+4$ and $x+4<10$

a) Write the solution set using set-builder notation.

b) Write the solution set using interval notation.

c) Graph the solution set.

Objective 2 Solve compound inequalities involving *or*.

Review this example for Objective 2:

3. $4x+5\geq -19$ or $6x+2\geq -10$

a) Write the solution set using set-builder notation.

b) Write the solution set using interval notation.

c) Graph the solution set.

Isolate the variable term in each inequality. Subtract 5 from both sides of the first inequality and subtract 2 from both sides of the second inequality.

$4x\geq -24$ or $6x\geq -12$

Solve for x in each inequality. Divide both sides of the first inequality by 4 and divide both sides of the second inequality by 6.

$x\geq -6$ or $x\geq -2$

Since $x\geq -6$ covers all of $x\geq -2$ and more, the union of the two inequalities is written as $x\geq -6$.

a) Set-builder notation: $\{x\,|\,x\geq -6\}$

b) Interval notation: $[-6, \infty)$

c) Graph:

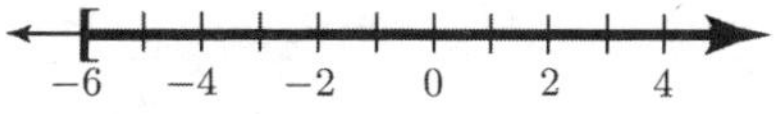

Practice this exercise:

3. $9x+9\geq -45$ or $3x+1\geq -5$

a) Write the solution set using set-builder notation.

b) Write the solution set using interval notation.

c) Graph the solution set.

Name:
Instructor:

Date:
Section:

ADDITIONAL EXERCISES

Objective 1 Solve compound inequalities involving *and.*

For extra help, see Examples 1–3 on pages 549–551 of your text and the Section 9.4 lecture video.

a) *Write the solution set using set-builder notation.*
b) *Write the solution set using interval notation.*
c) *Graph the solution set.*

1. $x < 2$ and $x > -2$

2. $x + 8 \geq 6$ and $x + 7 \leq 9$

3. $-4 < 2 - 2x \leq 10$

4. $-3 < x + 1 < 2$

Objective 2 Solve compound inequalities involving *or.*

For extra help, see Example 4 on pages 552–553 of your text and the Section 9.4 lecture video.

a) *Write the solution set using set-builder notation.*
b) *Write the solution set using interval notation.*
c) *Graph the solution set.*

5. $5x - 1 > -16$ or $5x - 1 \leq 5$

6. $x > -4$ or $x \leq 4$

7. $8x + 1 \geq -71$ or $5x + 3 \geq -12$

8. $x \geq -6$ or $x \geq 8$

9. $-2x - 2 < -10$ or $-2x - 2 > 10$

Name: Date:
Instructor: Section:

Chapter 9 LINEAR EQUATIONS AND INEQUALITIES IN ONE VARIABLE

9.5 Equations Involving Absolute Value

Learning Objectives
1 Solve equations involving absolute value.

GUIDED EXAMPLES AND PRACTICE

Objective 1 Solve equations involving absolute value.

Review these examples for Objective 1:

1. Solve using the absolute value property.
$|6x-3|=15$

Separate the equation into a positive case and a negative case.
$6x-3=15$ or $6x-3=-15$
Solve each case separately. Isolate the variable term.
$6x=18$ or $6x=-12$
Solve for x.
$x=3$ or $x=-2$
The solutions are $x=3$ and $x=-2$.

2. Isolate the absolute value; then use the absolute value property.
$|x-7|-4=20$

Add 4 to each side of the equation to isolate the absolute value.
$|x-7|=24$
Separate the equation into a positive case and a negative case.
$x-7=24$ or $x-7=-24$
Solve each case separately. Isolate the variable term.
$x=31$ or $x=-17$
The solutions are $x=31$ and $x=-17$.

Practice these exercises:

1. Solve using the absolute value property. $|x|=9$

2. Isolate the absolute value; then use the absolute value property.
$|x+4|+3=18$

Name: Date:
Instructor: Section:

3. Solve by separating the two absolute values into equal and opposite cases.
$|9x+4| = |x-13|$

Separate the absolute value equation into two equations.
$9x+4 = x-13$ or $9x+4 = -(x-13)$

Solve each case separately. Distribute the negative sign in the second equation.
$9x+4 = x-13$ or $9x+4 = -x+13$
Isolate the variable term.
$8x = -17$ or $10x = 9$
Solve for x.
$x = -\frac{17}{8}$ or $x = \frac{9}{10}$

The solutions are $x = -\frac{17}{8}$ and $x = \frac{9}{10}$.

3. Solve by separating the two absolute values into equal and opposite cases.
$|3x+8| = |x-11|$

ADDITIONAL EXERCISES

Objective 1 Solve equations involving absolute value.
For extra help, see Examples 1–3 on pages 559–561 of your text and the Section 9.5 lecture video.
Solve using the absolute value property.

1. $|x| = -19$

2. $|x-6| = 14$

3. $|4x-4| = 0$

4. $|4x-3| = 6$

5. $\left|\frac{1}{2}x+2\right| = 4$

Isolate the absolute value; then use the absolute value property.

6. $|6x|-3 = 33$

7. $|7x+3|-8 = -2$

8. $12-|x+7| = 4$

9. $|x+8|+2 = 31$

Solve by separating the two absolute values into equal and opposite cases.

10. $|x-15| = |15-x|$

11. $|3x+8| = |x-17|$

Name: Date:
Instructor: Section:

Chapter 9 LINEAR EQUATIONS AND INEQUALITIES IN ONE VARIABLE

9.6 Inequalities Involving Absolute Value

Learning Objectives
1 Solve absolute value inequalities involving *less than.*
2 Solve absolute value inequalities involving *greater than.*

GUIDED EXAMPLES AND PRACTICE

Objective 1 Solve absolute value inequalities involving *less than.*

Review this example for Objective 1:

1. Graph the solution set. Show set-builder and interval notation.
$|x+5|+8<28$

Subtract 8 from both sides to isolate the absolute value.
$|x+5|<20$
Rewrite as a compound inequality.
$x+5<20$ and $x+5>-20$
or $-20<x+5<20$.
Subtract 5 from all parts of the inequality.
$-25<x<15$
Graph:

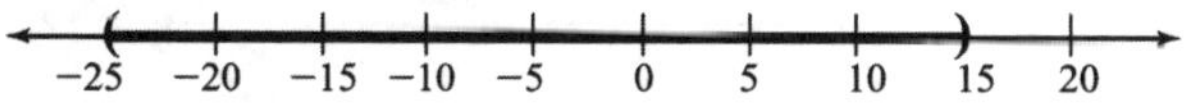

Set-builder notation: $\{x \mid -25<x<15\}$

Interval notation: $(-25, 15)$

Practice this exercise:

1. Graph the solution set. Show set-builder and interval notation.
$|x+5| \le 4$

Name: Date:
Instructor: Section:

Objective 2 Solve absolute value inequalities involving *greater than*.

Review this example for Objective 2:

2. Solve. Graph the solution set and write the solution set in both set-builder and interval notation.

$|x+19|>9$

Rewrite as a compound inequality.

$x+19>9$ or $x+19<-9$

Subtract 19 from both sides of each inequality.

$x>-10$ or $x<-28$

Graph:

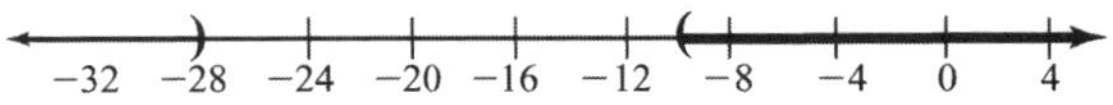

Set-builder notation: $\{x \mid x<-28 \text{ or } x>-10\}$

Interval notation: $(-\infty,-28)\cup(-10,\infty)$

Practice this exercise:

2. Graph the solution set. Show set-builder and interval notation.

$5-|x+3|>9$

ADDITIONAL EXERCISES

Objective 1 Solve absolute value inequalities involving *less than*.

For extra help, see Example 1 on pages 563–564 of your text and the Section 9.6 lecture video.

a) Write the solution set using set-builder notation.

b) Write the solution set using interval notation.

c) Graph the solution set.

1. $|x|<5$

2. $|x+2|+8\le 11$

3. $2|x|-2\le 4$

4. $|x|-3<-1$

Name: Date:
Instructor: Section:

Objective 2 **Solve absolute value inequalities involving *greater than*.**

For extra help, see Example 2 on pages 565–566 of your text and the Section 9.6 lecture video.

a) *Write the solution set using set-builder notation.*
b) *Write the solution set using interval notation.*
c) *Graph the solution set.*

5. $|x| > 2$

6. $|5x - 5| > -23$

7. $|x + 7| > 6$

8. $3|x| + 4 \geq 10$

9. $9 - |x + 3| > -1$

Name: Date:
Instructor: Section:

Chapter 10 GRAPHING LINEAR EQUATIONS AND INEQUALITIES

10.1 The Rectangular Coordinate System

Learning Objectives

1 Determine the coordinates of a given point.
2 Plot points in the coordinate plane.
3 Determine the quadrant for a given ordered pair.
4 Determine whether the graph of a set of data points is linear.

GUIDED EXAMPLES AND PRACTICE

Objective 1 Determine the coordinates of a given point.

Review this example for Objective 1:

1. Write the coordinates for each point.

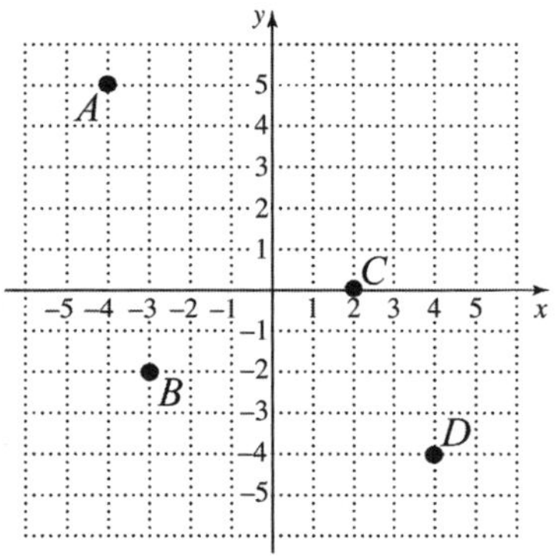

A: $(-4, 5)$, B: $(-3, -2)$, C: $(2, 0)$, D: $(4, -4)$

Practice this exercise:

1. Write the coordinates for each point.

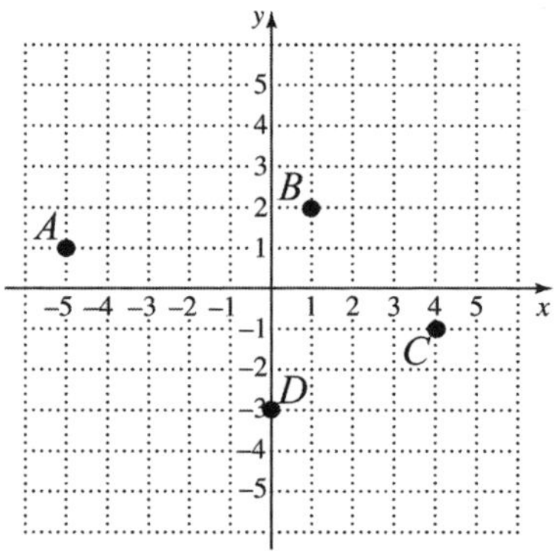

Name: Date:
Instructor: Section:

Objective 2 Plot points in the coordinate plane.

Review this example for Objective 2:

2. Plot and label the points indicated by the coordinate pairs.

$(3, -4)$, $(-4, 2)$, $(0, 5)$,

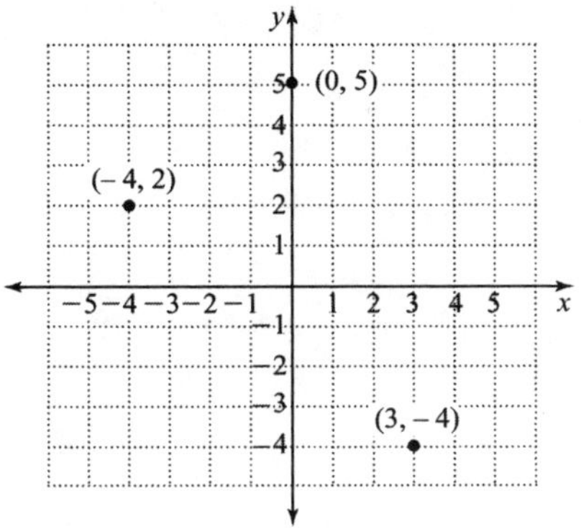

Practice this exercise:

2. Plot and label the points indicated by the coordinate pairs.

$(-4, -1)$, $(2, 5)$, $(-2, -4)$, $(5, 0)$

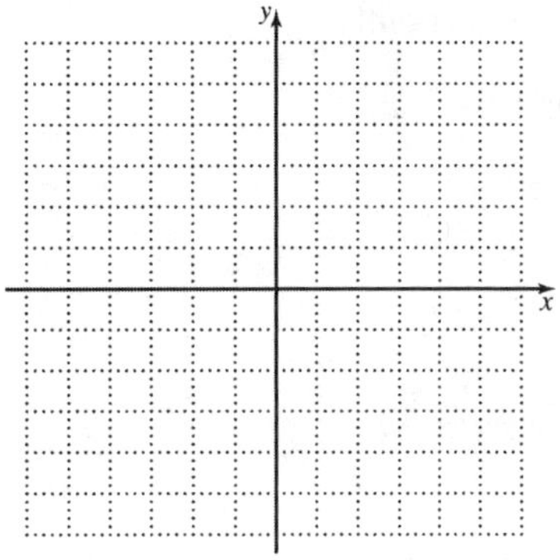

Objective 3 Determine the quadrant for a given ordered pair.

Review this example for Objective 3:

3. State the quadrant in which each point is located. If the point is on an axis, state which axis.

$(5.4, -255)$

$(+, -)$ means the point is in quadrant IV.

Practice this exercise:

3. State the quadrant in which each point is located. If the point is on an axis, state which axis.

$(-7.2, 112)$

Name: Date:
Instructor: Section:

Objective 4 Determine whether the graph of a set of data points is linear.

Review this example for Objective 4:

4. Determine if the graph is linear or nonlinear. (day, length of leaf of grass): $(1, 2)$, $(2, 2.2)$, $(3, 3)$, $(3, 3.5)$

Plot the set of data points.

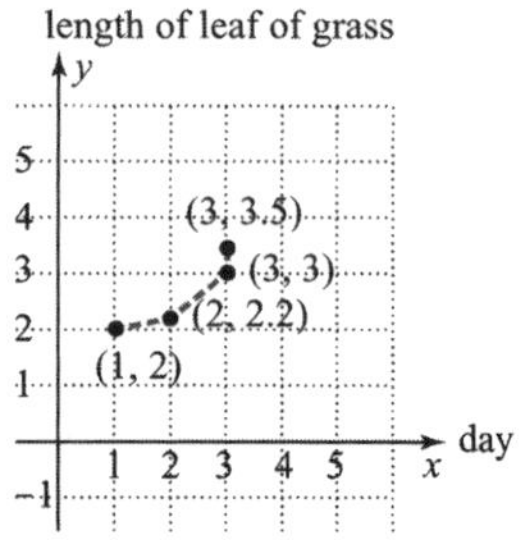

The points do not form a straight line, so they are nonlinear.

Practice this exercise:

4. Determine if the graph is linear or nonlinear. (age of tree, number of growth rings): $(1, 15)$, $(2, 20)$, $(10, 30)$, $(20, 100)$

ADDITIONAL EXERCISES

Objective 1 Determine the coordinates of a given point.

For extra help, see Example 1 on page 583 of your text and the Section 10.1 lecture video.

Write the coordinates for each point.

1.

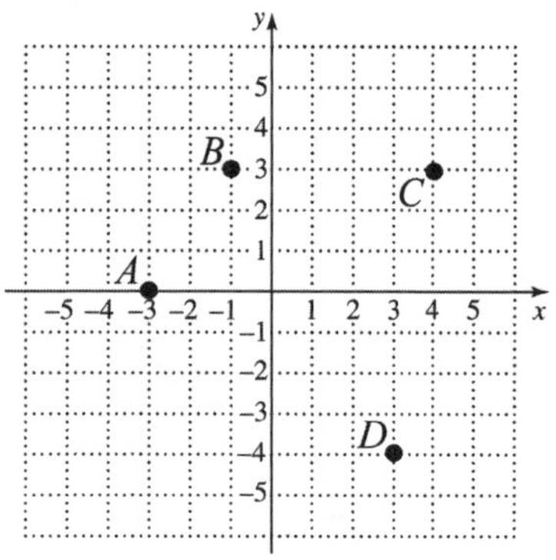

Name: Date:
Instructor: Section:

Objective 2 Plot points in the coordinate plane.
For extra help, see Example 2 on pages 583–584 of your text and the Section 10.1 lecture video.
Plot and label the points indicated by the coordinate pairs.

2. $(-2, 5)$, $(0, 1)$, $(1, -5)$, $(5, 5)$

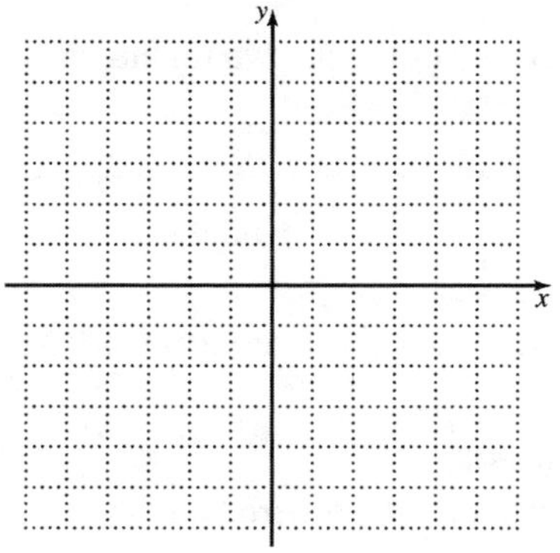

Objective 3 Determine the quadrant for a given ordered pair.
For extra help, see Example 3 on pages 584–585 of your text and the Section 10.1 lecture video.
State the quadrant in which the point is located. If the point lies on an axis, state which axis.

3. $(0, 708)$

4. $(-65, 301)$

5. $(-34, 0)$

6. $(45, 267)$

7. $(-58, -0.26)$

Objective 4 Determine whether the graph of a set of data points is linear.
For extra help, see Example 4 on page 585 of your text and the Section 10.1 lecture video.
Determine whether the set of points is linear or nonlinear.

8. (age of child, height):
$(2, 32)$, $(3, 36)$, $(4, 40)$, $(5, 44)$

9. (age of child, ounces of milk):
$(1, 8)$, $(3, 6)$, $(5, 6)$, $(7, 10)$

Name: Date:
Instructor: Section:

Chapter 10 GRAPHING LINEAR EQUATIONS AND INEQUALITIES

10.2 Graphing Linear Equations

Learning Objectives
1 Determine whether a given pair of coordinates is a solution to a given equation with two unknowns.
2 Find solutions for an equation with two unknowns.
3 Graph linear equations.

GUIDED EXAMPLES AND PRACTICE

Objective 1 Determine whether a given pair of coordinates is a solution to a given equation with two unknowns.

Review this example for Objective 1:

1. Determine whether the ordered pair is a solution for the equation.
$(2, -4)$; $9x-9y=15$

Replace x with 2 and y with -4 to see if the equation is true.

$$9x-9y=15$$
$$9(2)-9(-4)\stackrel{?}{=}15$$
$$18+36\stackrel{?}{=}15$$
$$54\neq 15$$

The ordered pair is not a solution for the equation.

Practice this exercise:

1. Determine whether the ordered pair is a solution for the equation.
$(-3, -5)$; $5x+3y=15$

Objective 2 Find solutions for an equation with two unknowns.

Review this example for Objective 2:

2. Find three solutions for the equation $3x+2y=6$.

To find a solution, we replace one of the variables with a chosen value then solve for the value of the other variable. There are an infinite number of correct solutions for a given equation in two variables.

Practice this exercise:

2. Find three solutions for the equation $y=-3x+2$.

Name:
Instructor:

Date:
Section:

For the first solution, we will choose x to be 0.

$$\begin{aligned} 3x+2y&=6 \\ 3(0)+2y&=6 \\ 2y&=6 \\ y&=3 \end{aligned}$$

Solution: $(0, 3)$

For the second solution, we will choose x to be 2.

$$\begin{aligned} 3x+2y&=6 \\ 3(2)+2y&=6 \\ 6+2y&=6 \\ 2y&=0 \\ y&=0 \end{aligned}$$

Solution: $(2, 0)$

For the third solution, we will choose y to be -3.

$$\begin{aligned} 3x+2y&=6 \\ 3x+2(-3)&=6 \\ 3x-6&=6 \\ 3x&=12 \\ x&=4 \end{aligned}$$

Solution: $(4, -3)$

Objective 3 Graph linear equations.

Review this example for Objective 3:

3. Use the three solutions in Example 2 to graph the equation $3x+2y=6$.

We plot each solution as a point in the rectangular coordinate system and then connect the points to form a straight line.
Solutions: $(0, 3), (2, 0), (4, -3)$

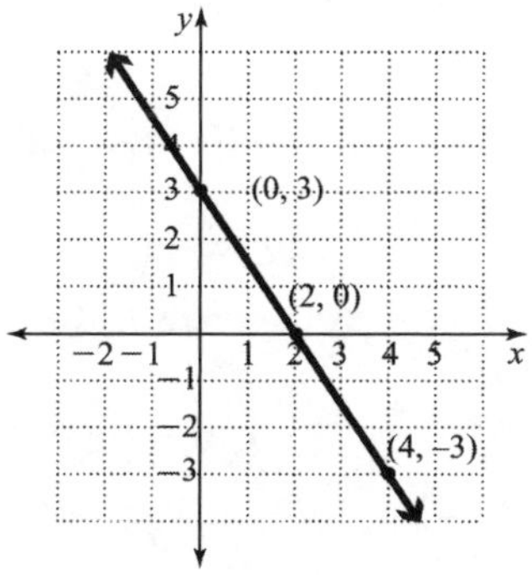

Practice this exercise:

3. Use the three solutions in Practice 2 to graph the equation $y=-3x+2$.

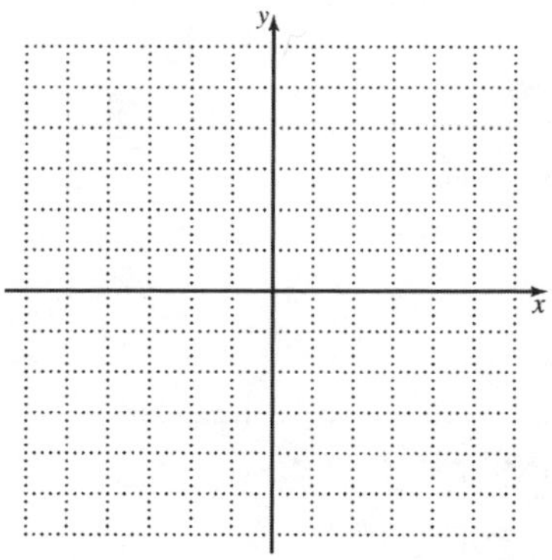

Name: Date:
Instructor: Section:

ADDITIONAL EXERCISES

Objective 1 Determine whether a given pair of coordinates is a solution to a given equation with two unknowns.

For extra help, see Example 1 on page 590 of your text and the Section 10.2 lecture video.

Determine whether the given ordered pair is a solution for the equation.

1. $(2, 6); y = 3x + 11$

2. $\left(0, \frac{5}{6}\right); 4x + 6y = 5$

Objective 2 Find solutions for an equation with two unknowns.

For extra help, see Example 2 on page 591 of your text and the Section 10.2 lecture video.

Find three solutions for the given equation.

3. $y = 4x$

4. $y = -\frac{1}{4}x + 3$

5. $4x - 5y = 20$

Objective 3 Graph linear equations.

For extra help, see Examples 3–5 on pages 592–594 of your text and the Section 10.2 lecture video.

Graph.

6. $y = 4x$

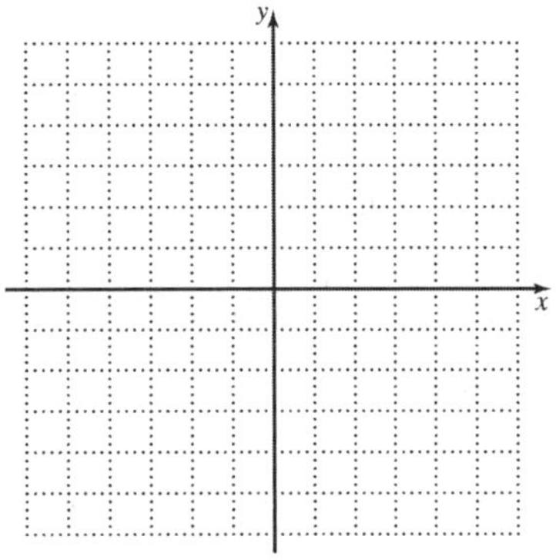

7. $y = -\frac{1}{4}x + 3$

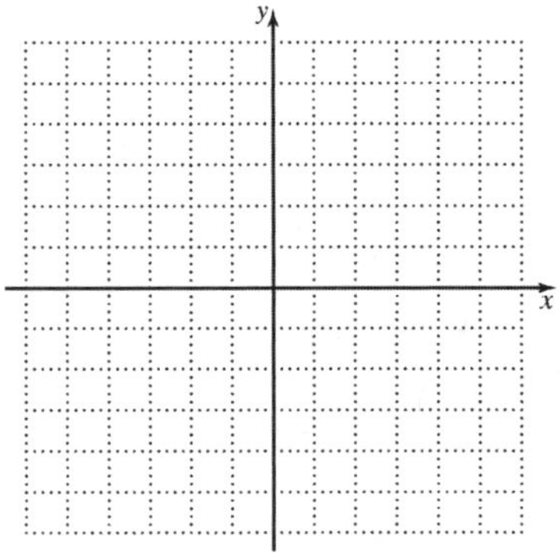

8. $4x - 5y = 20$

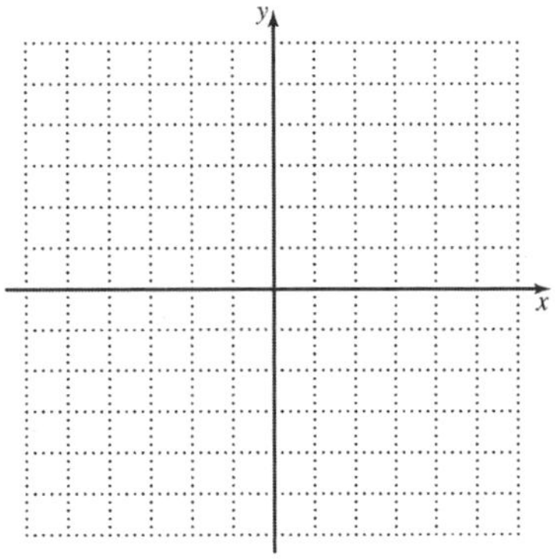

Name: Date:
Instructor: Section:

Solve.

9. A businesswoman buys a new computer for \$1600. For each year that it is in use, she can deduct its depreciated value. The equation $c = -200n + 1600$ gives the value after n years of use.

a) Find the value of the computer after 2 years.

b) In how many years will the computer be worth half of its initial value?

c) After how many years will the computer be worth \$0?

d) Graph the equation with n along the horizontal axis and c along the vertical axis. Because n and c are nonnegative, the graph is restricted to the first quadrant only.

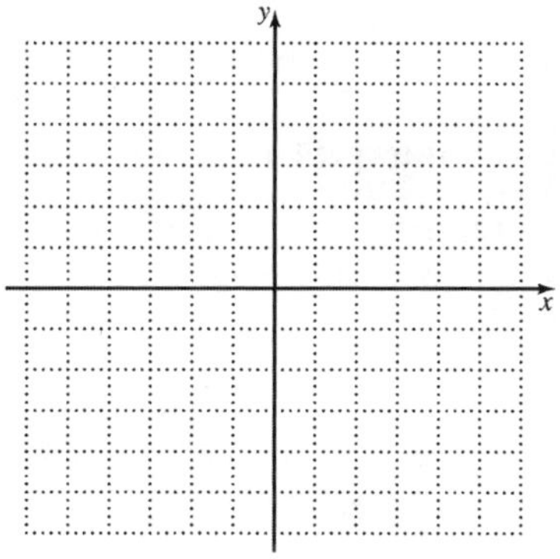

Name: Date:
Instructor: Section:

Chapter 10 GRAPHING LINEAR EQUATIONS AND INEQUALITIES

10.3 Graphing Using Intercepts

Learning Objectives
1 Given an equation, find the coordinates of the x- and y-intercepts.
2 Graph linear equations using intercepts.

Key Terms
Use the vocabulary terms listed below to complete each statement in Exercises 1 and 2.

***x*-intercept** ***y*-intercept**

1. The ________________ is a point where a graph intersects the y-axis.

2. A point where a graph intersects the x-axis is called the ________________.

GUIDED EXAMPLES AND PRACTICE

Objective 1 Given an equation, find the coordinates of the *x*- and *y*-intercepts.

Review this example for Objective 1:

1. Find the x- and y-intercepts for $4x+2y=8$.

For the x-intercept, replace y with 0 and solve for x.

$$4x+2y=8$$
$$4x+2(0)=8$$
$$4x=8$$
$$x=2$$

x-intercept: $(2, 0)$

For the y-intercept, replace x with 0 and solve for y.

$$4x+2y=8$$
$$4(0)+2y=8$$
$$2y=8$$
$$y=4$$

y-intercept: $(0, 4)$

Practice this exercise:

1. Find the x- and y-intercepts for $6x+5y=30$.

Name: Date:
Instructor: Section:

Objective 2 Graph linear equations using intercepts.

Review this example for Objective 2:

2. Graph $x+7y=14$ using the x- and y-intercepts.

Find the x-intercept.

$$x+7y=14$$
$$x+7(0)=14$$
$$x=14$$

x-intercept: $(14, 0)$

Find the y-intercept.

$$x+7y=14$$
$$0+7y=14$$
$$y=2$$

y-intercept: $(0, 2)$

Plot the x- and y-intercepts and draw a line through the two points.

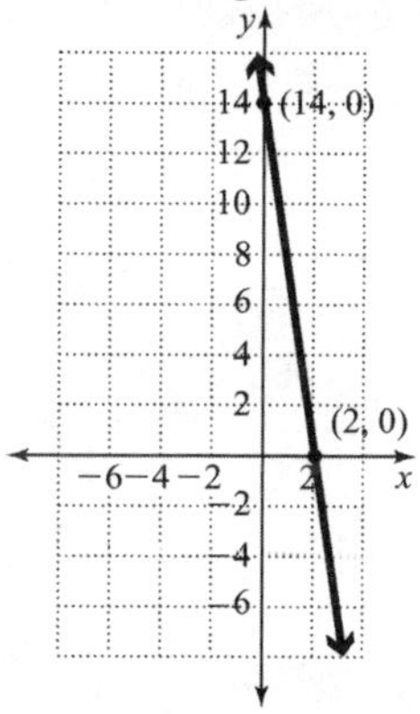

Practice this exercise:

2. Graph $x-5y=10$ using the x- and y-intercepts.

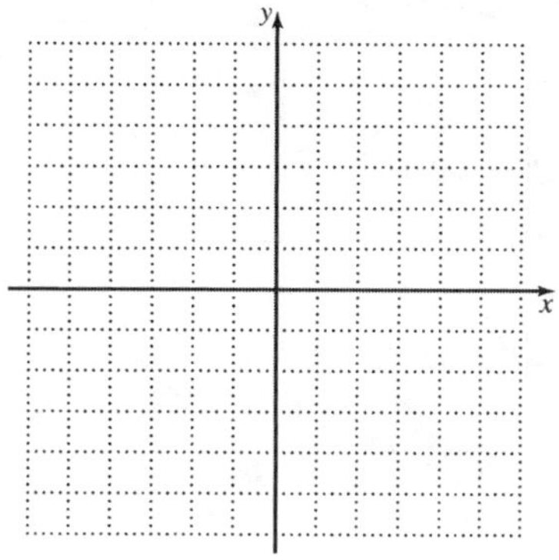

ADDITIONAL EXERCISES

Objective 1 Given an equation, find the coordinates of the x- and y-intercepts.

For extra help, see Examples 1–4 on pages 600–602 of your text and the Section 10.3 lecture video.

Find the x- *and* y-*intercepts.*

1. $y=2x+5$

2. $x-4=0$

Name: Date:
Instructor: Section:

Objective 2 Graph linear equations using intercepts.

For extra help, see Examples 5–6 on pages 602–604 of your text and the Section 10.3 lecture video.

Graph using the x- *and* y-*intercepts.*

3. $y = 2x$

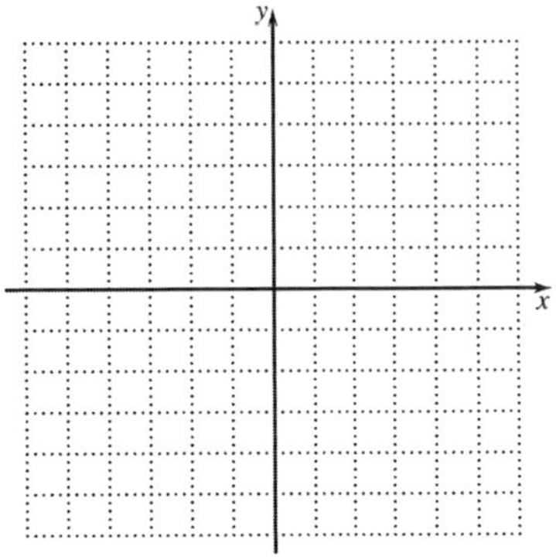

4. $2x - 2 = y$

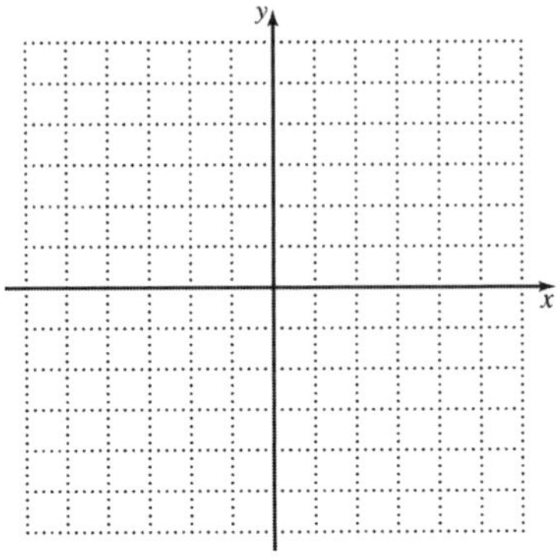

5. $2x + y = 4$

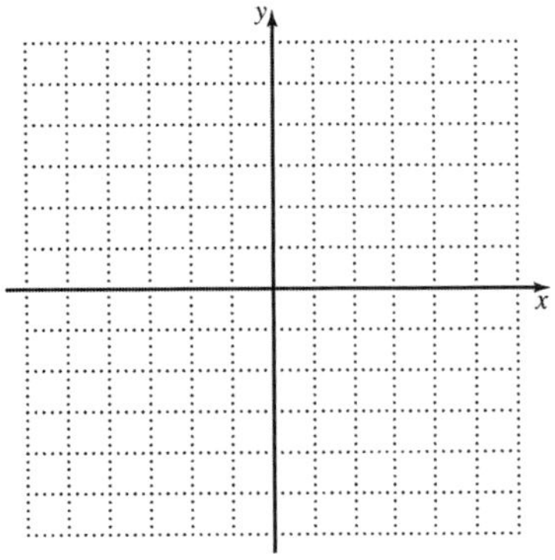

6. $y + 3 = 0$

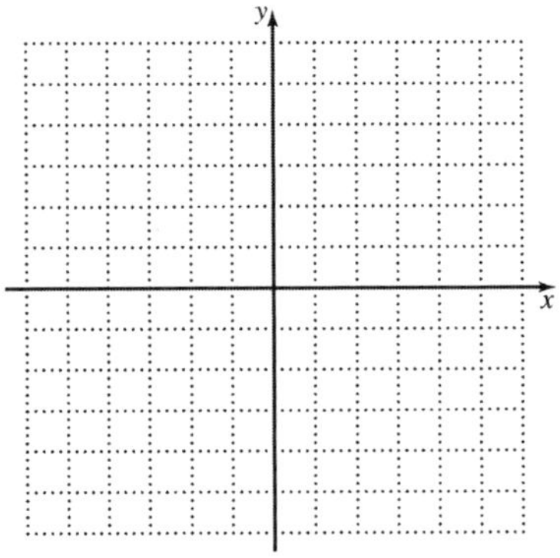

7. $x - 1 = 0$

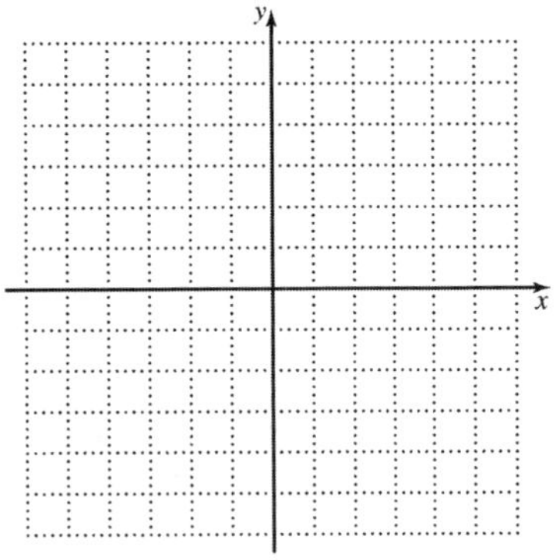

Name: Date:
Instructor: Section:

Chapter 10 GRAPHING LINEAR EQUATIONS AND INEQUALITIES

10.4 Slope-Intercept Form

Learning Objectives
1 Compare lines with different slopes.
2 Graph equations in slope–intercept form.
3 Find the slope of a line given two points on the line.
4 Use slope–intercept form to write the equation of a line.

Key Terms
Use the vocabulary term listed below to complete the statement in Exercise 1.

slope

1. The ____________________ is the ratio of the vertical change to the horizontal change between two points on a line.

GUIDED EXAMPLES AND PRACTICE

Objective 1 Compare lines with different slopes.

Review this example for Objective 1:

1. Graph each equation on the same grid. For each equation, compare the slopes, y-intercepts, and their effects on the graphs.

$y=\frac{1}{2}x$, $y=\frac{1}{4}x$, $y=2x$

Notice that $b=0$ for all equations, so the equations are of the form $y=mx$ and the graphs all have x- and y-intercepts at the point $(0, 0)$. Choose three different values for x and calculate the y-values for each equation.

If x is	and $y=\frac{1}{2}x$, then y is	and $y=\frac{1}{4}x$, then y is	and $y=2x$, then y is
0	0	0	0
1	$\frac{1}{2}$	$\frac{1}{4}$	2
2	1	$\frac{1}{2}$	4

Practice this exercise:

1. Graph each equation on the same grid. For each equation, compare the slopes, y-intercepts, and their effects on the graphs.

$y=\frac{1}{4}x$, $y=x$, $y=4x$

Name:
Instructor:

Date:
Section:

Plot the points on the graph.

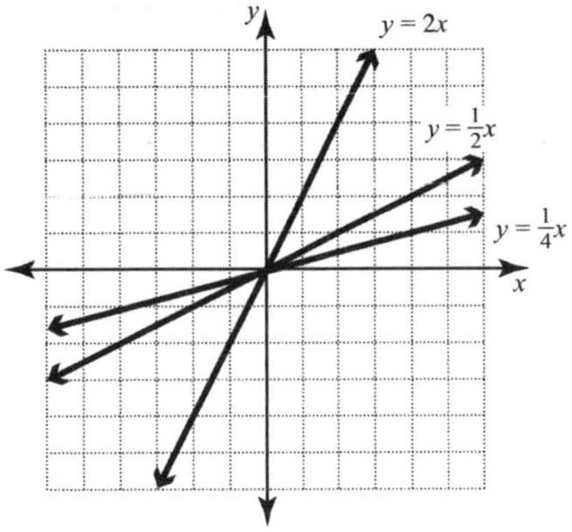

Objective 2 Graph equations in slope–intercept form.

Review this example for Objective 2:

2. Determine the slope and the y-intercept. Then graph the equation.
$3x + y = 6$

Write the equation of the line in slope-intercept form, $y = mx + b$.

$3x + y = 6$

$y = -3x + 6$

The slope is –3 and the y-intercept is 6.

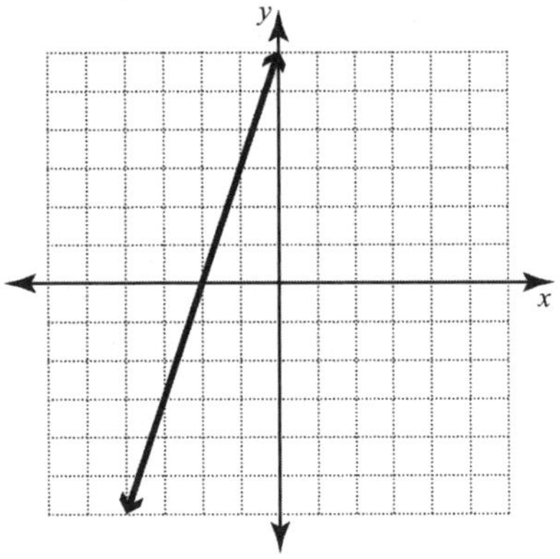

Practice this exercise:

2. Determine the slope and the y-intercept. Then graph the equation.
$2x + 5y = 10$

Name: Date:
Instructor: Section:

Objective 3 Find the slope of a line given two points on the line.

Review this example for Objective 3:

3. Find the slope of the line through the given points.
$(4, 5)$, $(7, 9)$

The slope the vertical distance over the horizontal distance.

$$\frac{\text{Vertical distance}}{\text{Horiziontal distance}} = \frac{y_2 - y_1}{x_2 - x_1}$$
$$= \frac{9-5}{7-4}$$
$$= \frac{4}{3}$$

The slope of the line through $(4, 5)$ and $(7, 9)$ is $\frac{4}{3}$.

Practice this exercise:

3. Find the slope of the line through the given points.
$(2, 3)$, $(8, 4)$

Objective 4 Use slope–intercept form to write the equation of a line.

Review this example for Objective 4:

4. Write the equation of the line in slope-intercept form given the slope and the coordinates of the y-intercept.
$m = 2$; $\left(0, -\frac{1}{4}\right)$

Recall that when x is 0, y is equal to the y-intercept.
Use the slope-intercept form of the equation of a line.
$y = mx + b$
$y = 2x - \frac{1}{4}$

Practice this exercise:

4. Write the equation of the line in slope-intercept form given the slope and the coordinates of the y-intercept.
$m = 4$; $\left(0, -\frac{1}{2}\right)$

Name: Date:
Instructor: Section:

ADDITIONAL EXERCISES

Objective 1 Compare lines with different slopes.

For extra help, see Examples 1–3 on pages 609–611 of your text and the Section 10.4 lecture video.

Graph each set of equations on the same grid. For each set of equations, compare the slopes, y-intercepts, and their effects on the graphs.

1. $y=-\frac{2}{3}x$, $y=-x$, $y=-\frac{3}{2}x$

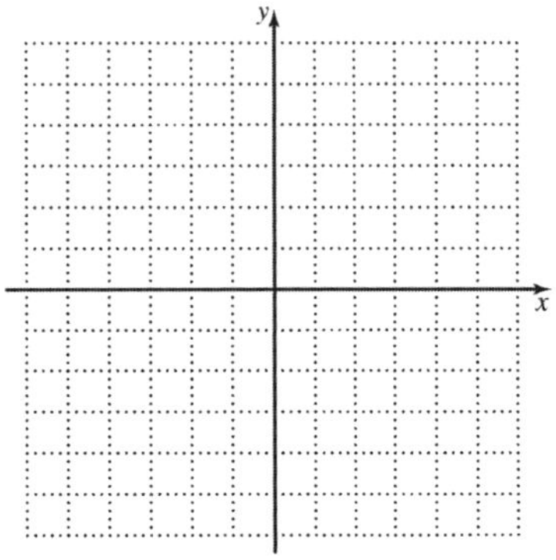

Objective 2 Graph equations in slope–intercept form.

For extra help, see Examples 4 and 5 on pages 612–613 of your text and the Section 10.4 lecture video.

Determine the slope and the y*-intercept. Then graph the equation.*

2. $2x+y=4$

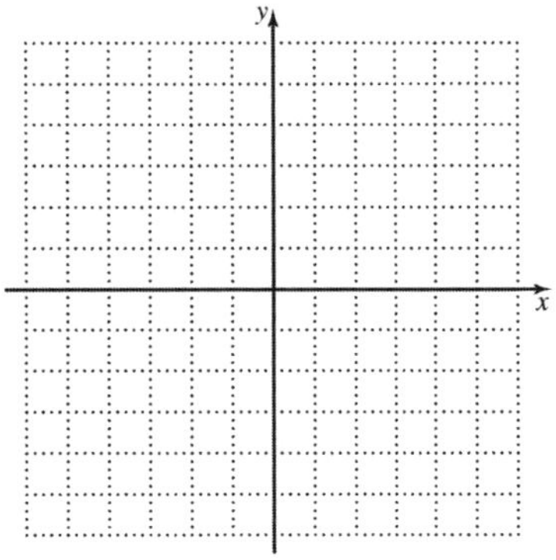

Name: Date:
Instructor: Section:

Objective 3 Find the slope of a line given two points on the line.
For extra help, see Examples 6 and 7 on page 614 of your text and the Section 10.4 lecture video.
Find the slope of the line through the given points.

3. $(-4, 2), (3, 5)$

4. $(1, -7), (8, 0)$

5. $(5, 4), (5, -9)$

Objective 4 Use slope–intercept form to write the equation of a line.
For extra help, see Example 8 on page 615 of your text and the Section 10.4 lecture video.
Write the equation of the line in slope-intercept form given the slope and the coordinates of the y-intercept.

6. $m = -0.4$; $(0, 2.3)$

Name: Date:
Instructor: Section:

Chapter 10 GRAPHING LINEAR EQUATIONS AND INEQUALITIES

10.5 Point-Slope Form

Learning Objectives
1 Use point–slope form to write the equation of a line.
2 Write the equation of a line parallel to a given line.
3 Write the equation of a line perpendicular to a given line.

GUIDED EXAMPLES AND PRACTICE

Objective 1 Use point–slope form to write the equation of a line.

Review this example for Objective 1:

1. Write the equation of the line in slope-intercept form with the given slope passing through the given point.
$m=-5;\ (3,11)$

Use the point slope form, then isolate y to write the slope-intercept form.

$y-y_1=m(x-x_1)$

$y-11=-5(x-3)$

$y-11=-5x+15$

$y=-5x+26$

Practice this exercise:

1. Write the equation of the line in slope-intercept form with the given slope passing through the given point.
$m=-8;\ (7,9)$

Objective 2 Write the equation of a line parallel to a given line.

Review this example for Objective 2:

2. Write the equation of the line that passes through the given point and is parallel to the given line.
$(0,3); y=4x-6$

a) Write the equation in slope-intercept form.

Remember that parallel lines have equal slopes. The point $(0,3)$ is where the line intersects the y-axis. So, the y-intercept is 3.

$y=mx+b$

$y=4x+3$

Practice this exercise:

2. Write the equation of the line that passes through the given point and is parallel to the given line.
$(0,2); y=5x-8$

a) Write the equation in slope-intercept form.

b) Write the equation in the form $Ax+By=C$, where A, B, and C are integers and $A>0$.

Name: Date:
Instructor: Section:

b) Write the equation in the form $Ax + By = C$, where A, B, and C are integers and $A > 0$.

First, isolate the constant term on the right side of the equation.

$y = 4x + 3$

$-4x + y = 3$

Next, make $A > 0$ by multiplying both sides of the equation by -1.

$-1(-4x + y) = -1(3)$

$4x - y = -3$

Objective 3 Write the equation of a line perpendicular to a given line.

Review this example for Objective 3:

3. Write the equation of the line that passes through the given point and is perpendicular to the given line.

$(2,8); y = \frac{1}{4}x + 7$

a) Write the equation in slope-intercept form.

$y = \frac{1}{4}x + 7$ is in the slope-intercept form. From the equation we know that the slope is $\frac{1}{4}$. Thus the slope of the perpendicular line is -4.

Next, use the point slope form to write the equation of the perpendicular line.

$y - y_1 = m(x - x_1)$

$y - 8 = -4(x - 2)$

$y - 8 = -4x + 8$

$y = -4x + 16$

b) Write the equation in the form $Ax + By = C$, where A, B, and C are integers and $A > 0$.

Isolate the constant term on the right side of the equation.

$y = -4x + 16$

$4x + y = 16$

Practice this exercise:

3. Write the equation of the line that passes through the given point and is perpendicular to the given line.

$(2,10); y = \frac{1}{2}x + 5$

a) Write the equation in slope-intercept form.

b) Write the equation in the form $Ax + By = C$, where A, B, and C are integers and $A > 0$.

Name: Date:
Instructor: Section:

ADDITIONAL EXERCISES

Objective 1 Use point–slope form to write the equation of a line.

For extra help, see Examples 1–4 on pages 622–624 of your text and the Section 10.5 lecture video.

Write the equation of the line in slope-intercept form with the given slope passing through the given point.

1. $m=\frac{3}{2};\ (2,\ -4)$

2. $m=-\frac{1}{2};\ (-5,\ -8)$

3. $m=\frac{1}{4};\ (2,4)$

Write the equation of the line through the given points in the form Ax + By = C, where A, B, and C are integers and A > 0.

4. $(4,0),\ (0,5)$

5. $(2,\ -1),\ (0,\ -5)$

6. $(5,3),\ (1,10)$

7. $(4,\ -8),\ (-8,7)$

Objective 2 Write the equation of a line parallel to a given line.

For extra help, see Example 5 on page 624 of your text and the Section 10.5 lecture video.

Write the equation of the line that passes through the given point and is parallel to the given line.

8. **a)** Write the equation in slope-intercept form.
b) Write the equation in the form $Ax + By = C$, where A, B, and C are integers and $A > 0$.
$(2,-2);4x-7y=6$

Objective 3 Write the equation of a line perpendicular to a given line.

For extra help, see Examples 6–7 on pages 625–626 of your text and the Section 10.5 lecture video.

Write the equation of the line that passes through the given point and is perpendicular to the given line.

9. **a)** Write the equation in slope-intercept form.
b) Write the equation in the form $Ax + By = C$, where A, B, and C are integers and $A > 0$.
$(2,9);y=-5x+3$

Determine whether the given lines are parallel, perpendicular, or neither.

10. $y=7x+2$
$y=7x-2$

11. $x=3$
$y=-9$

Name:
Instructor:

Date:
Section:

Chapter 10 GRAPHING LINEAR EQUATIONS AND INEQUALITIES

10.6 Graphing Linear Inequalities

Learning Objectives	
1	Determine whether an ordered pair is a solution for a linear inequality with two variables.
2	Graph linear inequalities.

GUIDED EXAMPLES AND PRACTICE

Objective 1 Determine whether an ordered pair is a solution for a linear inequality with two variables.

Review this example for Objective 1:

1. Determine whether the ordered pair is a solution for the linear inequality.
$(-3,10); y < -5x - 4$

Substitute the x- and y-coordinates into the inequality.

$y < -5x - 4$

$10 \overset{?}{<} -5(-3) - 4$

$10 \overset{?}{<} 15 - 4$

$10 < 11$

Since $10 < 11$ is a true statement, $(-3, 10)$ is a solution for the linear inequality.

Practice this exercise:

1. Determine whether the ordered pair is a solution for the linear inequality.
$(-5,8); y < -7x - 2$

Objective 2 Graph linear inequalities.

Review this example for Objective 2:

2. Graph the linear inequality.
$y \geq -6x + 3$

Remember that solutions for the inequality $y \geq -6x + 3$ satisfy $y = -6x + 3$ and $y > -6x + 3$.

First find the x- and y-intercepts.

x-intercept:

$y = -6x + 3$

$0 = -6x + 3$

$6x = 3$

$x = \frac{1}{2}$

y-intercept:

$y = -6x + 3$

$y = -6(0) + 3$

$y = 3$

x-intercept: $\left(\frac{1}{2}, 0\right)$ y-intercept: $(0, 3)$

Practice this exercise:

2. Graph the linear inequality.
$y \geq -8x + 4$

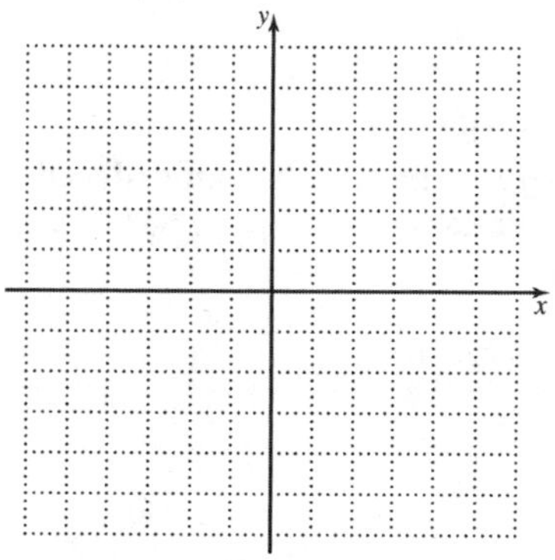

Name:
Instructor:

Date:
Section:

Graph $y = -6x + 3$.

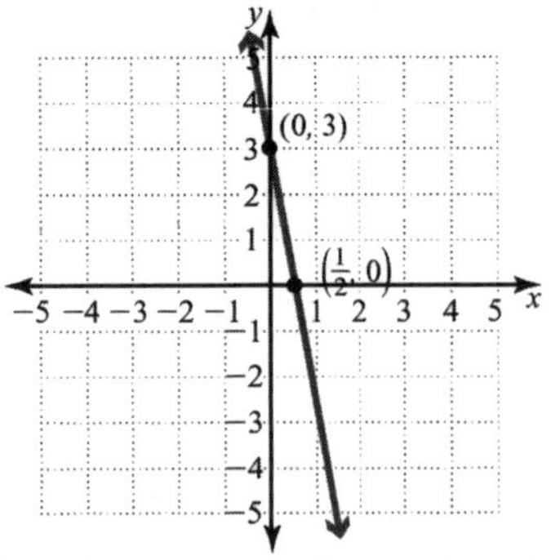

The "greater than" part of the inequality consists of ordered pairs that satisfy $y > -6x + 3$.

Select the point $(1, 1)$ and substitute the coordinates in the inequality.

$y > -6x + 3$

$1 > -6(1) + 3$

$1 > -3$

$1 > -3$ is a true statement. Thus, the region on the same side of the boundary $y = -6x + 3$ as $(1, 1)$ contains all of the ordered pair solutions for $y > -6x + 3$.

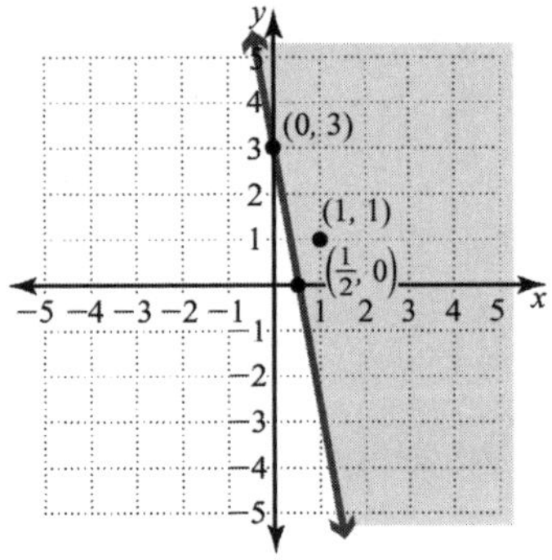

ADDITIONAL EXERCISES

Objective 1 Determine whether an ordered pair is a solution for a linear inequality with two variables.

For extra help, see Example 1 on page 631 of your text and the Section 10.6 lecture video.

Determine whether the ordered pair is a solution for the linear inequality.

1. $(-9,2); 5x + y \leq -6$

2. $(-2,9); 2x + y > 5$

Name: Date:
Instructor: Section:

Objective 2 Graph linear inequalities.
For extra help, see Example 2 on pages 633–635 of your text and the Section 10.6 lecture video.
Graph the linear inequality.

3. $y \le \frac{1}{5}x$

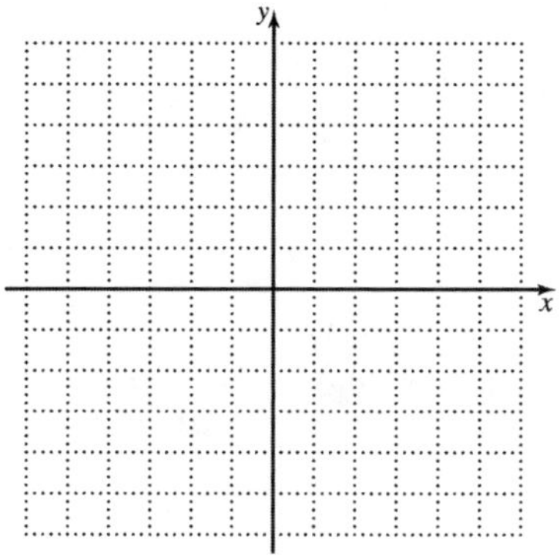

4. $3x + y < 1$

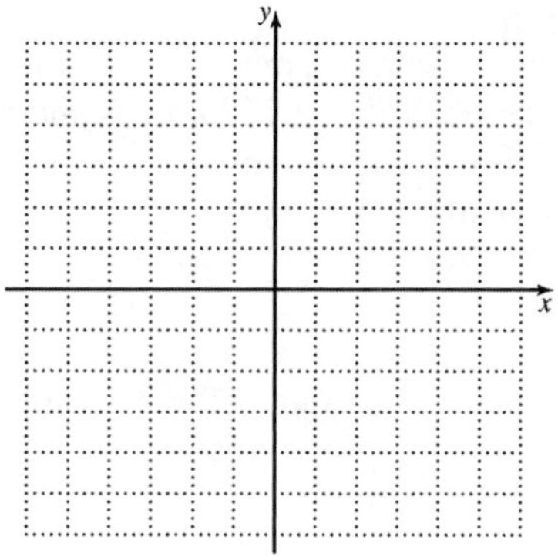

5. $y + 6x > 0$

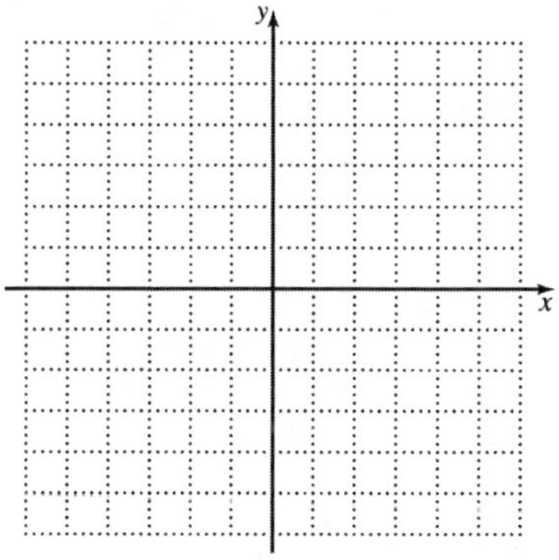

Solve and then graph.

6. Jose sells flowers for \$6 apiece and vases for \$15 apiece. Write an inequality that shows the possible combinations of x flowers and y vases that Jose needs to sell in order to earn at least \$120. Then graph the inequality.

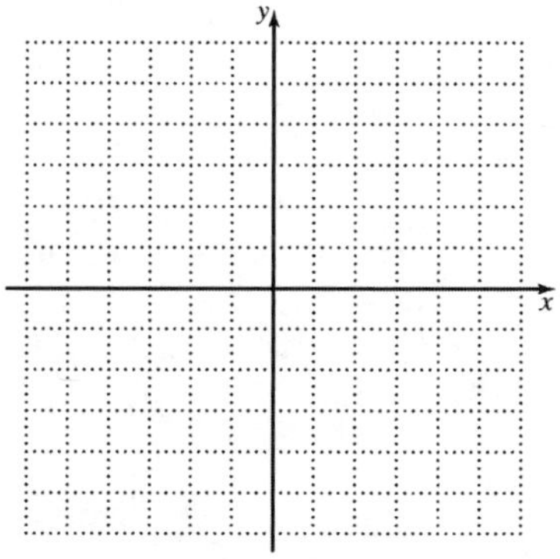

Name: Date:
Instructor: Section:

Chapter 10 GRAPHING LINEAR EQUATIONS AND INEQUALITIES

10.7 Introduction to Functions and Function Notation

Learning Objectives

1 Identify the domain and range of a relation.
2 Identify functions and their domains and ranges.
3 Find the value of a function.
4 Graph linear functions.
5 Graph absolute value functions.

Key Terms

Use the vocabulary terms listed below to complete each statement in Exercises 1–4.

relation **domain** **range** **function**

1. The ____________ is the set of all input values (x-values) for a relation.

2. A set of ordered pairs is called a(n) ____________.

3. A relation in which every value in the domain is paired with (or assigned to) exactly one value in the range is called a(n) ____________.

4. The ____________ is the set of all output values (y-values) for a relation.

GUIDED EXAMPLES AND PRACTICE

Objective 1 Identify the domain and range of a relation.

Review this example for Objective 1:

1. Determine the domain and range of the relation.

$\{(-5, -20), (-1, -4), (0, 0), (2, 8), (3, 12)\}$

The domain is the set containing all of the x-values.

$\{-5, -1, 0, 2, 3\}$

The range is the set containing all of the y-values.

$\{-20, -4, 0, 8, 12\}$

Practice this exercise:

1. Determine the domain and range of the relation.

$\{(7, 2), (24, -9), (38, 2), (7, 8), (50, 8)\}$

Name:
Instructor:

Date:
Section:

Objective 2 Identify functions and their domains and ranges.

Review this example for Objective 2:

2. Determine whether the relation is a function and explain your answer.

$\{(2,1),\ (-1,1),\ (2,0)\}$

Remember that a function is a relation in which every value in the domain is paired with exactly one value in the range. Since the value 2 in the domain is paired with both 1 and 0, this relation is not a function.

Practice this exercise:

2. Determine whether the relation is a function and explain your answer.

$\{(5,4),\ (2,4),\ (5,3)\}$

Objective 3 Find the value of a function.

Review this example for Objective 3:

3. Find the value of the function.

$f(x) = x^2 - x + 6$

Replace x with the given value and solve for $f(x)$.

a) $f(1)$

$f(1) = 1^2 - 1 + 6$

$f(1) = 6$

b) $f(-3)$

$f(-3) = (-3)^2 - (-3) + 6$

$f(-3) = 9 + 3 + 6$

$f(-3) = 18$

c) $f(5)$

$f(5) = (5)^2 - 5 + 6$

$f(5) = 25 + 1$

$f(5) = 26$

Practice this exercise:

3. Find the value of the function.

$f(x) = x^2 - x + 10$

a) $f(1)$

b) $f(-5)$

c) $f(0)$

Name: Date:
Instructor: Section:

Objective 4 Graph linear functions.

Review this example for Objective 4:

4. Graph.

$f(x)=\frac{3}{7}x+2$

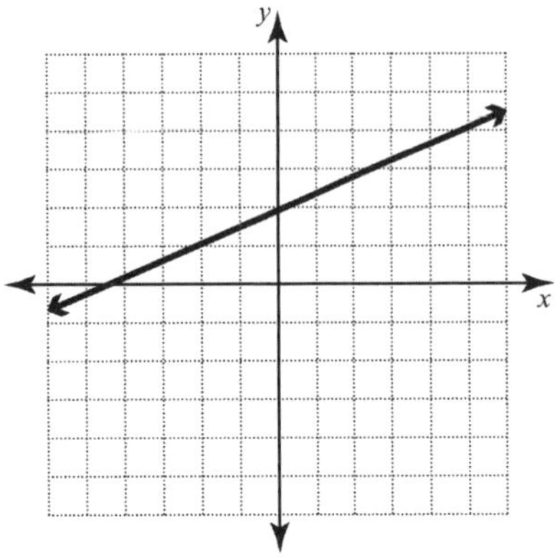

Practice this exercise:

4. Graph.

$f(x)=\frac{4}{5}x+3$

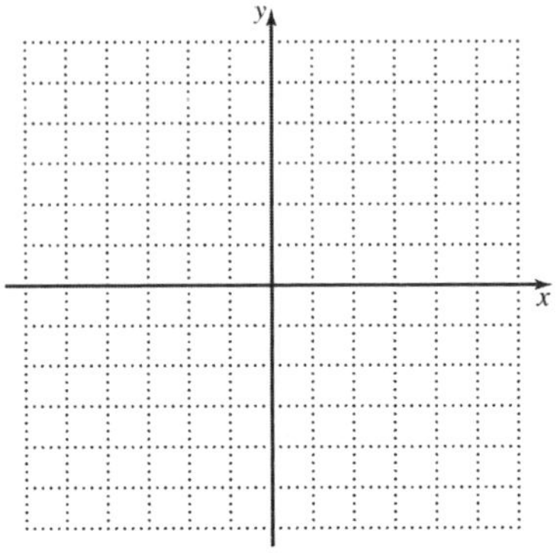

Objective 5 Graph absolute value functions.

Review these examples for Objective 5:

5. Graph.

$f(x)=|x|$

Create a table of ordered pairs. Keep in mind that x is an absolute value, so any negative values of x become positive for $f(x)$.

x	$f(x)$
−2	2
−1	1
0	0
1	1
2	2

Practice these exercises:

5. Graph.

$f(x)=|x|-2$

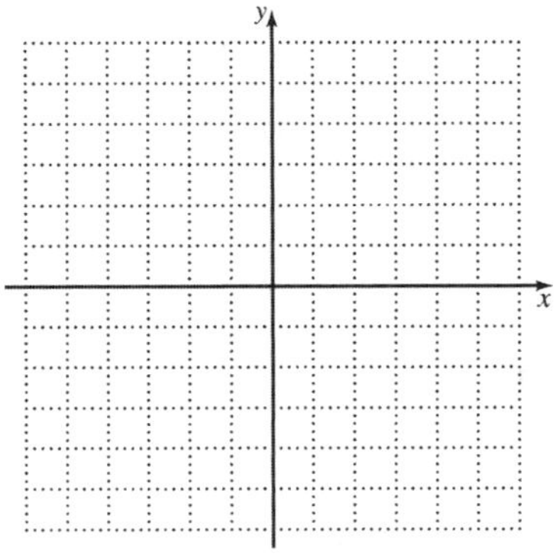

Name: Date:
Instructor: Section:

Plot each point on a graph and connect them.

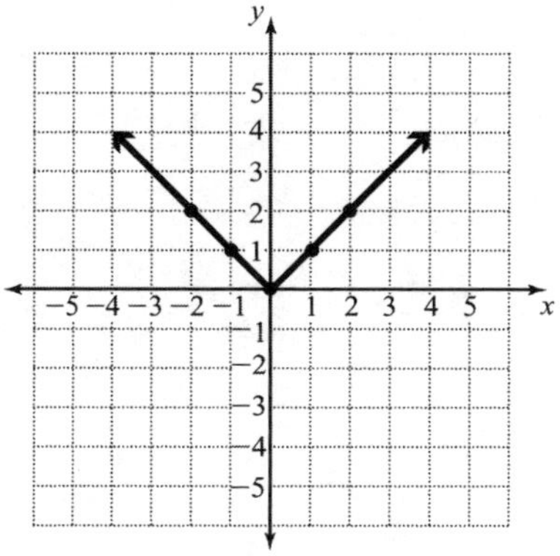

ADDITIONAL EXERCISES

Objective 1 Identify the domain and range of a relation.

For extra help, see Examples 1–2 on pages 640–641 of your text and the Section 10.7 lecture video.

Determine the domain and range of the relation.

1. $\{(6, 4), (16, -2), (35, 4), (6, 2), (59, 2)\}$

Objective 2 Identify functions and their domains and ranges.

For extra help, see Examples 3–4 on pages 642–645 of your text and the Section 10.7 lecture video.

Determine whether the relation is a function and explain your answer.

2. $\{(7, 5), (-2, -4), (2, -1), (3, -7)\}$

Objective 3 Find the value of a function.

For extra help, see Examples 5–6 on pages 646–647 of your text and the Section 10.7 lecture video.

Find the value of the function.

3. $f(x) = \sqrt{x+7}$

a. $f(0)$

b. $f(-10)$

c. $f(b)$

Name: Date:
Instructor: Section:

Use the given graph to find the value of the function.

4. $f(-3)$

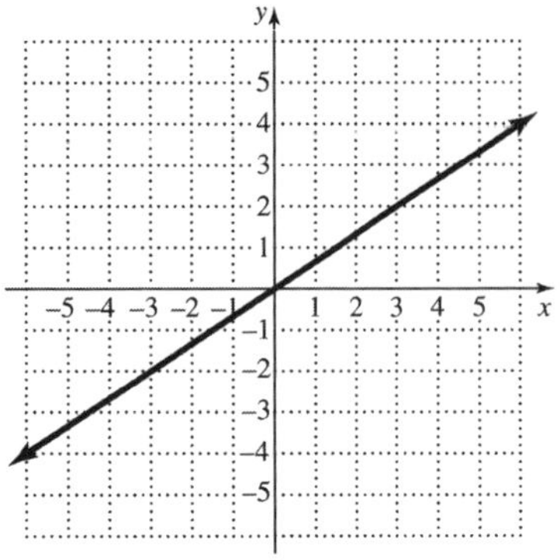

Objective 4 Graph linear functions.
For extra help, see Example 7 on pages 647–648 of your text and the Section 10.7 lecture video.
Graph.

5. Graph the function.
$f(x) = -\frac{3}{2}x - 2$

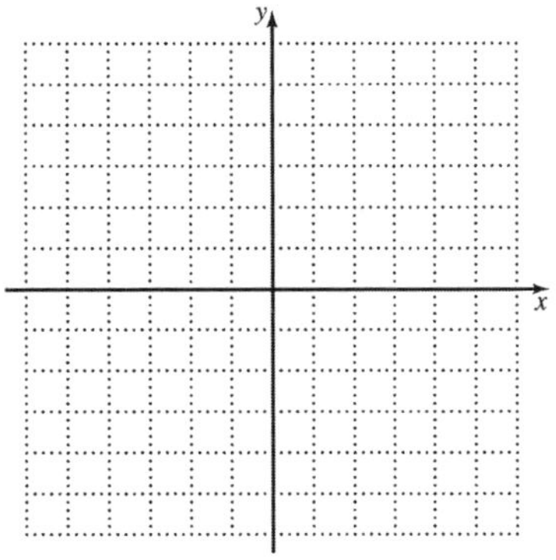

Objective 5 Graph absolute value functions.
For extra help, see Examples 4–8 on page 648–649 of your text and the Section 10.7 lecture video.
Graph.

6. $f(x) = |x+5| - 4$

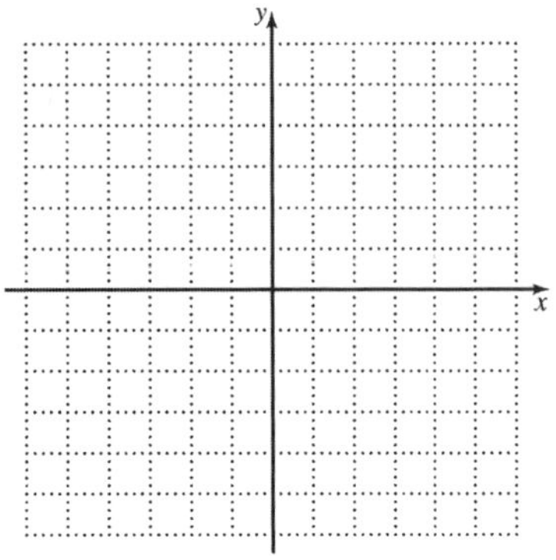

7. $f(x) = |x| + 1$

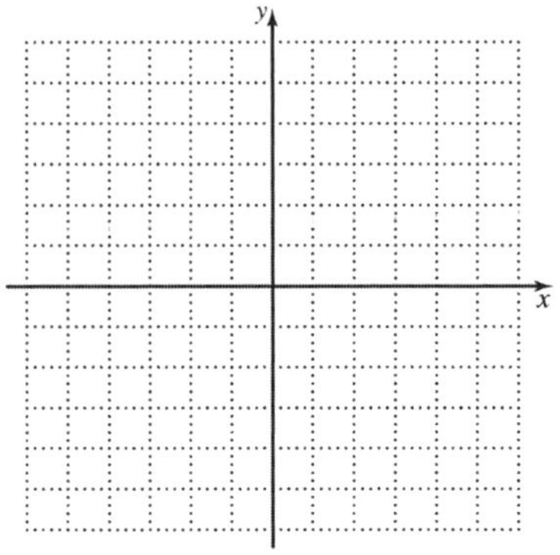

Name: Date:
Instructor: Section:

8. $f(x)=|x-3|+4$

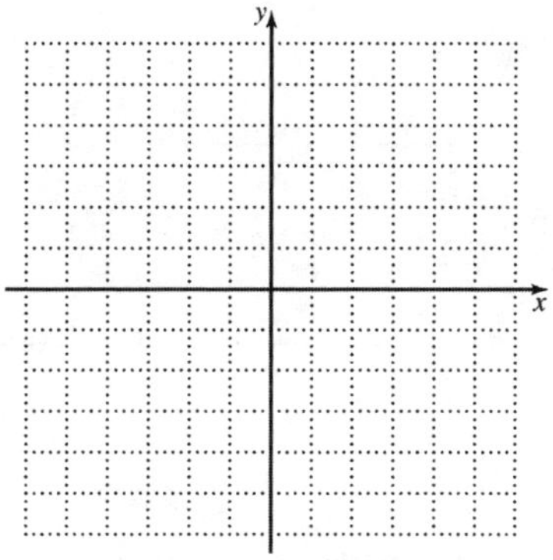

Name: Date:
Instructor: Section:

Chapter 11 SYSTEMS OF LINEAR EQUATIONS AND INEQUALITIES

11.1 Solving Systems of Linear Equations Graphically

Learning Objectives
1 Determine whether an ordered pair is a solution for a system of equations.
2 Solve a system of linear equations graphically.
3 Classify systems of linear equations in two unknowns.

Key Terms
Use the vocabulary terms listed below to complete each statement in Exercises 1–4.

system of equations
consistent system of equations
solution for a system of equations
inconsistent system of equations

1. Some systems of equations have no solutions. This is called a(n) __________________.

2. $\begin{cases} x + y = 3 \\ 2x+3y = 8 \end{cases}$ is called a(n)__________________, or a group of two or more equations.

3. A(n) __________________ is an ordered set of numbers that makes all equations in the system true.

4. A system of equations that has at least one solution is called a(n) ________________.

GUIDED EXAMPLES AND PRACTICE

Objective 1 Determine whether an ordered pair is a solution for a system of equations

Review this example for Objective 1:

1. Determine whether the given ordered pair is a solution to the given system of equations.

$(-1, 4)$; $\begin{cases} x+y=3 \\ y=4x-2 \end{cases}$

Plug in $x=-1$ and $y=4$ into each equation.

Equation 1:	Equation 2:
$x+y=3$	$y=4x-2$
$-1+4\overset{?}{=}3$	$4\overset{?}{=}4(-1)-2$
$3=3$	$4\neq-6$

Because $(-1, 4)$ does not satisfy both equations, it is not a solution for the system.

Practice this exercise:

1. Determine whether the given ordered pair is a solution to the given system of equations.

$(-4, -2)$; $\begin{cases} 3x+5y=-22 \\ 4x+2y=-20 \end{cases}$

Name: Date:
Instructor: Section:

Objective 2 Solve a system of linear equations graphically.

Review this example for Objective 2:

2. Solve the system of linear equations graphically.

$$\begin{cases} y=-x+3 \\ 2x-y=6 \end{cases}$$

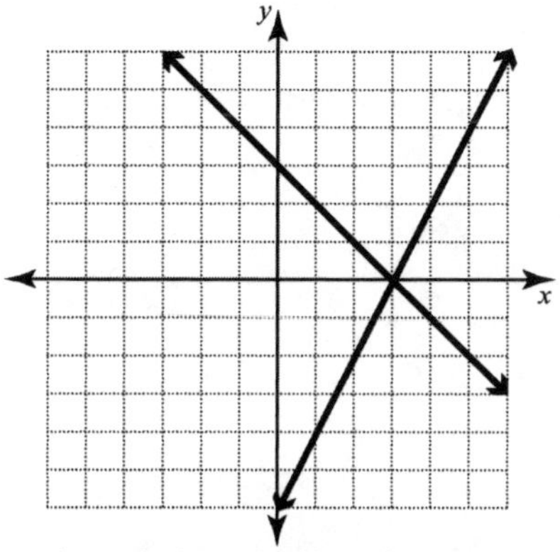

In the solution, the lines intersect at a single point, (3, 0). To check the answer, the point of intersection should satisfy both equations. Since this point satisfies both equations, the graph is correct.

Practice this exercise:

2. Solve the system of linear equations graphically.

$$\begin{cases} x+y=10 \\ x-y=-2 \end{cases}$$

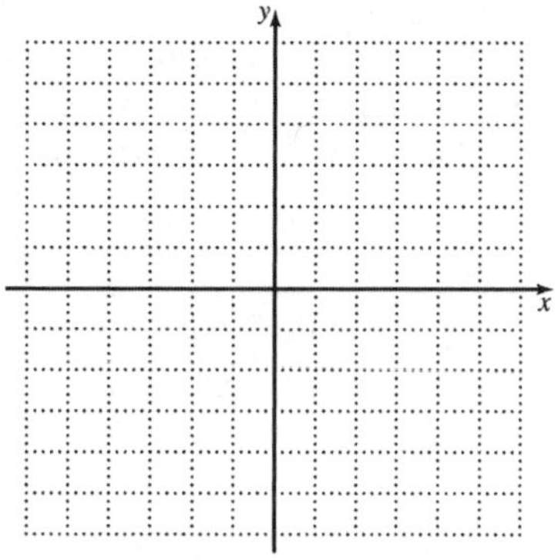

Objective 3 Classify systems of linear equations in two unknowns.

Review this example for Objective 3:

3. **a)** Determine whether the system of equations is consistent with independent equations, consistent with dependent equations, or inconsistent.

b) Determine how many solutions the system has.

$$\begin{cases} x-y=2 \\ 9x-3y=45 \end{cases}$$

a) Write each equations in standard form:

$y=x-2$

$y=3x-15$

Looking at the equations in standard form, we can see that they have different slopes and different y-intercepts. Therefore, the system is consistent (it has a solution) with independent equations (different graphs).

b) The system has one solution.

Practice this exercise:

3. **a)** Determine whether the system of equations is consistent with independent equations, consistent with dependent equations, or inconsistent.

b) Determine how many solutions the system has.

$$\begin{cases} x-y=6 \\ 6x-6y=37 \end{cases}$$

Name: Date:
Instructor: Section:

ADDITIONAL EXERCISES

Objective 1 Determine whether an ordered pair is a solution for a system of equations.
For extra help, see Example 1 on page 674 of your text and the Section 11.1 lecture video.
Determine whether the given ordered pair is a solution to the given system of equations.

1. $(7, 4); \begin{cases} 3x - 2y = 13 \\ 4x - 9y = 11 \end{cases}$

2. $\left(\frac{1}{2}, \frac{3}{4}\right); \begin{cases} 10x - 4y = 1 \\ 14x - 8y = 2 \end{cases}$

Objective 2 Solve a system of linear equations graphically.
For extra help, see Example 2 on page 676 of your text and the Section 11.1 lecture video.
Solve the system of linear equations graphically.

3. $\begin{cases} y = 2x - 2 \\ 3x + y = 3 \end{cases}$

4. $\begin{cases} 3x - y = 3 \\ 2x + y = 2 \end{cases}$

5. $\begin{cases} 2x - 6y = 6 \\ 3x - 9y = -6 \end{cases}$

6. $\begin{cases} 4x - 5y = 2 \\ 10y = 8x - 4 \end{cases}$

Objective 3 Classify systems of linear equations in two unknowns.
For extra help, see Example 3 on page 678 of your text and the Section 11.1 lecture video.
a) Determine whether the system of equations is consistent with independent equations, consistent with dependent equations, or inconsistent.
b) Determine how many solutions the system has.

7. $\begin{cases} 7x + y = 4 \\ 35x + 5y = 20 \end{cases}$

8. $\begin{cases} 3x + y = 13 \\ 4x + 5y = -1 \end{cases}$

Name: Date:
Instructor: Section:

Chapter 11 SYSTEMS OF LINEAR EQUATIONS AND INEQUALITIES

11.2 Solving Systems of Linear Equations by Substitution; Applications

Learning Objectives
1 Solve systems of linear equations using substitution.
2 Solve applications involving two unknowns using a system of equations.

GUIDED EXAMPLES AND PRACTICE

Objective 1 Solve systems of linear equations using substitution.

Review this example for Objective 1:

1. Solve the system of linear equations using substitution.

$$\begin{cases} x - y = 3 \\ 7x + 3y = -39 \end{cases}$$

Isolate x in the first equation.

$x = y + 3$

Substitute x in the second equation.

$$7(y+3) + 3y = -39$$
$$7y + 21 + 3y = -39$$
$$10y = -60$$
$$y = -6$$

Substitute y in the first equation to solve for x.

$x = y + 3$

$x = -6 + 3$

$x = -3$

The solution for the system of equations is $(-3, -6)$.

Practice this exercise:

1. Solve the system of linear equations using substitution.

$$\begin{cases} x + y = 4 \\ y = x - 2 \end{cases}$$

Name: Date:
Instructor: Section:

Objective 2 Solve applications involving two unknowns using a system of equations.

Review this example for Objective 2:

2. Solve.
There were 334 tickets sold for a basketball game. Student tickets cost \$1.50 and non-student tickets cost \$2.00. The total amount of money collected was \$527.00. How many of each kind of ticket were sold?

Let x represent the number of student tickets sold and y represent the number of non-student tickets sold.

$$\begin{cases} x+y=334 \\ 1.5x+2y=527 \end{cases}$$

Solve for x in the first equation.

$x=334-y$

Substitute x in the second equation.

$$\begin{aligned} 1.5(334-y)+2y&=527 \\ 501-1.5y+2y&=527 \\ 501+0.5y&=527 \\ 0.5y&=26 \\ y&=52 \end{aligned}$$

Substitute y in the first equation to solve for x.

$x=334-y$

$x=334-52$

$x=282$

The solution for the system of equations is $(282, 52)$. So, 282 student tickets and 52 non-student tickets were sold.

Practice this exercise:

2. Solve.
A motel clerk counts his \$1 and \$10 bills at the end of a day. He finds that he has a total of 45 bills having a combined monetary value of \$162. Find the number of bills of each denomination that he has.

ADDITIONAL EXERCISES

Objective 1 Solve systems of linear equations using substitution.

For extra help, see Examples 1–5 on pages 683–687 of your text and the Section 11.2 lecture video.

Solve the system of equations using substitution. Note that some systems may be inconsistent or consistent with dependent equations.

1. $\begin{cases} 9x+2y=-55 \\ x=17-6y \end{cases}$

2. $\begin{cases} x-9y=62 \\ 4y-7x=-80 \end{cases}$

3. $\begin{cases} 3x+y=24 \\ y=3x \end{cases}$

4. $\begin{cases} x-2=y \\ 4x-4y=8 \end{cases}$

Name: Date:
Instructor: Section:

5. $\begin{cases} 2x-5y=5 \\ 2x+5y=5 \end{cases}$

6. $\begin{cases} 4x+5y=14 \\ 2x+3y=10 \end{cases}$

7. $\begin{cases} x-9=y \\ 9x-9y=45 \end{cases}$

8. $\begin{cases} 9x+3y=21 \\ 2x=18-4y \end{cases}$

Objective 2 Solve applications involving two unknowns using a system of equations.
For extra help, see Examples 6–8 on pages 687–689 of your text and the Section 11.2 lecture video.
Solve.

9. The sum of two numbers is 35. One number is 9 more than the other. Find the numbers.

10. Two angles are supplementary. One angle is 3° smaller than twice the other. Find the measures of the angles.

11. The Everton College store paid \$1665 for an order of 45 calculators. The store paid \$9 for each scientific calculator. The others, all graphing calculators, cost the store \$54 each. How many of each type of calculator was ordered?

12. A train leaves a station and travels North at 75 kilometers per hour. Two hours later, a second train leaves on a parallel track and travels North at 115 kilometers per hour. How far from the station will they meet?

Name: Date:
Instructor: Section:

Chapter 11 SYSTEMS OF LINEAR EQUATIONS AND INEQUALITIES

11.3 Solving Systems of Linear Equations by Elimination; Applications

Learning Objectives
1 Solve systems of linear equations using elimination.
2 Solve applications using elimination.

GUIDED EXAMPLES AND PRACTICE

Objective 1 Solve systems of linear equations using elimination.

Review this example for Objective 1:

1. Solve the systems of equations using the elimination method.

$$\begin{cases} 4x - 2y = 38 \\ 5x + 3y = -2 \end{cases}$$

The system is already in standard form and there are no fractions or decimals. Multiply both equations by numbers that create a pair of terms that are additive inverses.

$$4x - 2y = 38 \xrightarrow{\text{Multiply by 3.}} 12x - 6y = 114$$

$$5x + 3y = -2 \xrightarrow{\text{Multiply by 2.}} 10x + 6y = -4$$

Our new system of equations:

$$\begin{cases} 12x - 6y = 114 \\ 10x + 6y = -4 \end{cases}$$

Add the equations to eliminate y since the y terms are additive inverses.

$22x = 110$

Solve for x.

$x = 5$

Substitute the value for x into either one of the original equations in the system.

$$\begin{aligned} 4x - 2y &= 38 \\ 4(5) - 2y &= 38 \\ 20 - 2y &= 38 \\ -2y &= 18 \\ y &= -9 \end{aligned}$$

The solution for the system of equations is $(5, -9)$.

Practice this exercise:

1. Solve the systems of equations using the elimination method.

$$\begin{cases} x + y = 3 \\ x - y = 5 \end{cases}$$

Name: Date:
Instructor: Section:

Objective 2 Solve applications using elimination.

Review this example for Objective 2:

2. Solve.

A pharmacist has a 40% acid solution and a 25% acid solution. How many liters of each must be mixed to form 105 liters of a 27% acid solution?

There are three acid solutions in this problem: the 40% acid solution, the 25% acid solution, and the new 27% acid solution that is created by mixing the 40% and 25% acid solutions. Find the volume of the 40% acid solution and the volume of the 25% acid solution. Choose x and y to represent these unknown amounts, respectively. Since both of the acid solution volumes are unknown, write an equation that describes the total volume.

$x+y=105$

Next, write an equation that describes the relationship between the volumes of acid solution.

$0.4x+0.25y=28.35$

The system of equations is below. Solve the system using the elimination method.

$$\begin{cases} x+y=105 \\ 0.4x+0.25y=28.35 \end{cases}$$

Multiply the first equation by -0.4 to eliminate the x variable.

$$\begin{cases} -0.4x-0.4y=-42 \\ 0.4x+0.25y=28.35 \end{cases}$$

Add the equations to eliminate x.

$-0.15y=-13.65$

Divide to solve for y

$y=91$

Now substitute 91 for y into the first equation.

$x+y=105$

$x+91=105$

$x=14$

The solution for the system of equations is (14, 91). So, 14 liters of a 40% acid solution and 91 liters of a 25% solution are mixed to form the 27% acid solution.

Practice this exercise:

2. Solve.

If a plane can travel 490 miles per hour with the wind, and 410 miles per hour against the wind, find the speed of the plane in still air.

Name: Date:
Instructor: Section:

ADDITIONAL EXERCISES

Objective 1 Solve systems of linear equations using elimination.

For extra help, see Examples 1–5 on pages 694–698 of your text and the Section 11.3 lecture video.

Solve the system of equations using the elimination method. Note that some systems may be inconsistent or consistent with dependent equations.

1. $\begin{cases} x+y=15 \\ 8x-y=57 \end{cases}$

2. $\begin{cases} 2x+5y=14 \\ 9x-5y=-102 \end{cases}$

3. $\begin{cases} \frac{5}{2}x+\frac{5}{3}y=15 \\ \frac{1}{4}x+\frac{1}{3}y=2 \end{cases}$

4. $\begin{cases} 0.5x=-45.5+3y \\ -\frac{6}{5}x+y=\frac{19}{5} \end{cases}$

Objective 2 Solve applications using elimination.

For extra help, see Examples 6–8 on pages 698–701 of your text and the Section 11.3 lecture video.

Solve.

5. Two angles are supplementary. One is $90°$ more than twice the other. Find the measures of the angles.

6. The perimeter of a rectangular rug is 28 feet. The length is 2 feet longer than the width. Find the dimensions of the rug.

7. Mr. White invested $29,000 in two accounts, one yielding 6% interest and the other yielding 10%. If he received a total of $2100 in interest at the end of the year, how much did he invest in each account?

8. Sheila Grant has twice as much money invested at 5% simple annual interest as she does at 4%. If her yearly income from the two investments is $119, how much does she have at each rate?

9. One canned juice drink is 25% orange juice; another is 10% orange juice. How many liters of each should be mixed together in order to get 15 liters that is 13% orange juice?

10. A store mixes Kenyan coffee, worth $12 per kilogram, and Turkish coffee, worth $15 per kilogram. The mixture is to sell for $13 per kilogram. Find how much of each should be used to make a 588-kilogram mixture.

11. A mother wants to invest $15,000 for her son's future education. She invests a portion of the money in a bank certificate of deposit (CD account), which earns 4%, and the remainder in a savings bond that earns 7%. If the total interest earned after one year is $900, how much money was invested in the CD account?

12. A chocolate factory makes dark chocolate that is 33% fat, and white chocolate that is 47% fat. How many kilograms of dark chocolate should be mixed with 700 kilograms of white chocolate to make a ripple blend that is 35% fat?

Name: Date:
Instructor: Section:

Chapter 11 SYSTEMS OF LINEAR EQUATIONS AND INEQUALITIES

11.4 Solving Systems of Linear Equations in Three Variables; Applications

Learning Objectives
1 Determine whether an ordered triple is a solution for a system of equations.
2 Understand the types of solution sets for systems of three equations.
3 Solve a system of linear equations using the elimination method.
4 Solve application problems that translate to a system of three linear equations.

GUIDED EXAMPLES AND PRACTICE

Objective 1 Determine whether an ordered triple is a solution for a system of equations.

Review this example for Objective 1:

1. Determine whether each ordered triple is a solution to the system of equations.

$$\begin{cases} x+y+z=3 \\ x+2y+2z=6 \\ 2x-3y+z=8 \end{cases};\ (2,1,0)$$

In all three equations, replace x with 2, y with 1, and z with 0.

Equation 1 $x+y+z=3$

$$2+1+0\stackrel{?}{=}3$$
$$3=3$$

Equation 2 $x+2y+2z=6$

$$(2)+2(1)+2(0)\stackrel{?}{=}6$$
$$4\neq 6$$

Equation 3 $2x-3y+z=8$

$$2(2)-3(1)+0\stackrel{?}{=}8$$
$$1\neq 8$$

Because $(2, 1, 0)$ does not satisfy all three equations in the system, it is not a solution for the system.

Practice this exercise:

1. Determine whether each ordered triple is a solution to the system of equations.

$$\begin{cases} x+y+z=-6 \\ x-2y-z=-9 \\ 2x+2y-z=-3 \end{cases};\ (-6, 3, -3)$$

Name: Date:
Instructor: Section:

Objective 2 **Understand the types of solution sets for systems of three equations.**

Review this example for Objective 2:

2. If the planes are as shown, how many solutions are there for the system of three equations?

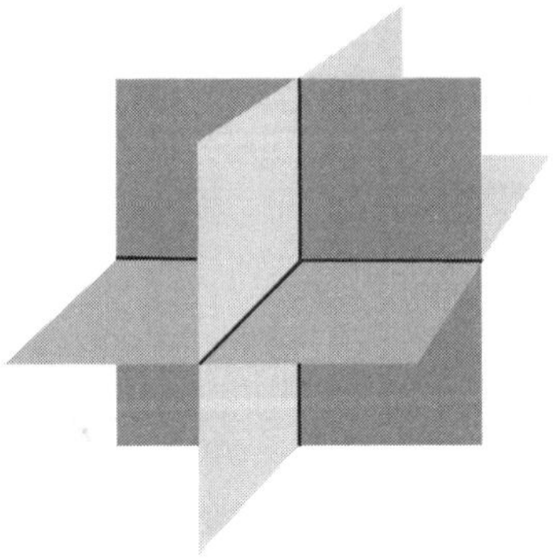

All three planes intersect at one single point in the center. So, there is one solution for the system.

Practice this exercise:

2. If the planes are as shown, how many solutions are there for the system of three equations?

Objective 3 **Solve a system of linear equations using the elimination method.**

Review this example for Objective 3:

3. Solve the system of equations using elimination.

$$\begin{cases} 2x+3y-z=-6 \\ -6x-2y-2z=-8 \\ x-y+2z=7 \end{cases}$$

Equation 1 $2x+3y-z=-6$

Equation 2 $-6x-2y-2z=-8$

Equation 3 $x-y+2z=7$

Choose a variable and eliminate that variable using two of the equations.
Eliminate x using equations 1 and 2.

$$2x+3y-z=-6 \xrightarrow{\text{Multiply by 3.}} 6x+9y-3z=-18$$

Now, add this equation to equation 2 to get $7y-5z=-26$.

Now, use equations 2 and 3 to eliminate x.

$$x-y+2z=7 \xrightarrow{\text{Multiply by 6.}} 6x-6y+12z=42$$

Add this equation to equation 2 to get $-8y+10z=34$.

Now, add the two equations together and solve for z.

$$7y-5z=-26 \xrightarrow{\text{Multiply by 2.}} 14y-10z=-52$$

Add $14y-10z=-52$ to $-8y+10z=34$ to get $6y=-18$.

Practice this exercise:

3. Solve the system of equations using elimination.

$$\begin{cases} x+y+z=-2 \\ 2x+5y+2z=-10 \\ -x+6y-3z=-16 \end{cases}$$

Name: Date:
Instructor: Section:

Solve for y.
$y = -3$
Now, substitute the value for y into either equation with two variables to solve for z.

$$\begin{aligned} -8y + 10z &= 34 \\ -8(-3) + 10z &= 34 \\ 10z &= 10 \\ z &= 1 \end{aligned}$$

Substitute the values for y and z into either of the original equations and solve for x.

$$\begin{aligned} 2x + 3y - z &= -6 \\ 2x + 3(-3) - 1 &= -6 \\ 2x - 10 &= -6 \\ 2x &= 4 \\ x &= 2 \end{aligned}$$

The solution for the system of equations is $(2, -3, 1)$.

Objective 4 Solve application problems that translate to a system of three linear equations.

Review this example for Objective 4:

3. Solve.
At a fair, John buys one soft drink, one candy apple, and one pretzel, all for \$5. Fred buys two soft drinks, three candy apples, and two pretzels for \$14. Carla buys one soft drink, two candy apples, and three pretzels for \$18. Find the price of one soft drink, one candy apple, and one pretzel.

Let x represent the cost of one soft drink, y the cost of one candy apple, and z the cost of one pretzel.

$x + y + z = 5$ (1)
$2x + 3y + 2z = 14$ (2)
$x + 2y + 3z = 10$ (3)

Choose to eliminate x from pairs of equations. In this example, start with equations 1 and 2.

$$x + y + z = 5 \xrightarrow{\text{Multiply by } -2.} -2x - 2y - 2z = -10$$

Now, add this equation to equation 2 to get $y = 4$.

Now, choose another pair of equations and eliminate z again. This time, choose equations 1 and 3.

$$x + y + z = 5 \xrightarrow{\text{Multiply by } -3.} -3x - 3y - 3z = -15$$

Now, add this equation to equation 3 to get $-2x - y = -5$.

Substitute to find x.

$$\begin{aligned} -2x - 4 &= -5 \\ -2x &= -1 \\ x &= 0.5 \end{aligned}$$

Practice this exercise:

3. Solve.
The sum of the measures of the angles in every triangle is $180°$. In triangle ABC, the measure of angle B is $26°$ more than three times the measure of angle A. The measure of angle C is $49°$ more than the measure of angle A. Find the measure of each angle.

Name: Date:
Instructor: Section:

Substitute the values of x and y to find z. Use equation 1.

$$x + y + z = 5$$
$$0.5 + 4 + z = 5$$
$$z = 1.5$$

Therefore, a soft drink costs \$0.50, a candy apple costs \$4.00 and a pretzel costs \$0.50.

ADDITIONAL EXERCISES

Objective 1 Determine whether an ordered triple is a solution for a system of equations.
For extra help, see Example 1 on page 705 of your text and the Section 11.4 lecture video.
Determine whether the ordered triple is a solution of the system.

1. $\begin{cases} x + y + z = 4 \\ x - 2y - z = 5 \\ 3x + 2y - z = 15 \end{cases}$; $(5, 1, -2)$

2. $\begin{cases} x + y + z = 6 \\ x + z = 8 \\ 3x + 2y = 16 \end{cases}$; $(4, 2, 0)$

Objective 2 Understand the types of solution sets for systems of three equations.
For extra help, see the Section 11.4 lecture video.
If the planes are as shown, how many solutions are there for the system of three equations?

3.

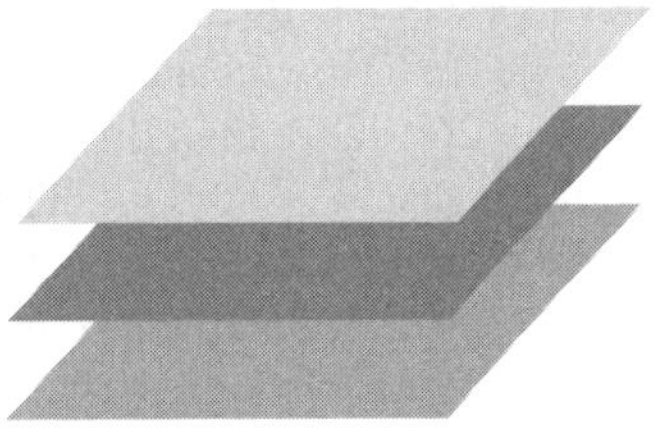

Objective 3 Solve a system of linear equations using the elimination method.
For extra help, see Examples 2–3 on pages 707–709 of your text and the Section 11.4 lecture video.
Solve the system of equations.

4. $\begin{cases} 4x + 3y + z = 4 \\ x - 3y + 2z = 9 \\ 11x - 2y + 3z = 51 \end{cases}$

5. $\begin{cases} 7x - y + z = -25 \\ 2x + 2y - 3z = 6 \\ x - 3y + 2z = -11 \end{cases}$

6. $\begin{cases} 5x + 5y - 10z = 7 \\ 3x + y - z = 5 \\ -x - y + 2z = 3 \end{cases}$

7. $\begin{cases} -3x + 3y - 9z = -12 \\ x - y + 3z = 4 \\ 5x - 5y + 15z = 20 \end{cases}$

Name: Date:
Instructor: Section:

8. $\begin{cases} x = -y - z - 5 \\ 2x + 5y + 2z = 2 \\ -x + 7y - 3z = 47 \end{cases}$

9. $\begin{cases} \frac{7}{2}x + y - z = 0 \\ 2y - 4z = -72 \\ -0.4x + 0.2y = -3.6 \end{cases}$

10. $\begin{cases} 2x - 3y + 4z = 4 \\ x + 2y + z = 14 \\ 5y - z = 18 \end{cases}$

Objective 4 Solve application problems that translate to a system of three linear equations.
For extra help, see Example 4 on page 709-710 of your text and the Section 11.4 lecture video.
Translate to a system of three equations; then solve.

11. A basketball team recently scored a total of 80 points on a combination of 2-point field goals, 3-point field goals, and 1-point foul shots. Altogether, the team made 45 baskets and 19 more 2-point field goals than foul shots. How many shots of each kind were made?

Name: Date:
Instructor: Section:

Chapter 11 SYSTEMS OF LINEAR EQUATIONS AND INEQUALITIES

11.5 Solving Systems of Linear Equations Using Matrices

Learning Objectives
1 Write a system of equations as an augmented matrix.
2 Solve a system of linear equations by transforming its augmented matrix to row echelon form.
3 Solve applications problems using matrices.

Key Terms
Use the vocabulary terms listed below to complete each statement in Exercises 1–3.

matrix **augmented matrix** **row echelon form**

1. A matrix made up of the coefficients and the constant terms of a system is called a(n) __________________. The constant terms are separated from the coefficients by a dashed vertical line.

2. $\left[\begin{array}{cc|c} 1 & 5 & -2 \\ 0 & 1 & 2 \end{array}\right]$ is in __________________ because it is an augmented matrix whose coefficient portion has 1's on the diagonal from upper left to lower right and has 0's below the 1's.

3. The rectangular array of numbers: $\begin{bmatrix} 7 & 1 \\ 2 & 3 \end{bmatrix}$ is an example of a(n)__________________.

GUIDED EXAMPLES AND PRACTICE

Objective 1 Write a system of equations as an augmented matrix.

Review these examples for Objective 1:

1. Write the system of equations as an augmented matrix.

$$\begin{cases} 2x+2y-2z=2 \\ 12x-5y+3z=-1 \\ 4x-y+z=4 \end{cases}$$

Think of the dashed line as the equal sign of each equation.

$$\left[\begin{array}{ccc|c} 2 & 2 & -2 & 2 \\ 12 & -5 & 3 & -1 \\ 4 & -1 & 1 & 4 \end{array}\right]$$

Practice these exercises:

1. Write the system of equations as an augmented matrix.

$$\begin{cases} -x+6y=5 \\ 3x-5y=7 \end{cases}$$

Name: Date:
Instructor: Section:

Objective 2 Solve a system of linear equations by transforming its augmented matrix to row echelon form.

Review this example for Objective 2:

2. Solve the following linear systems by transforming their augmented matrices into row echelon form.

$$\begin{cases} -15x - 3y = -54 \\ 6x + y = 22 \end{cases}$$

The augmented matrix is:

$$\left[\begin{array}{cc|c} -15 & -3 & -54 \\ 6 & 1 & 22 \end{array}\right]$$

To get a 1 in row 1, column 1, multiply row 1 by $-\frac{1}{15}$.

$$-\frac{1}{15}R_1 \to \left[\begin{array}{cc|c} 1 & \frac{1}{5} & \frac{18}{5} \\ 6 & 1 & 22 \end{array}\right]$$

To get a 0 in row 2, column 1, multiply the new row 1 by -6 and add it to row 2.

$$-6R_1 + R_2 \to \left[\begin{array}{cc|c} 1 & \frac{1}{5} & \frac{18}{5} \\ 0 & -\frac{1}{5} & \frac{2}{5} \end{array}\right]$$

To get a 1 in row 2, column 2, multiply row 2 by –5.

$$-5R_2 \to \left[\begin{array}{cc|c} 1 & \frac{1}{5} & \frac{18}{5} \\ 0 & 1 & -2 \end{array}\right]$$

This matrix represents the new system of equations:

$$\begin{cases} x + \frac{1}{5}y = \frac{18}{5} \\ y = -2 \end{cases}$$

Substituting for y, solve for x.

$$x + \frac{1}{5}y = \frac{18}{5}$$

$$x + \frac{1}{5}(-2) = \frac{18}{5}$$

$$x = \frac{20}{5}$$

$$x = 4$$

The solution is $(4, -2)$.

Practice this exercise:

2. Solve the following linear systems by transforming their augmented matrices into row echelon form.

$$\begin{cases} x + y = 2 \\ x - y = 6 \end{cases}$$

Name: Date:
Instructor: Section:

Objective 3 Solve applications problems using matrices.

Review this example for Objective 3:

3. Solve.

Jeff invests \$3000 in two different accounts. The first account has an annual percentage rate (APR) of 3%. The second account has an APR of 5%. If the total interest earned after one year is \$110, what principal was invested in each account?

Let x represent the amount invested at 3% and y represent the amount invested at 5%. This gives the following system of equations:

$$\begin{cases} x+y=3000 \\ 0.03x+0.05y=110 \end{cases}$$

The augmented matrix representing this system of equations is:

$$\left[\begin{array}{cc|c} 1 & 1 & 3000 \\ 0.03 & 0.05 & 110 \end{array}\right]$$

Now, put the augmented matrix into row echelon form. To do this, first get a 0 in the second row, first column. There is currently 0.03 there, so multiply row 1 by −0.03 and add it to row 2.

$$-0.03R_1 + R_2 \rightarrow \left[\begin{array}{cc|c} 1 & 1 & 3000 \\ 0 & 0.02 & 20 \end{array}\right]$$

Finally, get all 1's in the diagonal from left to right by multiplying the second row by $\frac{1}{0.02}$.

$$\frac{1}{0.02}R_2 \rightarrow \left[\begin{array}{cc|c} 1 & 1 & 3000 \\ 0 & 1 & 1000 \end{array}\right]$$

The system of equations represented by this matrix is:

$$\begin{cases} x+y=3000 \\ y=1000 \end{cases}$$

Since $y=1000$, substitute the value of y in for x.

$$x+y=3000$$
$$x+1000=3000$$
$$x-2000$$

So, Jeff invested \$2000 at 3% and \$1000 at 5%.

Practice this exercise:

3. Solve.

Two rivers have a combined length of 5920 miles. The longer of the two rivers is 220 miles longer than the shorter. What is the length of each river?

Name: Date:
Instructor: Section:

ADDITIONAL EXERCISES

Objective 1 Write a system of equations as an augmented matrix.

For extra help, see Example 1 on page 714 of your text and the Section 11.5 lecture video.

Write the augmented matrix for the system of equations.

1. $\begin{cases} 2x-8y+3z=-5 \\ 5x-4y+2z=9 \\ 4y-6z=-9 \end{cases}$

2. $\begin{cases} x-y=-7 \\ x+2y+5z=16 \\ 5y-3z=-1 \end{cases}$

Objective 2 Solve a system of linear equations by transforming its augmented matrix to row echelon form.

For extra help, see Examples 2–3 on pages 715–717 of your text and the Section 11.5 lecture video.

Given the matrices in row echelon form, find the solution for the system.

3. $\left[\begin{array}{cc|c} 1 & 4 & -3 \\ 0 & 1 & 7 \end{array}\right]$

4. $\left[\begin{array}{ccc|c} 1 & 1 & -1 & 11 \\ 0 & 1 & -1 & 2 \\ 0 & 0 & 1 & 1 \end{array}\right]$

Complete the indicated row operation.

5. Replace R_1 in $\left[\begin{array}{cc|c} 16 & 12 & 4 \\ 0 & 3 & 9 \end{array}\right]$ with $\frac{1}{4}R_1$.

6. Replace R_2 in $\left[\begin{array}{ccc|c} 1 & 7 & 4 & 0 \\ -6 & 8 & -1 & 4 \\ 5 & 7 & 0 & 2 \end{array}\right]$ with $6R_1+R_2$.

Solve by transforming the augmented matrix into row echelon form.

7. $\begin{cases} -25x-3y=-116 \\ 10x+y=47 \end{cases}$

8. $\begin{cases} x+y-z=6 \\ 2x-y+z=-12 \\ x-4y+3z=-46 \end{cases}$

9. $\begin{cases} -35x-2y+z=-97 \\ 20x+y=57 \\ -40x-2y+z=-112 \end{cases}$

Solve using a matrix on a graphing calculator.

10. $\begin{cases} 3x+y=19 \\ 5x+y=29 \end{cases}$

11. $\begin{cases} -21x-2y+z=-73 \\ 12x+y=44 \\ -24x-2y+z=-85 \end{cases}$

12. $\begin{cases} 5x+y-4z=-12 \\ 5x+5y+z=-20 \\ -3x-5y+5z=16 \end{cases}$

Name: Date:
Instructor: Section:

Objective 3 Solve applications problems using matrices.

For extra help, see Example 4 on page 718 of your text and the Section 11.5 lecture video.

Translate the problem to a system of equations; then solve using matrices.

13. Michelle and Deena went hopping for clothes. Michelle bought 1 T-shirt and 2 belts for \$20. Deena bought 2 T-shirts and 6 belts for \$52. How much does a T-shirt cost? How much does a belt cost?

Name: Date:
Instructor: Section:

Chapter 11 SYSTEMS OF LINEAR EQUATIONS AND INEQUALITIES

11.6 Solving Systems of Linear Inequalities

Learning Objectives
1 Graph the solution set of a system of linear inequalities.
2 Solve applications involving a system of linear inequalities.

GUIDED EXAMPLES AND PRACTICE

Objective 1 Graph the solution set of a system of linear inequalities.

Review this example for Objective 1:

1. Graph the solution set for the system of inequalities.

$$\begin{cases} 3x + y \geq 2 \\ x \leq 2 \end{cases}$$

Graph the inequalities on the same grid. Use a solid line with $\leq$ or $\geq$.
The solution set for this system contains all ordered pairs in the region of overlap together with all ordered pairs on the portion of both lines that touch the region of overlap.

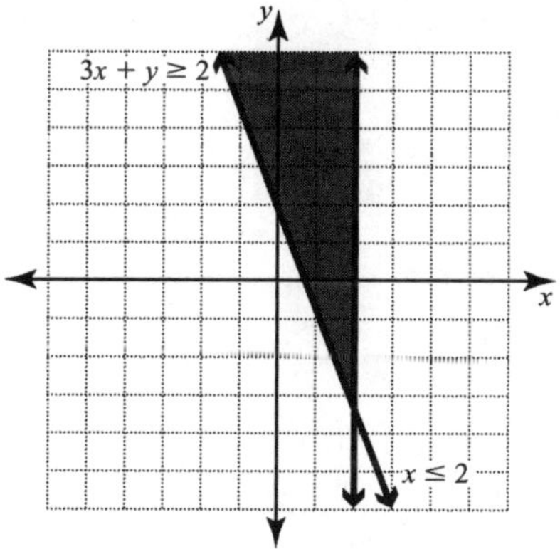

Practice this exercise:

1. Graph the solution set for the system of inequalities.

$$\begin{cases} 5x + 3y < 15 \\ x - y < 5 \end{cases}$$

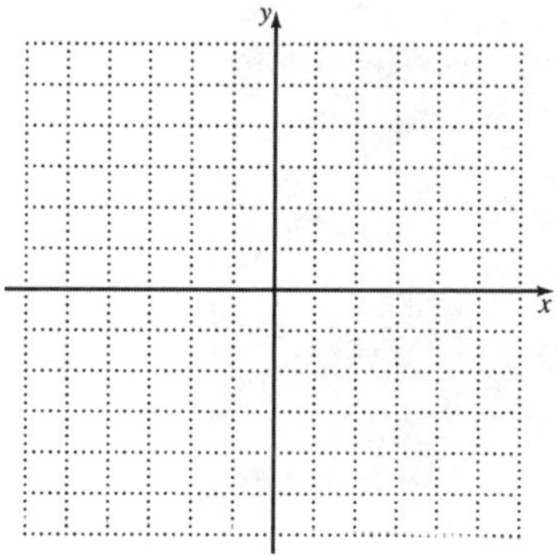

Name: Date:
Instructor: Section:

Objective 2 Solve applications involving a system of linear inequalities.

Review this example for Objective 2:

2. Solve.
A company sells a regular version of software for $34.99 and a deluxe version for $49.99. The company gives a bonus to any salesperson who sells at least 55 units and has at least $2500 in total sales. Write a system of inequalities that describes the requirements a salesperson must meet in order to receive the bonus. Solve the system by graphing.

Let x represent the number of regular units sold and y represent the number of deluxe units. The system of equations represented is:

$$\begin{cases} x+y \geq 55 \\ 34.99x+49.99y \geq 2500 \end{cases}$$

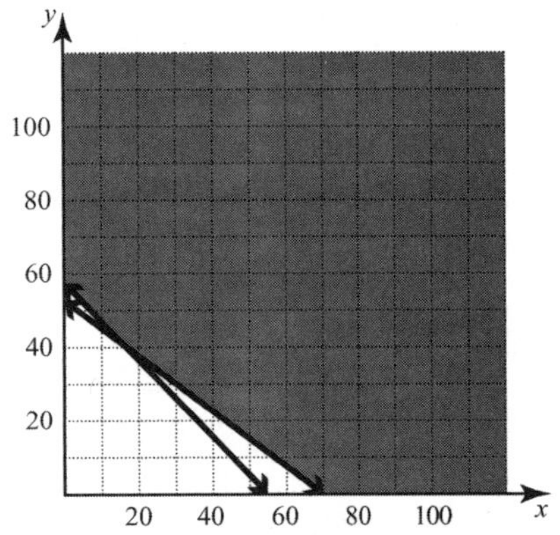

Practice this exercise:

2. Solve.
A computer manufacturer sells a standard model DVD player and a top-of-the-line DVD player. The standard model is $25 and the top-of-the-line model is $50. The manufacturer sells at least 75 DVD players in a day and makes a profit of at least $2100. Write a system of equations and solve the system by graphing.

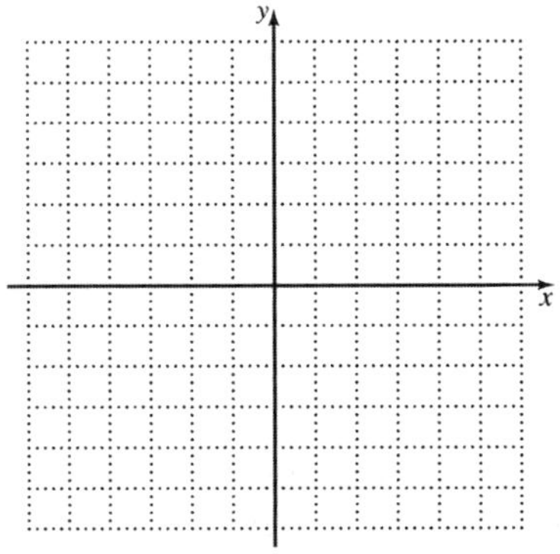

Name: Date:
Instructor: Section:

ADDITIONAL EXERCISES

Objective 1 Graph the solution set of a system of linear inequalities.

For extra help, see Examples 1 and 2 on page 725 of your text and the Section 11.6 lecture video.

Graph the solution set for the system of inequalities.

1. $\begin{cases} 2x+5y\geq 10 \\ 3x-2y\leq 6 \end{cases}$

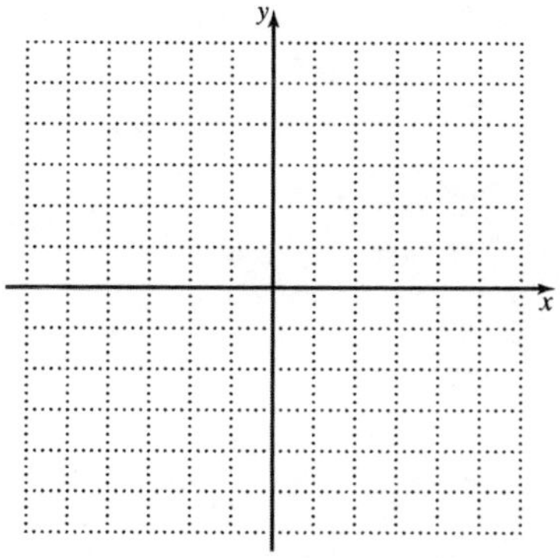

2. $\begin{cases} y>x \\ y<-x+3 \end{cases}$

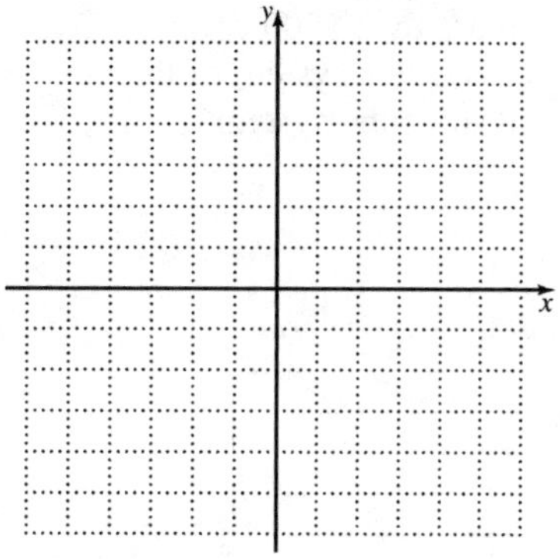

3. $\begin{cases} 2x>-3y \\ 2x+3y\leq -6 \end{cases}$

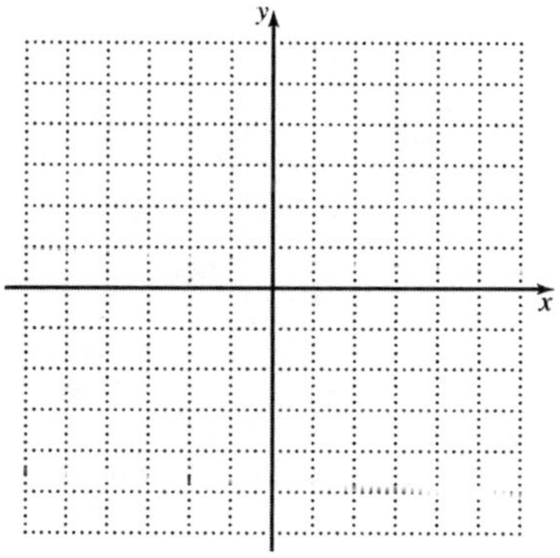

4. $\begin{cases} x+y\leq 2 \\ x-y\leq 4 \end{cases}$

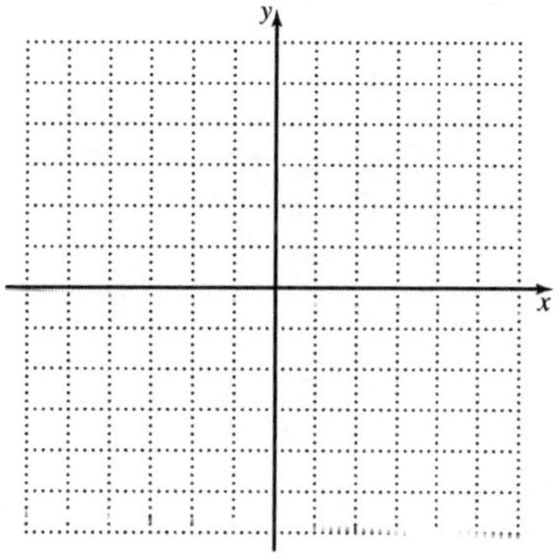

5. $\begin{cases} y<3x+1 \\ 3x-y\leq 5 \end{cases}$

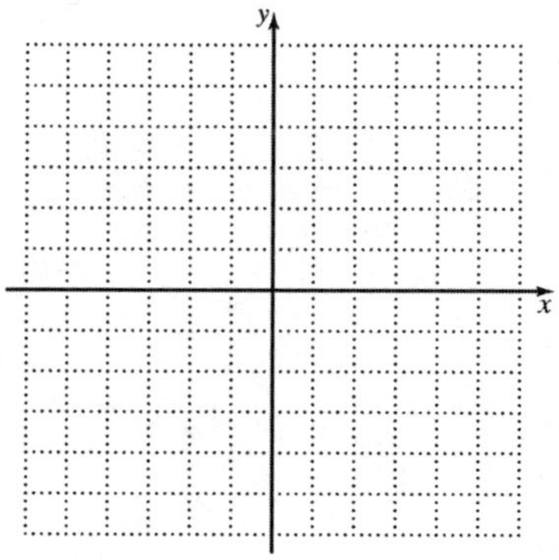

6. $\begin{cases} 4x+4y\geq 16 \\ 2x-4y<8 \end{cases}$

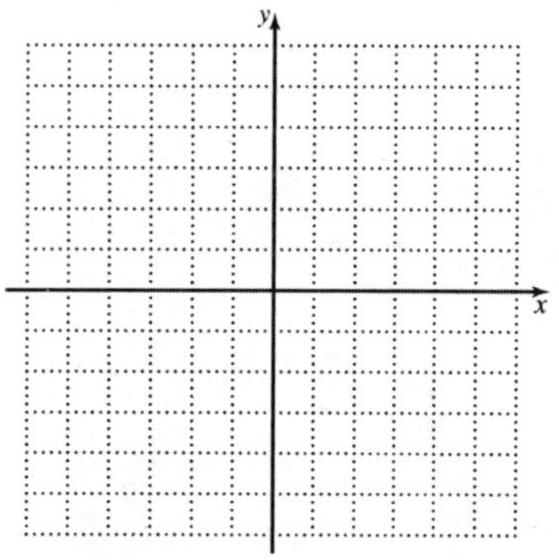

Name: Date:
Instructor: Section:

Objective 2 Solve applications involving a system of linear inequalities.
For extra help, see Example 3 on page 726 of your text and the Section 11.6 lecture video.
Solve.

7. A local circus charges an adult admission and a child admission. The adult admission price is $12 and the individual price is $4. Everyday, at least 50 people attend the circus. However, the circus never collects more than $500 for admission sales. Write a system of linear inequalities and solve the system by graphing.

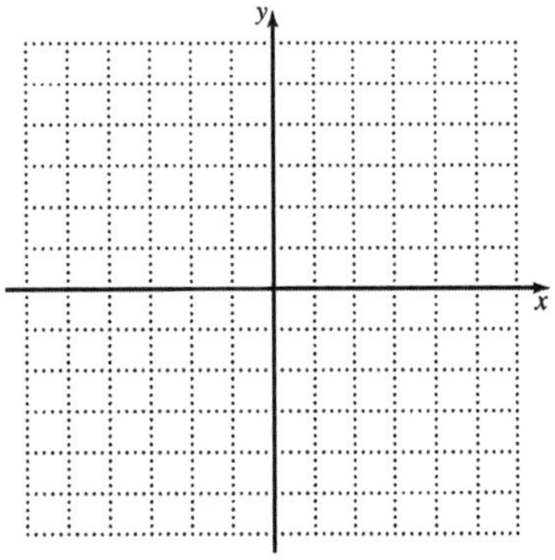

Name:
Instructor:

Date:
Section:

Chapter 12 POLYNOMIALS

12.1 Exponents and Scientific Notation

Learning Objectives
1 Evaluate exponential form with integer exponents.
2 Write scientific notation in standard form.
3 Write standard form numbers in scientific notation.

Key Terms
Use the vocabulary terms listed below to complete each statement in Exercise 1.

scientific notation

1. A number expressed in the form $a \times 10^n$ where a is a decimal number with $1 \le |a| \le 10$ and n is an integer is in ___________________.

GUIDED EXAMPLES AND PRACTICE

Objective 1 Evaluate exponential form with integer exponents.

Review these examples for Objective 1:

1. Evaluate the exponential form.
$(-2)^3$

$$(-2)^3 = (-2)(-2)(-2) = -8$$

2. Rewrite the expression with positive exponents; then if the expression is numeric, evaluate it.
$(-3)^{-5}$

$$(-3)^{-5} = \frac{1}{(-3)(-3)(-3)(-3)(-3)} = -\frac{1}{243}$$

Practice these exercises:

1. Evaluate the exponential form.
-7^2

2. Rewrite the expression with positive exponents; then if the expression is numeric, evaluate it.
6^{-2}

Objective 2 Write scientific notation in standard form.

Review this example for Objective 2:

3. Write the number in standard form.
2.13×10^4

Multiplying 2.13 by 10^4 means that the decimal will move four places to the right.
$2.13 \times 10^4 = 21{,}300$

Practice this exercise:

3. Write the number in standard form.
4.12×10^5

Name: Date:
Instructor: Section:

Objective 3 Write standard form numbers in scientific notation.

Review this example for Objective 3:

4. Write the number in scientific notation.
0.00000706

Place the decimal point between the 6 and 0 digits so that you have 7.06, which is a decimal number greater than 1 but less than 10. Because there are six decimal places to the left of the new decimal position, the exponent is −6.

$0.00000706 = 7.06 \times 10^{-6}$

Practice this exercise:

4. Write the number in scientific notation.
Sales totals for the Automation Company were \$25,000,000.

ADDITIONAL EXERCISES

Objective 1 Evaluate exponential forms with integer exponents.

For extra help, see Examples 1–4 on pages 741–744 of your text and the Section 12.1 lecture video.

Evaluate the exponential form.

1. $\left(\frac{2}{5}\right)^3$

2. 9^0

3. $(-4)^3$

4. $-\left(\frac{1}{6}\right)^2$

Rewrite each expression with positive exponents; then if the expression is numeric, evaluate it.

5. $\left(\frac{4}{5}\right)^{-4}$

6. $(-3)^{-3}$

Objective 2 Write scientific notation in standard form.

For extra help, see Examples 5 and 6 on pages 745–746 of your text and the Section 12.1 lecture video.

Write the number in standard form.

7. 7.32×10^6

8. 6.11×10^{-5}

9. 1.43×10^{-7}

Objective 3 Write standard form numbers in scientific notation.

For extra help, see Examples 7 and 8 on page 747 of your text and the Section 12.1 lecture video.

Write the number in scientific notation.

10. It is estimated that in the year 2020 the population of the world will be 7,518,000,000.

11. One cubic orc is approximately equal to 0.0000000048 cubic balrogs.

Name: Date:
Instructor: Section:

Chapter 12 POLYNOMIALS

12.2 Introduction to Polynomials

Learning Objectives
1 Identify monomials.
2 Identify the coefficient and degree of a monomial.
3 Classify polynomials.
4 Identify the degree of a polynomial.
5 Evaluate polynomials.
6 Write polynomials in descending order of degree.
7 Combine like terms.

Key Terms
Use the vocabulary terms listed below to complete each statement in Exercises 1–8.

monomial	**degree of a polynomial**	**binomial**	**degree of a monomial**
polynomial	**coefficient of a monomial**	**trinomial**	**polynomial in one variable**

1. The ________________ is the greatest degree of any of the terms in a polynomial.

2. The numerical factor in a monomial is the ________________.

3. $9t^2 + 3.4t - 8.2$ is a(n) ________________ because it is a polynomial that contains three terms.

4. A(n) ________________ is defined as a monomial or a sum of monomials.

5. An expression that is a constant, a variable, or a product of a constant and variable(s) that are raised to whole number powers is a(n)________________.

6. $x^2 - 5x$ is a polynomial in which every variable term has the same variable. This is a(n) ________________.

7. A(n) ________________ is a polynomial containing two terms.

8. The sum of the exponents of all variables in a monomial is the ________________.

GUIDED EXAMPLES AND PRACTICE

Objective 1 Identify monomials.

Review this example for Objective 1:

1. Determine whether the expression is a monomial.
$2x^4$

Yes, $2x^4$ is a monomial because it is a product of a constant, 2, and a variable, x, which has a whole-number exponent of 4.

Practice this exercise:

1. Determine whether the expression is a monomial.
$-6x^2y^4$

Name: Date:
Instructor: Section:

Objective 2   Identify the coefficient and degree of a monomial.

Review this example for Objective 2:

2. Identify the coefficient and degree of each monomial.

$-2t^{15}$

The coefficient is -2 because it is the numerical factor. The degree is 15 because it is the exponent for the single variable t.

Practice this exercise:

2. Identify the coefficient and degree of each monomial.

$6t^7$

Objective 3   Classify polynomials.

Review this example for Objective 3:

3. Indicate whether the expression is a monomial, binomial, or trinomial; has no special polynomial name; or is not a polynomial.

$-2a^3+b$

This is a binomial because it contains two terms.

Practice this exercise:

3. Indicate whether the expression is a monomial, binomial, or trinomial; has no special polynomial name; or is not a polynomial.

x^2-2x+1

Objective 4   Identify the degree of a polynomial.

Review this example for Objective 4:

4. Identify the degree of the polynomial.

$2x^3+11x^8-11x^2+4x-1$

The degree is 8 because it is the greatest degree of all of the terms.

Practice this exercise:

4. Identify the degree of the polynomial.

$x^5-6x+x^7-5x^6$

Objective 5   Evaluate polynomials.

Review this example for Objective 5:

5. Evaluate the polynomial using the given values.

$x^3-4x^2+2;\ \ x=-2$

$$\begin{aligned}(-2)^3-4(-2)^2+2&=(-8)-4(4)+2\\&=-8-16+2\\&=-24+2\\&=-22\end{aligned}$$

Practice this exercise:

5. Evaluate the polynomial using the given values.

$3x^2-3x+1;\ \ x=4$

Name: Date:
Instructor: Section:

Objective 6 Write polynomials in descending order of degree.

Review this example for Objective 6:

6. Write the polynomial in descending order of degree.

$11x^2 + 4x - x^3 - 5 + 2x^4$

Rearrange the terms so that the highest degree is first, then the next highest degree, and so on:

$2x^4 - x^3 + 11x^2 + 4x - 5$

Practice this exercise:

6. Write the polynomial in descending order of degree.

$x^5 + x + 8x^3 + 7 + 6x^2$

Objective 7 Combine like terms.

Review this example for Objective 7:

7. Combine like terms and write the resulting polynomial in descending order of degree.

$11r - 8r^2 + 22d^3 + 9r - 5d^3 + 88r^2 + 16d^3$

Collect like terms as needed.

$11r + 9r - 8r^2 + 88r^2 + 22d^3 - 5d^3 + 16d^3$

Combine like terms.

$20r + 80r^2 + 33d^3$

Rearrange the terms so they are in descending order of degree.

$33d^3 + 80r^2 + 20r$

Practice this exercise:

7. Combine like terms and write the resulting polynomial in descending order of degree.

$9b^5 + b^2 - b^3 - 6b^5 - 7b^2$

ADDITIONAL EXERCISES

Objective 1 Identify monomials.

For extra help, see Example 1 on pages 752–753 of your text and the Section 12.2 lecture video.

Determine whether the expression is a monomial.

1. $-8x^{-3}y^2$

2. 40

Objective 2 Identify the coefficient and degree of a monomial.

For extra help, see Example 2 on page 753 of your text and the Section 12.2 lecture video.

Identify the coefficient and degree of each monomial.

3. mn^2

4. $\frac{3}{4}x^2y^4$

Objective 3 Classify polynomials.

For extra help, see Example 3 on page 754 of your text and the Section 12.2 lecture video.

Indicate whether the expression is a monomial, binomial, or trinomial or has no special polynomial name. If the expression is a polynomial, give the degree.

5. 68

6. $3p^3m^4 - 5p^2m^3 + 5p^3m - 5pm^3 + 4p^4m^3$

Name: Date:
Instructor: Section:

Objective 4 Identify the degree of a polynomial.
For extra help, see Example 4 on pages 754–755 of your text and the Section 12.2 lecture video.
Identify the degree of each polynomial.

7. $1.2x^5 - 7.3x^3 + 4.5x + 9.5$

8. $2x^9 - 7x^6 + 3x^3 + 900$

Objective 5 Solve equations containing an unknown addend.
For extra help, see Examples 5 and 6 on page 755 of your text and the Section 12.2 lecture video.
Evaluate the polynomial using the given values.

9. $a^3 - 3a^2 + 4a + 7$; $a = 5$

10. A 12-ounce beverage can has a height of 7.3 inches and a radius of 1.6 inches. The polynomial $2\pi rh + 2\pi r^2$ describes the surface area of a cylindrical can. Evaluate the polynomial for $h = 7.3$ inches and $r = 1.6$ inches to find the surface area of the can. Round the result to the nearest tenth.

Objective 6 Write polynomials in descending order of degree.
For extra help, see Example 7 on page 756 of your text and the Section 12.2 lecture video.
Write the polynomial in descending order of degree.

11. $z + 875 + 6z^3 + z^2$

12. $m^3 + m^5 + 34m + 2m^{10}$

Objective 7 Combine like terms.
For extra help, see Examples 8 and 9 on pages 756–757 of your text and the Section 5.2 lecture video.
Combine like terms and write the resulting polynomial in descending order of degree.

13. $11a^2p + 4p^2 - 33a^2p - 6p^2$

14. $2mn + m^2 + 3m^2n + 2n^2 + 4mn + n^2$

Name: Date:
Instructor: Section:

Chapter 12 POLYNOMIALS

12.3 Adding and Subtracting Polynomials

Learning Objectives
1 Add polynomials.
2 Subtract polynomials.

GUIDED EXAMPLES AND PRACTICE

Objective 1 Add polynomials.

Review these examples for Objective 1:

1. Add and write the resulting polynomial in descending order of degree.
$(3y^4-5y^3+2y-7)+(y^4+2y^3-y^2+6y+1)$

Combine like terms. Combining in order of degree places the resulting polynomial in descending order of degree.
$=3y^4-5y^3+2y-7+y^4+2y^3-y^2+6y+1$
$=3y^4+y^4-5y^3+2y^3-y^2+2y+6y-7+1$
$=4y^4-3y^3-y^2+8y-6$

2. Write an expression in simplest form for the perimeter of the triangle.

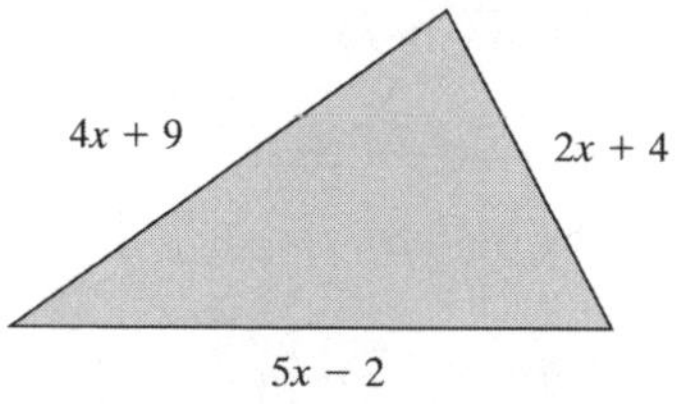

Since the perimeter is the total distance around the shape, we need to add the polynomials to represent the perimeter.
Perimeter = side + side + side
$=(4x+9)+(2x+4)+(5x-2)$
$=4x+9+2x+4+5x-2$
$=4x+2x+5x+4+9-2$
$=11x+11$

To check, choose a value for x and evaluate the original expressions for the side lengths, determine the corresponding numeric perimeter, and evaluate the perimeter expression using the same value for x to verify that you get the same numeric perimeter.

Practice these exercises:

1. Add and write the resulting polynomial in descending order of degree.
$(7x+8)+(-8x+6)$

2. Write an expression for the perimeter in simplest form.

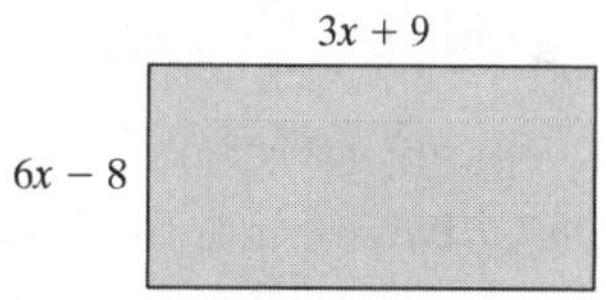

Name: Date:
Instructor: Section:

Objective 2 Subtract polynomials.

Review this example for Objective 2:

3. Subtract and write the resulting polynomial in descending order of degree.

$(-11m^4 - 27m^3 + 6m - 8) - (7m^4 - 3m^3 + 9m + 1)$

Change the minus sign between the polynomials in parentheses to a plus sign, and change all signs in the subtrahend.

$= -11m^4 - 27m^3 + 6m - 8$

$= -7m^4 + 3m^3 - 9m - 1$

Combine like terms.

$= -11m^4 - 7m^4 - 27m^3 + 3m^3$

$= +6m - 9m - 8 - 1$

$= -18m^4 - 24m^3 - 3m - 9$

Practice this exercise:

3. Subtract and write the resulting polynomial in descending order of degree.

$(4x + 5) - (2x + 9)$

ADDITIONAL EXERCISES

Objective 1 Add polynomials.

For extra help, see Examples 1–3 on pages 763–764 of your text and the Section 12.3 lecture video.

Add and write the resulting polynomial in descending order of degree.

1. $(5x + 9y) + (-6x + 8y)$

2. $(-9x + 4) + (x^2 + x - 6)$

3. $(9x^2 - 6x + 15) + (7x^2 + 8x - 89)$

4. $(2 + 8x + 3x^2 + 7x^3) +$
$(9 - 8x + 3x^2 - 7x^3)$

5. $(-2x^4 - 3x^3 - 3x^2 + x - 14) +$
$(-3x^4 + 9x^3 + x^2 + 14x + 7)$

6. $(6x^5 - 2x^3 + x^2 - x - 2) +$
$(-4x^4 + 7x^3 + x^2 + x + 9)$

Write an expression for the perimeter in simplest form.

7. A rectangle with width $5x - 10$ and length $2x + 4$

Objective 2 Subtract polynomials.

For extra help, see Examples 4 and 5 on page 766 of your text and the Section 12.3 lecture video.

Subtract and write the resulting polynomial in descending order of degree.

8. $(-3v^2 + 9v + 7) - (7v^2 + 11v - 2)$

9. $(-4a^3 - 3a^2 + 6a + 9) -$
$(-4a^3 + 9a^2 + 10a - 4)$

10. $(-y^5 + 3y^4 + 6y^2 - y - 15) -$
$(-3y^5 + y^2 - y - 5)$

11. $(4w^2 + 9wt - 2t^2) -$
$(6w^2 - 7wt + 16t^2)$

Name: Date:
Instructor: Section:

Chapter 12 POLYNOMIALS

12.4 Exponent Rules and Multiplying Monomials

Learning Objectives
1 Multiply monomials.
2 Multiply numbers in scientific notation.
3 Simplify a monomial raised to a power.

GUIDED EXAMPLES AND PRACTICE

Objective 1 Multiply monomials.

Review these examples for Objective 1:

1. Multiply.
$\left(5n^3\right)\left(8n^6\right)$

Multiply the coefficients and add the exponents of the like bases.
$\left(5n^3\right)\left(8n^6\right) = 5 \cdot 8\left(n^{3+6}\right)$
$= 40n^9$

2. Write an expression in the simplest form for the volume of the box shown.

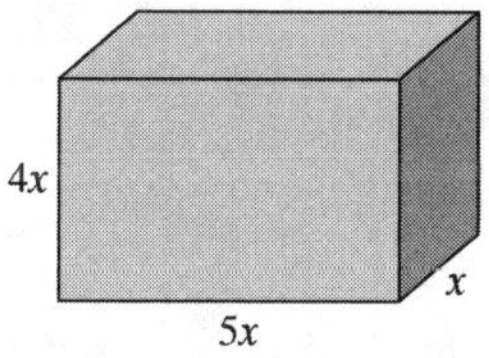

The volume of a box is found by multiplying length, width, and height.
$V = lwh$
$V = (5x)(4x)(x)$
$V = 20x^3$
The expression for the volume is $20x^3$.

Practice these exercises:

1. Multiply.
$v^3 \cdot v^6$

2. Write an expression in simplest form for the area of the figure.

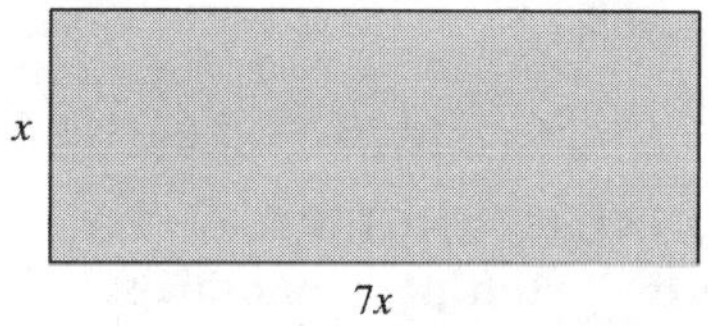

Name: Date:
Instructor: Section:

Objective 2 Multiply numbers in scientific notation.

Review this example for Objective 2:

3. Multiply and write the answers in scientific notation.

$(2.7\times10^2)(7.5\times10^5)$

Multiply 2.7 and 7.5; then add the exponents for base 10.

$$(2.7\times10^2)(7.5\times10^5)=2.7\times7.5\times10^{2+5}$$
$$=20.25\times10^7$$
$$=2.025\times10^8$$

Practice this exercise:

3. Multiply and write the answers in scientific notation.

$(3.1\times10^6)(2.4\times10^2)$

Objective 3 Simplify a monomial raised to a power.

Review this example for Objective 3:

4. Simplify.

$(2x^3z^4)^6$

Write the coefficient, 2, raised to the sixth power and multiply each variable's exponent by 6.

$(2x^3z^4)^6=2^6x^{3\cdot6}z^{4\cdot6}$

Simplify.

$64x^{18}z^{24}$

Practice this exercise:

4. Simplify.

$(c^2)^8$

ADDITIONAL EXERCISES

Objective 1 Multiply monomials.

For extra help, see Examples 1–3 on pages 771–773 of your text and the Section 12.4 lecture video.

Multiply.

1. $7^9\cdot7^3$

2. $7r^2\cdot5r$

3. $(-5x^7y^7)(4x^6y^2)$

4. $(3qr)(6q^2r^4)(2q^5)$

5. $(5xy)(4x^3y^4)(3x^2)$

6. $(27ab^2)\left(\frac{1}{27}a^3b^2\right)(3a^3)$

Write an expression in simplest form for the area of the figure.

7.

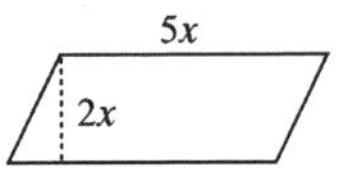

Name: Date:
Instructor: Section:

Objective 2 Multiply numbers in scientific notation.
For extra help, see Example 4 on page 774 of your text and the Section 12.4 lecture video.
Multiply and write your answer in scientific notation.

8. $\left(7.56\times10^{7}\right)\left(3.48\times10^{-4}\right)$

9. $\left(2.31\times10^{4}\right)\left(2.51\times10^{-2}\right)$

Objective 3 Simplify a monomial raised to a power.
For extra help, see Examples 5 and 6 on pages 775–776 of your text and the Section 12.4 lecture video.
Simplify.

10. $\left(-x^{7}\right)^{4}$

11. $\left(\frac{1}{3}x^{2}y\right)^{3}$

12. $\left(-r^{4}s\right)^{3}\left(-r^{5}s^{4}\right)^{2}$

13. $(3a)^{2}\left(a^{3}b\right)(-5ab)^{2}$

Name: Date:
Instructor: Section:

Chapter 12 POLYNOMIALS

12.5 Multiplying Polynomials; Special Products

Learning Objectives
1. Multiply a polynomial by a monomial.
2. Multiply binomials.
3. Multiply polynomials.
4. Determine the product when given special polynomial factors.

Key Terms
Use the vocabulary term listed below to complete the statement in Exercise 1.

conjugates

1. Binomials that differ only in the sign separating the terms are called ________________.

GUIDED EXAMPLES AND PRACTICE

Objective 1 Multiply a polynomial by a monomial.

Review this example for Objective 1:

1. Multiply the polynomial by the monomial.
$5x(6x^2-8x+3)$

Multiply each term in the polynomial by $5x$.
$5x\cdot 6x^2-5x\cdot 8x+5x\cdot 3$
Simplify.
$(5\cdot 6)x\cdot x^2-(5\cdot 8)x\cdot x+5\cdot 3x$
Combine like terms.
$30x^3-40x^2+15x$

Practice this exercise:

1. Multiply the polynomial by the monomial.
$-5x(x-3)$

Objective 2 Multiply binomials.

Review this example for Objective 2:

2. Multiply the binomials. (Use FOIL.)
$(8x+1)(6x-4)$

Multiply each term in $8x+1$ by each term in $6x-4$. (Think FOIL.)
$48x^2+6x-32x-4$
Combine like terms and simplify.
$48x^2-26x-4$

Practice this exercise:

2. Multiply the binomials. (Use FOIL.)
$(b+11)(b-9)$

Name: Date:
Instructor: Section:

Objective 3 Multiply polynomials.

Review this example for Objective 3:

3. Multiply the polynomials.

$(x-4)(x^2-3x+1)$

Multiply each term in $x-4$ by each term in x^2-3x+1.

$x(x^2-3x+1)-4(x^2-3x+1)$

Distribute.

$x^3-3x^2+x-4x^2+12x-4$

Combine like terms.

$x^3-7x^2+13x-4$

Practice this exercise:

3. Multiply the polynomials.

$(x^2+x+7)(x-7)$

Objective 4 Determine the product when given special polynomial factors.

Review these examples for Objective 4:

4. State the conjugate of the given polynomial.

$7b-9$

Since conjugates differ only in the sign separating the terms, reverse the sign to find the conjugate.

$7b+9$

5. Multiply the polynomials.

$(b+4)(b-4)$

Multiply each term in $b+4$ by each term in $b-4$. (Think FOIL.)

$b^2+4b-4b-16=b^2-16$

Practice these exercises:

4. State the conjugate of the given polynomial.

$a+7$

5. Multiply the polynomials.

$(s+6)(s-6)$

ADDITIONAL EXERCISES

Objective 1 Multiply a polynomial by a monomial.

For extra help, see Example 1 on page 782 of your text and the Section 12.5 lecture video.

Multiply the polynomial by the monomial.

1. $4x(4x^2-9x+9)$

2. $-4y^4(9y^2+4y-6)$

Name: Date:
Instructor: Section:

Objective 2 Multiply binomials.

For extra help, see Examples 2–5 on pages 783–785 of your text and the Section 12.5 lecture video.

Multiply the binomials. (Use FOIL.)

3. $(3x-1)(4x+3)$

4. $(4v-5f)(4v+4f)$

Objective 3 Multiply polynomials.

For extra help, see Example 6 on pages 785–786 of your text and the Section 12.5 lecture video.

Multiply the polynomial by the monomial.

5. $(9x+5)(2x^2+2x+4)$

6. $(r^2+r-1)(r^2+7r-9)$

Objective 4 Determine the product when given special polynomial factors.

For extra help, see Examples 7–8 on pages 786–788 of your text and the Section 12.5 lecture video.

State the conjugate of the given binomial.

7. $\frac{1}{2}b-2c$

Multiply using the rules for special products.

8. $(3r+5)(3r-5)$

9. $(5x+4y)^2$

10. $(y-5)^2$

Name: Date:
Instructor: Section:

Chapter 12 POLYNOMIALS

12.6 Exponent Rules and Dividing Polynomials

Learning Objectives
1 Divide exponential forms with the same base.
2 Divide numbers in scientific notation.
3 Divide monomials.
4 Divide a polynomial by a monomial.
5 Use long division to divide polynomials.
6 Simplify expressions using rules of exponents.

GUIDED EXAMPLES AND PRACTICE

Objective 1 Divide exponential forms with the same base.

Review this example for Objective 1:

1. Divide and write the result with a positive exponent.

$\frac{g^{-6}}{g^{-3}}$

Subtract the exponents and keep the same base.

$\frac{g^{-6}}{g^{-3}} = g^{-6-(-3)}$

Rewrite the subtraction as addition.

g^{-6+3}

Simplify.

g^{-3}

Write with a positive exponent.

$\frac{1}{g^{3}}$

Practice this exercise:

1. Divide and write the result with a positive exponent.

$\frac{z^{-8}}{z^{-1}}$

Name: Date:
Instructor: Section:

Objective 2 Divide numbers in scientific notation.

Review this example for Objective 2:

2. Divide and write the result in scientific notation.

$$\frac{4.5\times10^{-6}}{2.5\times10^{-3}}$$

The decimal factors and powers of 10 can be separated into a product of two fractions. This allows us to calculate the decimal division and divide the powers of 10 separately.

$$\frac{4.5\times10^{-6}}{2.5\times10^{-3}}=\frac{4.5}{2.5}\times\frac{10^{-6}}{10^{-3}}$$

Divide decimal numbers and subtract exponents.

$$1.8\times10^{-3}$$

Practice this exercise:

2. Divide and write the result in scientific notation.

$$\frac{-18.2\times10^{-7}}{2.6\times10^{-4}}$$

Objective 3 Divide monomials.

Review this example for Objective 3:

3. Divide the monomials.

$$\frac{-8a^7b^4c}{2ab^5c}$$

Divide coefficients and subtract exponents of like bases.

$$-\frac{8}{2}\cdot a^{7-1}\cdot b^{4-5}\cdot c^{1-1}=-4\cdot a^6\cdot b^{-1}\cdot c^0$$

Write with positive exponents.

$$-\frac{4a^6}{b}$$

Practice this exercise:

3. Divide the monomials.

$$\frac{15a^4b^5}{-3ab}$$

Objective 4 Divide a polynomial by a monomial.

Review this example for Objective 4:

4. Divide the polynomial by the monomial.

$$\frac{32x^5-64x^7+8x^2}{8x^2}$$

Divide each term in the polynomial by the monomial.

$$\frac{32x^5}{8x^2}-\frac{64x^7}{8x^2}+\frac{8x^2}{8x^2}$$

Divide coefficients and subtract exponents of like bases.

$$4x^3-8x^5+1$$

Practice this exercise:

4. Divide the polynomial by the monomial.

$$\frac{7x+28y}{7}$$

Name: Date:
Instructor: Section:

Objective 5 Use long division to divide polynomials.

Review this example for Objective 5:

5. Use long division to divide the polynomials.

$$\frac{x^2+6x+10}{x+2}$$

Begin by dividing the first term in the dividend by the first term in the divisor.

$$\frac{x^2}{x}=x$$

$$\begin{array}{r} x \\ x+2\overline{)x^2+6x+10} \\ -(x^2+2x) \\ \hline 4x+10 \end{array}$$

Determine the next part of the quotient by dividing $4x$ by x, which is 4, and repeat the multiplication and subtraction steps.

$$\begin{array}{r} x+4 \\ x+2\overline{)x^2+6x+10} \\ -(x^2+2x) \\ \hline 4x+10 \\ -(4x+8) \\ \hline 2 \end{array}$$

Since we have a remainder of 2, we get

$$x+4+\frac{2}{x+2}.$$

Practice this exercise:

5. Use long division to divide the polynomials.

$$\frac{x^2+7x+6}{x+6}$$

Objective 6 Simplify expressions using rules of exponents.

Review this example for Objective 6:

6. Divide and write the result with a positive exponent.

$$\frac{(x^3)^{-2}}{x^6\cdot x^{-8}}$$

In the numerator, use the rule for raising a power to a power. In the denominator, use the product rule for exponents.

$$\frac{(x^3)^{-2}}{x^6\cdot x^{-8}}=\frac{x^{-6}}{x^{6+(-8)}}=\frac{x^{-6}}{x^{-2}}$$

Use the quotient rule for exponents.

$$x^{-6-(-2)}$$

Write the subtraction as an equivalent addition.

$$x^{-4}$$

Write with a positive exponent.

$$\frac{1}{x^4}$$

Practice this exercise:

6. Divide and write the result with a positive exponent.

$$\frac{m^{-2}\cdot m^2}{(m^3)^{-2}}$$

Name: Date:
Instructor: Section:

ADDITIONAL EXERCISES

Objective 1 Divide exponential forms with the same base.

For extra help, see Examples 1–2 on pages 793–794 of your text and the Section 12.6 lecture video.

Divide and write the result with a positive exponent.

1. $\dfrac{3^8}{3^5}$

2. $\dfrac{x^3}{x^{-2}}$

Objective 2 Divide numbers in scientific notation.

For extra help, see Example 3 on page 794 of your text and the Section 12.6 lecture video.

Divide and write the result in scientific notation.

3. $\dfrac{-11.6\times10^{-4}}{2.9\times10^{3}}$

4. $\dfrac{-28.2\times10^{-6}}{3.75\times10^{2}}$

Objective 3 Divide monomials.

For extra help, see Example 4 on page 795 of your text and the Section 12.6 lecture video.

Divide the monomials.

5. $\dfrac{-2st^3}{10s^4t}$

6. $\dfrac{4a^6b^3}{30a^2b^9}$

Objective 4 Divide a polynomial by a monomial.

For extra help, see Example 5 on page 796 of your text and the Section 12.6 lecture video.

Divide the polynomial by the monomial.

7. $\dfrac{9a^5-9a^3+9}{3a}$

8. $\dfrac{18z^6+12z^4+6}{6z^3}$

Objective 5 Use long division to divide polynomials.

For extra help, see Examples 6–8 on pages 797–800 of your text and the Section 12.6 lecture video.

Use long division to divide polynomials.

9. $\dfrac{x^2-x-46}{x-7}$

10. $\dfrac{x^3+729}{x+9}$

11. $\dfrac{12y^2-31y+147}{3y+8}$

Objective 6 Simplify expressions using rules of exponents.

For extra help, see Example 9 on pages 800–801 of your text and the Section 12.6 lecture video.

Simplify. Write all answers with positive exponents.

12. $\left(\dfrac{4}{5}\right)^{-3}$

13. $\left(\dfrac{1}{6}\right)^{-1}$

Name: Date:
Instructor: Section:

Chapter 13 FACTORING

13.1 Greatest Common Factor and Factoring by Grouping

Learning Objectives

1 List all possible factors for a given number.
2 Find the greatest common factor of a set of numbers of monomials.
3 Write a polynomial as a product of a monomial GCF and a polynomial.
4 Factor by grouping.

Key Terms

Use the vocabulary terms listed below to complete each statement in Exercises 1 and 2.

factored form **greatest common factor (GCF)**

1. A number or an expression written as a product of factors is said to be in __________________.

2. The __________________ is the largest natural number that divides all given numbers with no remainder.

GUIDED EXAMPLES AND PRACTICE

Objective 1 List all possible factors for a given number.

Review this example for Objective 1:

1. List all natural number factors of 32.

To list all natural number factors, divide 32 by 1, 2, 3, and so on. Write each divisor and quotient pair as a product.

$1 \cdot 32$

$2 \cdot 16$

$4 \cdot 8$

The natural number factors of 32 are 1, 2, 4, 8, 16, and 32.

Practice this exercise:

1. List all natural number factors of 18.

Objective 2 Find the greatest common factor of a set of numbers of monomials.

Review this example for Objective 2:

2. Find the GCF of 42 and 126.

Factors of 42: 1, 2, 3, 6, 7, 14, 21, 42
Factors of 126: 1, 2, 3, 6, 7, 9, 14, 18, 21, 42, 63, 126

The greatest common factor of 42 and 126 is 42.

Practice this exercise:

2. Find the GCF of 40 and 100.

Name:	Date:
Instructor:	Section:

Objective 3 Write a polynomial as a product of a monomial GCF and a polynomial.

Review this example for Objective 3:

3. Factor $24xy^3 + 32y^2$.

First find the GCF of $24xy^3$ and $32y^2$. Write the prime factorization of each monomial, treating the variables like prime factors.

$24xy^3 = 2^3 \cdot 3 \cdot x \cdot y^3$

$32y^2 = 2^5 \cdot y^2$

GCF: $2^3 \cdot y^2 = 8y^2$

Write the polynomial $24xy^3 + 32y^2$ as the product of the GCF and the quotient of the polynomial and the GCF. Separate the terms and divide the terms by the GCF.

$$24xy^3 + 32y^2 = 8y^2\left(\frac{24xy^3 + 32y^2}{8y^2}\right)$$
$$= 8y^2\left(\frac{24xy^3}{8y^2} + \frac{32y^2}{8y^2}\right)$$
$$= 8y^2(3xy + 4)$$

Practice this exercise:

3. Factor $8m^3n^3 + 48m^3n^2$.

Objective 4 Factor by grouping.

Review this example for Objective 4:

4. Factor $2z^2 - 8z^4 + x - 4xz^2$.

Group pairs of terms and factor the GCF out of each pair.

$2z^2 - 8z^4 - x + 4xz^2$

$= (2z^2 - 8z^4) + (-x + 4xz^2)$

$= 2z^2(1 - 4z^2) - x(1 - 4z^2)$

$= (1 - 4z^2)(2z^2 - x)$

Practice this exercise:

4. Factor $sq + sz + fq + fz$.

ADDITIONAL EXERCISES

Objective 1 List all possible factors for a given number.

For extra help, see Example 1 on page 817 of your text and the Section 13.1 lecture video.

List all natural number factors of the given number.

1. 625

Name: Date:
Instructor: Section:

Objective 2 Find the greatest common factor of a set of numbers of monomials.
For extra help, see Examples 2–4 on pages 818–819 of your text and the Section 13.1 lecture video.
Find the GCF.

2. $32x, 20$

Objective 3 Write a polynomial as a product of a monomial GCF and a polynomial.
For extra help, see Examples 5–8 on pages 820–822 of your text and the Section 13.1 lecture video.
Factor by factoring out the negative of the GCF.

3. $-6z^2 + 14z$

Factor by factoring out the GCF.

4. $x^2 - 5x$

5. $5a^2y - 15ay$

6. $3x^7y^5 + 21x^5y^4 + 12xy$

Factor out the polynomial GCF.

7. $x(c-4) + t(c-4)$

8. $6m(7m-4) - 5(7m-4)$

Objective 4 Factor by grouping.
For extra help, see Example 9 on page 823 of your text and the Section 13.1 lecture video.
Factor by grouping.

9. $s^2 + 3s + 7s + 21$

10. $4z^2 + 8z - az - 2a$

11. $r^2 - 10tw + 2wr - 5tr$

Name: Date:
Instructor: Section:

Chapter 13 FACTORING

13.2 Factoring Trinomials of the Form $x^2 + bx + c$

Learning Objectives
1 Factor trinomials of the form $x^2 + bx + c$.
2 Factor out a monomial GCF; then factor the trinomial of the form $x^2 + bx + c$.

GUIDED EXAMPLES AND PRACTICE

Objective 1 Factor trinomials of the form $x^2 + bx + c$.

Review these examples for Objective 1:

1. Factor $t^2 + t - 12$.

We must find a pair of numbers whose product is -12 and whose sum is 1. Because the product is negative, the two numbers must have different signs. Because the sum is positive, the number with the greater absolute value will be positive.

Product	Sum
$(-1)(12) = -12$	$-1+12 = 11$
$(-2)(6) = -12$	$-2+6 = 4$
$(-3)(4) = -12$	$-3+4 = 1$

-3 and 4 form the correct combination, so these are the second terms in each binomial.

$$t^2 + t - 12 = (t-3)(t+4)$$

2. Factor $b^2 + 6bh - 16h^2$.

The variable h is in the last term, so think of the middle term $6bh$ as $6hb$ with "coefficient" $6h$. We must find a pair of terms whose product is $-16h^2$ and whose sum is $6h$.

Product	Sum
$(-h)(16h) = -16h^2$	$-h+16h = 15h$
$(-2h)(8h) = -16h^2$	$-2h+8h = 6h$
$(-4h)(4h) = -16h^2$	$-4h+4h = 0$

$-2h$ and $8h$ form the correct combination, so these are the second terms in each binomial.

$$b^2 + 6bh - 16h^2 = (b-2h)(b+8h)$$

Practice these exercises:

1. Factor $r^2 - 17r + 70$.

2. Factor $w^2 + 13wz + 42z^2$.

Name: Date:
Instructor: Section:

Objective 2 **Factor out a monomial GCF; then factor the trinomial of the form $x^2 + bx + c$.**

Review this example for Objective 2:

3. Factor $7z^6 - 28z^5 - 147z^4$.

Whenever factoring polynomials, the first step should be to look for a monomial GCF among the terms.
GCF: $7z^4$
Factor the trinomial to two binomials.
$7z^6 - 28z^5 - 147z^4 = 7z^4(z^2 - 4z - 21)$

We are looking for two numbers whose product is negative and whose sum is negative, so the numbers will have different signs and the number with the greater absolute value will be negative.

Product	Sum
$(1)(-21) = -21$	$1 + (-21) = -20$
$(3)(-7) = -21$	$3 + (-7) = -4$

3 and -7 form the correct combination, so these are the second terms in each binomial.

$7z^6 - 28z^5 - 147z^4 = 7z^4(z+3)(z-7)$

Practice this exercise:

3. Factor $2x^3 + 14x^2 + 20x$.

ADDITIONAL EXERCISES

Objective 1 **Factor trinomials of the form $x^2 + bx + c$.**

For extra help, see Examples 1–2 on pages 828–829 of your text and the Section 13.2 lecture video.

Fill in the missing values in the factors.

1. $x^2 + 15x + 44 = (x+4)(x+\_\_\_\_)$

2. $x^2 - 12x + 35 = (x-5)(x-\_\_\_\_)$

3. $x^2 - 6x - 55 = (x+5)(x-\_\_\_\_)$

4. $x^2 + 9x - 10 = (x-1)(x+\quad)$

Factor. If the polynomial is prime, so state.

5. $t^2 + 6t + 8$

6. $b^2 - 14b + 45$

7. $w^2 - w - 30$

8. $s^2 + 4s + 77$

9. $6c^2 - 60c + 126$

Name: Date:
Instructor: Section:

Factor the trinomials containing two variables. If the polynomial is prime, so state.

10. $b^2 - 6bx - 27x^2$

11. $s^2 - 8sz + 15z^2$

Objective 2 Factor out a monomial GCF; then factor the trinomial of the form $x^2 + bx + c$.
For extra help, see Example 3 on page 830 of your text and the Section 13.2 lecture video.
Factor completely.

12. $4x^7 - 8x^6 - 32x^5$

Name: Date:
Instructor: Section:

Chapter 13 FACTORING

13.3 Factoring Trinomials of the Form $ax^2 + bx + c$, where $a \neq 1$.

Learning Objectives

1 Factor trinomials of the form $ax^2 + bx + c$, where $a \neq 1$, by trial.

2 Factor trinomials of the form $ax^2 + bx + c$, where $a \neq 1$, by grouping.

GUIDED EXAMPLES AND PRACTICE

Objective 1 Factor trinomials of the form $ax^2 + bx + c$, where $a \neq 1$, by trial.

Review this example for Objective 1:

1. Factor $x^2 + 5x + 6$.

Write a pair of first terms whose product is ax^2.

$x^2 + 5x + 6 = (x + \_\_)(x + \_\_)$

Write a pair of last terms whose product is c. Possible combinations:

$x^2 + 5x + 6 = (x+1)(x+6)$

$x^2 + 5x + 6 = (x+2)(x+3)$

The second combination is correct since the product of 2 and 3 is 6 and the sum is 5. So, $x^2 + 5x + 6 = (x+2)(x+3)$.

Practice this exercise:

1. Factor $t^2 - 12t + 35$.

Objective 2 Factor trinomials of the form $ax^2 + bx + c$, where $a \neq 1$, by grouping.

Review this example for Objective 2:

2. Factor $32v^2 + 20v - 3$ by grouping.

In $32v^2 + 20v - 3$, $a = 32$, $b = 20$, and $c = -3$. So, $ac = -96$.

Now find two factors of the product ac whose sum is b. Because the product is negative, the factors have different signs. Because the sum is positive, the factor with the greater absolute value must be positive.

Practice this exercise:

2. Factor $4u^2 + 9u + 5$ by grouping.

Name: Date:
Instructor: Section:

Factors of ac	Sum of Factors of ac
$(-1)(96)=-96$	$-1+96=95$
$(-2)(48)=-96$	$-2+48=46$
$(-3)(32)=-96$	$-3+32=29$
$(-4)(24)=-96$	$-4+24=20$
$(-6)(16)=-96$	$-6+16=10$
$(-8)(12)=-96$	$-8+12=4$

-4 and 24 form the correct combination.

Now use the factors found in the table to write the middle term of the polynomial as a sum of two like terms.

$32v^2+20v-3=32v^2-4v+24v-3$

Factor the new polynomial by grouping.

$$\begin{aligned}32v^2-4v+24v-3&=\left(32v^2-4v\right)+\left(24v-3\right)\\&=4v(8v-1)+3(8v-1)\\&=(4v+3)(8v-1)\end{aligned}$$

ADDITIONAL EXERCISES

Objective 1 Factor trinomials of the form ax^2+bx+c, where $a\neq 1$, by trial.

For extra help, see Examples 1–2 on pages 835–836 of your text and the Section 13.3 lecture video.

Factor completely. If prime, so state.

1. t^2-5t+6

2. $b^2+5b+66$

3. $4a^2+21a+5$

4. $w^2-9wz+20z^2$

5. $a^2-16af+63f^2$

6. $4s^2-22s+10$

Objective 2 Factor trinomials of the form ax^2+bx+c, where $a\neq 1$, by grouping.

For extra help, see Example 3 on pages 837–838 of your text and the Section 13.3 lecture video.

Factor by grouping. If prime, so state.

7. $18v^2-85v+18$

8. $56u^2-122u+42$

Name: Date:
Instructor: Section:

9. $28b^2 + 17b - 3$

10. $25c^2 + 60c + 27$

11. $9a^2 + 18a + 8$

12. $30c^2 - 11cd - 30d^2$

13. $30u^3 + 145u^2 - 210u$

Name: Date:
Instructor: Section:

Chapter 13 FACTORING

13.4 Factoring Special Products

Learning Objectives
1 Factor perfect square trinomials.
2 Factor a difference of squares.
3 Factor a difference of cubes.
4 Factor a sum of cubes.

GUIDED EXAMPLES AND PRACTICE

Objective 1 Factor perfect square trinomials.

Review this example for Objective 1:

1. Factor $49v^2 - 56v + 16$.

Look at the terms in the trinomial $49v^2 - 56v + 16$ to determine whether the trinomial fits the form of a perfect square.

$49v^2 = (7v)^2$

$16 = (4)^2$

$56v = 2 \cdot (28) \cdot (v)$

The first and last terms of the trinomial are perfect squares and the middle term is equal to twice the product of the square roots of the first and last terms. So, $49v^2 - 56v + 16$ is a perfect square trinomial fitting the form $a^2 - 2ab + b^2$.

We can write the factored form as $(a-b)^2$, where $a = 7v$ and $b = 4$.

$49v^2 - 56v + 16 = (7v - 4)^2$

Practice this exercise:

1. Factor $25r^2 + 70rt + 49t^2$.

Name: Date:
Instructor: Section:

Objective 2 Factor a difference of squares.

Review this example for Objective 2:

2. Factor $16x^2 - 81$.

Look at the terms in the binomial $16x^2 - 81$ to determine whether the binomial is a difference of squares.

$16x^2 = (4x)^2$

$81 = (9)^2$

The two terms are both perfect squares. So, $16x^2 - 81$ is a difference of squares in the form $a^2 - b^2$.

We can write the factored form as $(a+b)(a-b)$, where $a = 4x$ and $b = 9$.

$16x^2 - 81 = (4x-9)(4x+9)$

Practice this exercise:

2. Factor $9a^2 - 121g^2$.

Objective 3 Factor a difference of cubes.

Review this example for Objective 3:

3. Factor $x^3 - 8$.

Look at the terms in the binomial $x^3 - 8$ to determine whether the binomial is a difference of cubes.

$x^3 = (x)^3$

$8 = (2)^3$

The two terms are both perfect cubes. So, $x^3 - 8$ is a difference of cubes in the form $a^3 - b^3$.

We can write the factored form as $(a-b)(a^2 + ab + b^2)$, where $a = x$ and $b = 2$.

$x^3 - 8 = (x-2)(x^2 + 2x + 4)$

Practice this exercise:

3. Factor $r^3 - 27$.

Name: Date:
Instructor: Section:

Objective 4 Factor a sum of cubes.

Review this example for Objective 4:

4. Factor $64y^3 + 125$.

Look at the terms in the binomial $64y^3 + 125$ to determine whether the binomial is a sum of cubes.

$64y^3 = (4y)^3$

$125 = (5)^3$

The two terms are both perfect cubes. So, $64y^3 + 125$ is a sum of cubes in the form $a^3 + b^3$.

We can write the factored form as $(a+b)(a^2 - ab + b^2)$, where $a = 4y$ and $b = 5$.

$64y^3 + 125 = (4y+5)(16y^2 - 20y + 25)$

Practice this exercise:

4. Factor $b^3 + 64$.

ADDITIONAL EXERCISES

Objective 1 Factor perfect square trinomials.

For extra help, see Examples 1–2 on pages 842–843 of your text and the Section 13.4 lecture video.

Factor the trinomials that are perfect squares. If the trinomial is not a perfect square, write the following: not a perfect square.

1. $s^2 + 12s + 36$

2. $v^2 - 16v + 64$

3. $25a^2 - 90a + 81$

4. $9c^2 - 24cg + 16g^2$

Objective 2 Factor a difference of squares.

For extra help, see Examples 3–4 on pages 843–844 of your text and the Section 13.4 lecture video.

Factor the binomials that are the difference of squares. If prime, so state.

5. $r^2 - 36$

6. $4c^2 - 121$

7. $25m^2 - \frac{36}{25}$

8. $45a^2 - 245$

Name: Date:
Instructor: Section:

Objective 3 Factor a difference of cubes.

For extra help, see Example 5 on page 844 of your text and the Section 13.4 lecture video.

Factor the difference of cubes.

9. $27x^3 - 64b^3$

10. $(y+z)^3 - 8$

Objective 4 Factor a sum of cubes.

For extra help, see Example 6 on page 845 of your text and the Section 13.4 lecture video.

Factor the sum of cubes.

11. $27x^3 + 64g^3$

Name: Date:
Instructor: Section:

Chapter 6 FACTORING

13.5 Strategies for Factoring

Learning Objectives
1 Factor polynomials.

GUIDED EXAMPLES AND PRACTICE

Objective 1 Factor polynomials.

Review these examples for Objective 1:

1. Factor $a^2+13a+42$.

First factor out any monomial GCF.
There is no monomial GCF in this polynomial.

There are three terms in this polynomial, so check to see if it is a perfect square.
a^2 is a perfect square, but 42 is not. So, the polynomial is not a perfect square.

Now consider the form of the trinomial.
It has the form ax^2+bx+c.
So, find two factors of *c* whose sum is *b* and write the factored form as (*a* + first number)(*a* + second number).
$a^2+13a+42=(a+7)(a+6)$

2. Factor $ac+ad+bc+bd$.

First factor out any monomial GCF.
There is no monomial GCF in this polynomial.

There are four terms in this polynomial, so try factoring by grouping.

$$\begin{aligned} ac+ad+bc+bd &= (ac+ad)+(bc+bd) \\ &= a(c+d)+b(c+d) \\ &= (a+b)(c+d) \end{aligned}$$

Practice these exercises:

1. Factor $r^2+10r+16$.

2. Factor $vw+vx+kw+kx$.

Name:
Instructor:

Date:
Section:

3. Factor $54x^3 - 128s^3$.

First factor out any monomial GCF.
The monomial GCF in this polynomial is 2.

There are two terms in the binomial factor of this polynomial, so check to see if the binomial is a difference of squares, a sum of cubes, or a difference of cubes. If so, use the appropriate rule to factor the polynomial completely.

$$54x^3 - 128s^3 = 2(27x^3 - 64s^3)$$
$$= 2\left[(3x)^3 - (4s)^3\right]$$
$$= 2(3x - 4s)(9x^2 + 12xs + 16s^2)$$

3. Factor $v^3 + 27$.

ADDITIONAL EXERCISES

Objective 1 Factor polynomials.

For extra help, see Example 1 on pages 849–851 of your text and the Section 13.5 lecture video.

Factor completely. If prime, so state.

1. $a(c-5) + z(c-5)$

2. $24b^2 - 294$

3. $t^2 + t - 30$

4. $t^2 - t - 30$

5. $b^2 - 25$

6. $125x^3 - 216a^3$

7. $54u^2 - 255u + 54$

8. $s^4 - 81$

9. $x^2 - 4x - 45$

10. $4v^2 - 36v + 81$

11. $x^2 - 6x$

12. $9r - 100r^3$

Name: Date:
Instructor: Section:

Chapter 13 FACTORING

13.6 Solving Quadratic Equations by Factoring

Learning Objectives
1 Use the zero-factor theorem to solve equations containing expressions in factored form.
2 Solve quadratic equations by factoring.
3 Solve problems involving quadratic equations.
4 Use the Pythagorean theorem to solve problems.

Key Terms
Use the vocabulary term listed below to complete the statement in Exercises 1.

quadratic equation in one variable

1. An equation that can be written in the form $ax^2 + bx + c = 0$, where a, b, and c are real numbers and $a \neq 0$, is a(n) ________________.

GUIDED EXAMPLES AND PRACTICE

Objective 1 Use the zero-factor theorem to solve equations containing expressions in factored form.

Review this example for Objective 1:

1. Solve $(x-6)(x+7)=0$.

Use the zero-factor theorem to solve.

$(x-6)(x+7)=0$

$x-6=0$ or $x+7=0$

$x=6$ $\quad x=-7$

Practice this exercise:

1. Solve $r(r+4)=0$.

Objective 2 Solve quadratic equations by factoring.

Review this example for Objective 2:

2. Solve $r^2-30r=-29$.

Write the equation in standard form, $ax^2+bx+c=0$.

$r^2-30r+29=0$

Factor the variable expression, and use the zero-factor theorem to solve.

$r^2-30r+29=0$

$(r-1)(r-29)=0$

$r-1=0$ or $r-29=0$

$r=1$ $\quad r=29$

Practice this exercise:

2. Solve $(v-5)^2=0$.

Name:
Instructor:

Date:
Section:

Objective 3 Solve problems involving quadratic equations.

Review this example for Objective 3:

3. The product of two consecutive even natural numbers is 288. Find the numbers.

Even natural numbers are 0, 2, 4, 6, . . .
Adding 2 to a given even natural number gives a consecutive even natural number.
The pattern is as follows:
First even natural number: x
Consecutive even natural number: $x+2$

The product of the two numbers is 288.
Multiply and solve for x.

$$x(x+2)=288$$
$$x^2+2x-288=0$$
$$(x-16)(x+18)=0$$
$$x-16=0 \quad \text{or} \quad x+18=0$$
$$x=16 \qquad\qquad x=-18$$

Since -18 is not a natural number, the first number is 16. So, the second number is $16+2=18$.

Practice this exercise:

3. The sum of the squares of two consecutive odd positive integers is 74. Find the integers.

Objective 4 Use the Pythagorean theorem to solve problems.

Review this example for Objective 4:

4. Find the length of the hypotenuse.

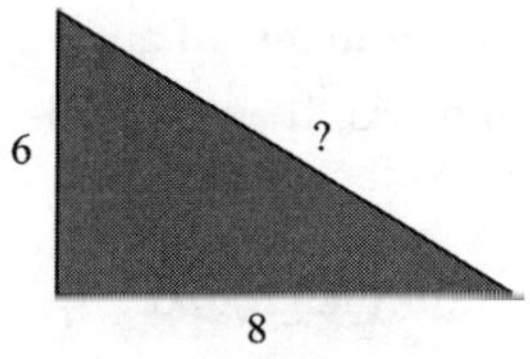

Use the Pythagorean theorem $a^2+b^2=c^2$, where the hypotenuse is c.

$$8^2+6^2=c^2$$
$$64+36=c^2$$
$$100=c^2$$
$$0=c^2-100$$
$$0=(c-10)(c+10)$$
$$c-10=0 \quad \text{or} \quad c+10=0$$
$$c=10 \qquad\qquad c=-10$$

Since the hypotenuse is a length, only positive solutions are sensible. So, the length of the hypotenuse is 10.

Practice this exercise:

4. Find the length of the hypotenuse.

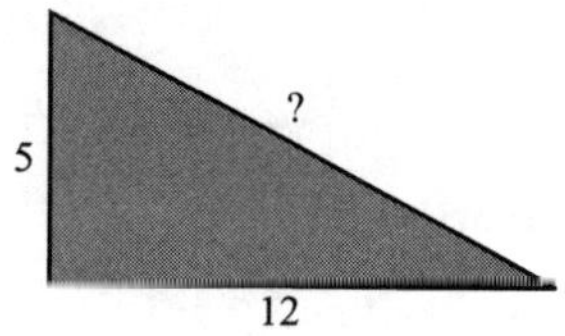

Name:
Instructor:

Date:
Section:

ADDITIONAL EXERCISES

Objective 1   Use the zero-factor theorem to solve equations containing expressions in factored form.
For extra help, see Examples 1–2 on pages 854–855 of your text and the Section 13.6 lecture video.
Solve using the zero-factor theorem.

1. $(a-25)(a+64)=0$

2. $(4t+9)(t+8)=0$

Objective 2   Solve quadratic equations by factoring.
For extra help, see Examples 3–5 on pages 856–857 of your text and the Section 13.6 lecture video.
Solve the quadratic equations.

3. $x^2-25x+24=0$

4. $w^2-6w+9=0$

5. $a^2=-12-7a$

6. $b^2=16$

7. $5w^2=26w+24$

8. $r(r-5)=36$

Objective 3   Solve problems involving quadratic equations.
For extra help, see Examples 6–8 on pages 858–859 of your text and the Section 13.6 lecture video.
Translate to an equation and then solve.

9. The length of the top of a rectangular table is 5 meters greater than the width. The area is 104 square meters. Find the dimensions of the table.

10. Use the formula $h=-16t^2+v_0t+h_0$, where h is the final height in feet, t is the time of travel in seconds, v_0 is the initial velocity in feet per second, and h_0 is the initial height in feet of an object traveling upward. If an object is thrown upward at 96 feet per second from a height of 4 feet, when will the object be 144 feet off the ground?

Objective 4   Use the Pythagorean theorem to solve problems.
For extra help, see Example 9 on page 860 of your text and the Section 13.6 lecture video.
Find the length of the hypotenuse.

11.

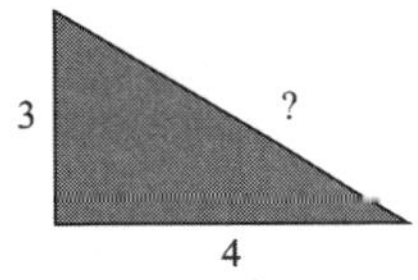

Name: Date:
Instructor: Section:

Chapter 13 FACTORING

13.7 Graphs of Quadratic Equations and Functions

Learning Objectives
1 Graph quadratic equations in the form $y = ax^2 + bx + c$.
2 Graph quadratic functions.

Key Terms
Use the vocabulary terms listed below to complete each statement in Exercises 1–3.

quadratic equation in two variables **axis of symmetry** **vertex of a parabola**

1. The lowest point on a parabola that opens up or the highest point on a parabola that opens down is called the ____________________.

2. An equation that can be written in the form $y = ax^2 + bx + c$, where a, b, and c are real numbers and $a \neq 0$, is a(n) ____________________.

3. The ____________________ is a line that divides a graph into two symmetrical halves.

GUIDED EXAMPLES AND PRACTICE

Objective 1 Graph quadratic equations in the form $y = ax^2 + bx + c$.

Review this example for Objective 1:
1. Graph $f(x) = 2x^2 - 8x + 5$.

Find enough ordered pairs to clearly see the graph. Then connect the points to form the parabola.

x	y
0	5
1	−1
2	−3
3	−1
4	5

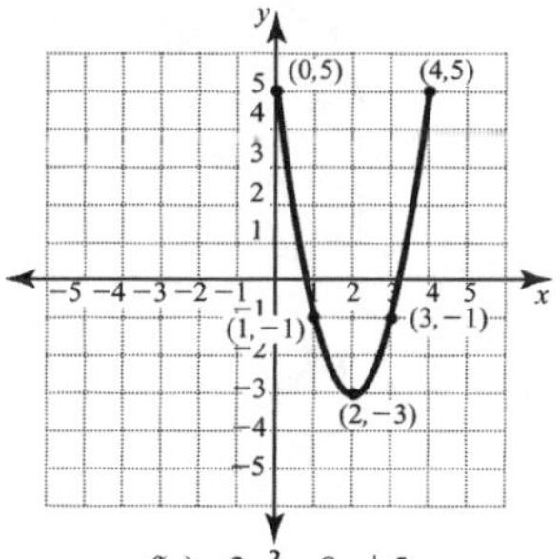

$f(x) = 2x^2 - 8x + 5$

Practice this exercise:
1. Graph $y = x^2 + 6x - 1$.

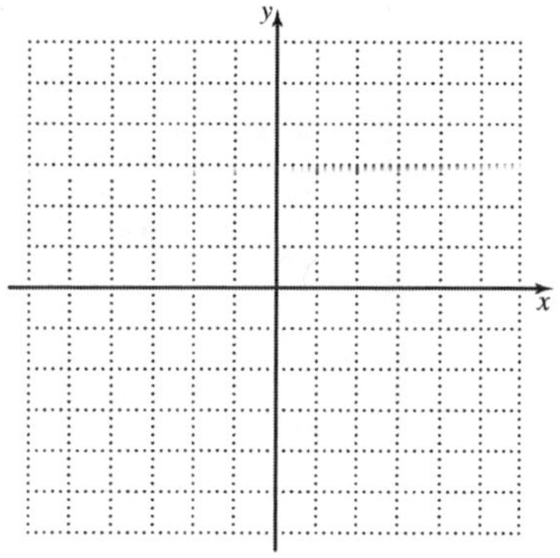

Name:　　　　　　　　　　　　　　　　　　Date:
Instructor:　　　　　　　　　　　　　　　　Section:

Objective 2　Graph quadratic functions.

Review this example for Objective 2:

2. Graph $f(x) = -x^2 + 7$.

Find enough ordered pairs to clearly see the graph. Then connect the points to form the parabola.

x	y
-2	3
-1	6
0	7
1	6
2	3

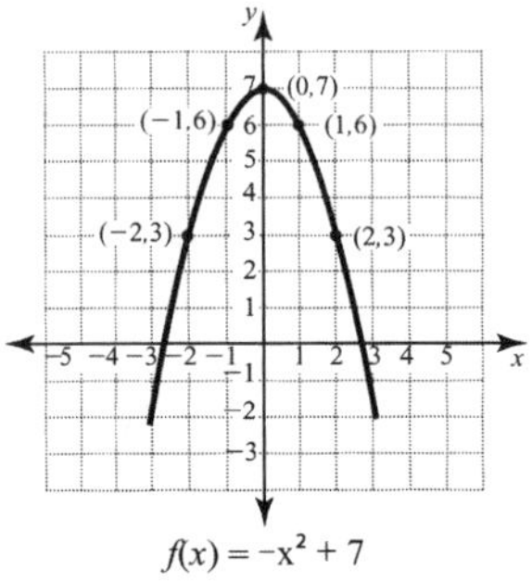

$f(x) = -x^2 + 7$

Practice this exercise:

2. Graph $f(x) = x^2 - 4x - 1$.

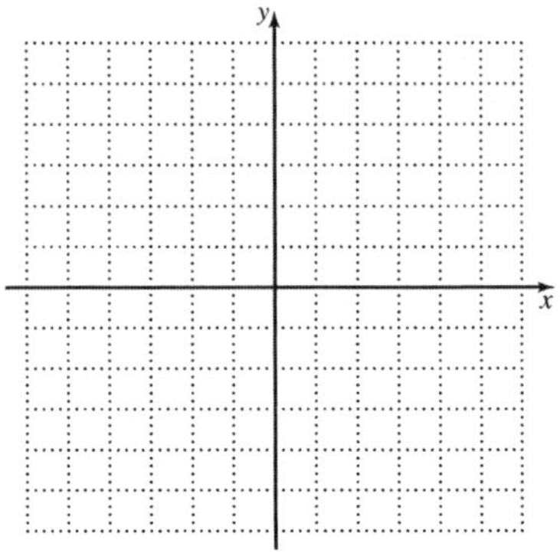

ADDITIONAL EXERCISES

Objective 1　Graph quadratic equations in the form $y = ax^2 + bx + c$.

For extra help, see Examples 1–2 on pages 866–867 of your text and the Section 13.7 lecture video.

Graph.

1. $y = x^2 - 9$

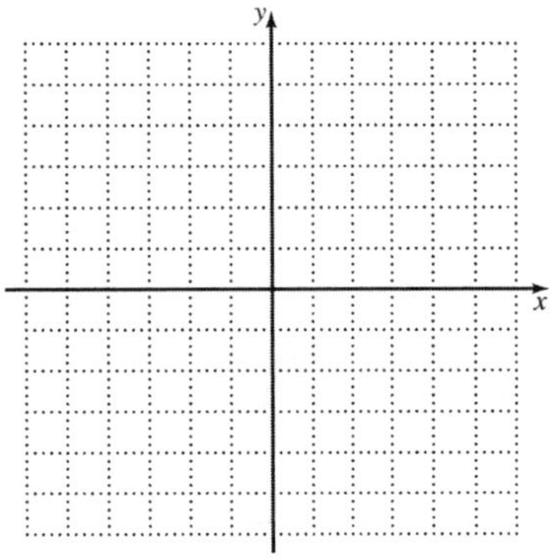

2. $y = 4 - x^2$

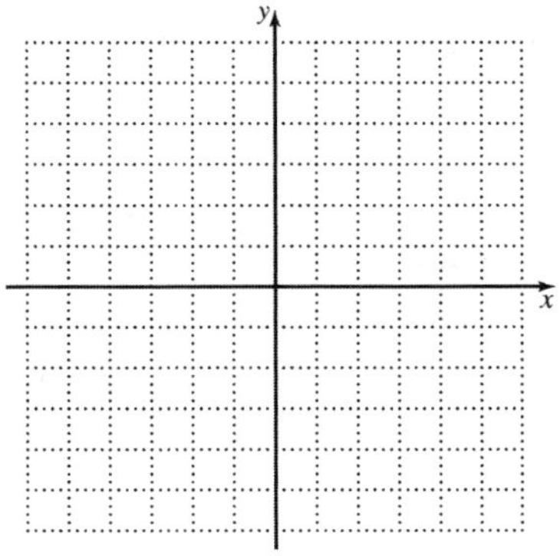

3. $y = -x^2 + 2x - 3$

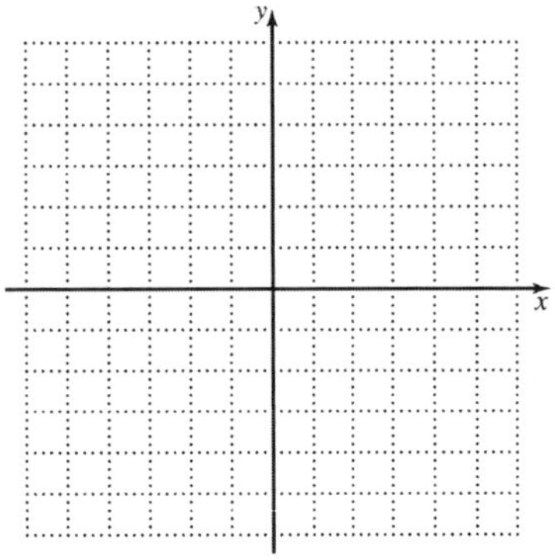

Name: Date:
Instructor: Section:

Objective 2 Graph quadratic functions.

For extra help, see Examples 3–4 on pages 868–869 of your text and the Section 13.7 lecture video.

Graph.

4. $f(x) = x^2 - x - 8$

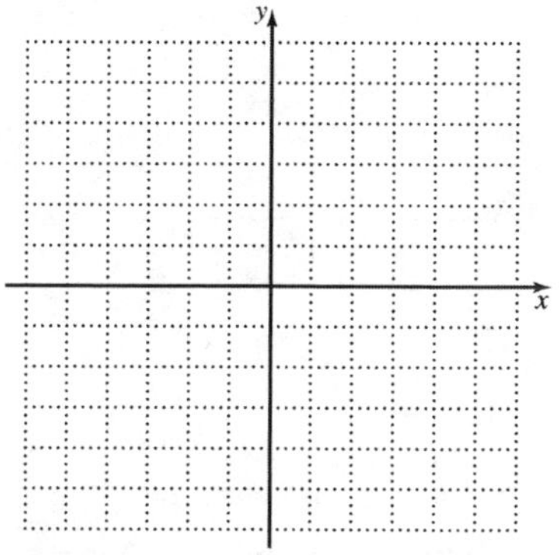

5. $f(x) = -x^2 - 8x - 10$

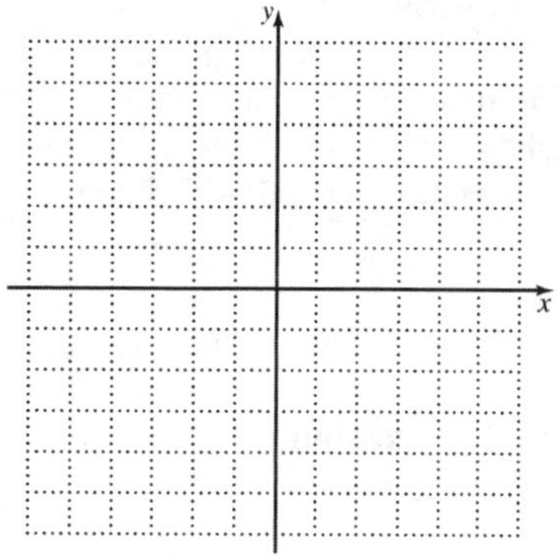

State whether the graph is the graph of a function. Give the domain and range.

6.

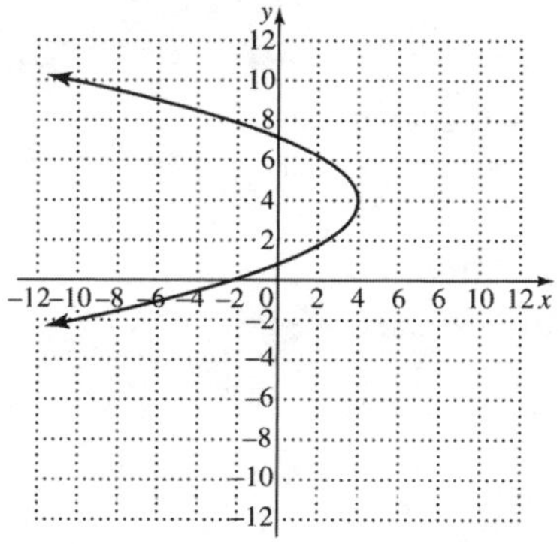

Name: Date:
Instructor: Section:

Chapter 14 RATIONAL EXPRESSIONS AND EQUATIONS

14.1 Simplifying Rational Expressions

Learning Objectives
1 Evaluate rational expressions.
2 Find numbers that cause a rational expression to be undefined.
3 Simplify rational expressions containing only monomials.
4 Simplify rational expressions containing multiterm polynomials.

Key Terms
Use the vocabulary term listed below to complete the statement in Exercise 1.

rational expression

1. A(n) expression that can be written in the form $\frac{P}{Q}$, where P and Q are polynomials and $Q \neq 0$ is referred to as a(n) ________________.

GUIDED EXAMPLES AND PRACTICE
Objective 1 Evaluate rational expressions.

Review this example for Objective 1:

1. Evaluate $\frac{7x+3}{9x+81}$

a) when $x=9$.
b) when $x=-9$.

a) Replace x with 9.
$\frac{7x+3}{9x+81}=\frac{7(9)+3}{9(9)+81}=\frac{66}{162}=\frac{11}{27}$

b) Replace x with -9.
$\frac{7x+3}{9x+81}=\frac{7(-9)+3}{9(-9)+81}=\frac{-60}{0}$; undefined

Practice this exercise:

1. Evaluate $\frac{8x-10}{3x}$

a) when $x=2$.
b) when $x=-3$.

Name: Date:
Instructor: Section:

Objective 2 **Find numbers that cause a rational expression to be undefined.**

Review this example for Objective 2:

2. Find every value for the variable that makes the expression undefined.

$$\frac{x}{x^2+4x-5}$$

Set the denominator equal to 0.

$x^2+4x-5=0$

Factor the polynomial.

$(x+5)(x-1)=0$

Use the zero-factor theorem.

$x+5=0$ or $x-1=0$

$x=-5$ $x=1$

The expression $\frac{x}{x^2+4x-5}$ is undefined if x is replaced by −5 or 1.

Practice this exercise:

2. Find every value for the variable that makes the expression undefined.

$$\frac{4}{x-5}$$

Objective 3 **Simplify rational expressions containing only monomials.**

Review this example for Objective 3:

3. Simplify.

$$\frac{1760a^7b^3}{15a^4b}$$

Factor the numerator and denominator completely.

$$\frac{1760a^7b^3}{15a^4b}=\frac{(352)(5)a^4a^3b^3}{(3)(5)a^4b}$$

Divide out the common factors and simplify.

$$\frac{(352)(\not{5})\not{a^4}a^3b^2\not{b}}{(3)(\not{5})\not{a^4}\not{b}}=\frac{352a^3b^2}{3}$$

Practice this exercise:

3. Simplify.

$$\frac{1250u^8z^{10}}{20u^4z^2}$$

Objective 4 **Simplify rational expressions containing multiterm polynomials.**

Review this example for Objective 4:

4. Simplify.

$$\frac{10z-40}{8z-32}$$

Factor the numerator and denominator completely.

$$\frac{10z-40}{8z-32}=\frac{2(5)(z-4)}{2(4)(z-4)}$$

Divide out the common factors and simplify.

$$\frac{\not{2}(5)\not{(z-4)}}{\not{2}(4)\not{(z-4)}}=\frac{5}{4}$$

Practice this exercise:

4. Simplify.

$$\frac{8a-24}{7a-21}$$

Name: Date:
Instructor: Section:

ADDITIONAL EXERCISES

Objective 1 Evaluate the rational expression.

For extra help, see Example 1 on pages 885–886 of your text and the Section 14.1 lecture video.

Evaluate the rational expression.

1. $\dfrac{(-5x)^2}{4x+12}$

a) when $x = 2$

b) when $x = -3$

2. $\dfrac{3x+1}{5x+25}$

a) when $x = 5$

b) when $x = -5$

Objective 2 Find numbers that cause a rational expression to be undefined.

For extra help, see Example 2 on page 886 of your text and the Section 14.1 lecture video.

Find every value for the variable that makes the expression undefined.

3. $\dfrac{3x}{x^2-16}$

4. $\dfrac{2x+4}{x^2-2x-3}$

Objective 3 Simplify rational expressions containing only monomials.

For extra help, see Examples 3–4 on pages 887–888 of your text and the Section 14.1 lecture video.

Simplify, if possible.

5. $\dfrac{820u^7z^3}{41u^5z^2}$

Objective 4 Simplify rational expressions containing multiterm polynomials.

For extra help, *see* Examples 5–8 on pages 889–891 of your text and the Section 14.1 lecture video.

Simplify, if possible.

6. $\dfrac{ak-av-dk+dv}{ak-av+dk-dv}$

7. $\dfrac{6v-12}{2-v}$

8. $\dfrac{a^3-c^3}{a^2-c^2}$

9. $\dfrac{r^2-49}{r^2-14r+49}$

10. $\dfrac{w^2+1}{w+1}$

11. $\dfrac{15(y-7)}{6(y-7)}$

Name: Date:
Instructor: Section:

Chapter 14 RATIONAL EXPRESSIONS AND EQUATIONS

14.2 Multiplying and Dividing Rational Expressions

Learning Objectives

1 Multiply rational expressions.
2 Divide rational expressions.
3 Convert units of measurement using dimensional analysis.

GUIDED EXAMPLES AND PRACTICE

Objective 1 Multiply rational expressions.

Review this example for Objective 1:

1. Multiply.

$$\frac{125-5z}{49}\cdot\frac{245}{7z-175}$$

Factor the numerators and denominators completely.

$$\frac{125-5z}{49}\cdot\frac{245}{7z-175}=\frac{5(25-z)}{(7)(7)}\cdot\frac{(7)(7)(5)}{-7(25-z)}$$

Divide out the common factors.

$$\frac{5(\cancel{25-z})}{(\cancel{7})(\cancel{7})}\cdot\frac{(\cancel{7})(\cancel{7})(5)}{-7(\cancel{25-z})}=\frac{5}{1}\cdot\frac{5}{-7}$$

Multiply the remaining numerator factors and denominator factors.

$$\frac{5}{1}\cdot\frac{5}{-7}=-\frac{25}{7}$$

Practice this exercise:

1. Multiply.

$$\frac{11}{m}\cdot\frac{m^5}{11}$$

Objective 2 Divide rational expressions.

Review this example for Objective 2:

2. Divide.

$$\frac{50x^9}{2y^9}\div\frac{625x^4}{10y^3}$$

Write an equivalent multiplication statement by changing the division sign to multiplication and changing the divisor to its reciprocal.

$$\frac{50x^9}{2y^9}\div\frac{625x^4}{10y^3}=\frac{50x^9}{2y^9}\cdot\frac{10y^3}{625x^4}$$

Factor the numerators and denominators completely.

$$\frac{50x^9}{2y^9}\cdot\frac{10y^3}{625x^4}=\frac{5(5)(2)x^5(x^4)}{2y^6(y^3)}\cdot\frac{(2)(5)y^3}{5(5)(5)(5)x^4}$$

Practice this exercise:

2. Divide.

$$\frac{3}{r}\div\frac{27}{r}$$

Divide out the common factors.

$$\frac{5(5)(2)x^5(x^4)}{2y^6(y^3)} \cdot \frac{(2)(5)y^3}{5(5)(5)(5)x^4} = \frac{x^5}{y^6} \cdot \frac{(2)}{5}$$

Multiply the remaining numerator factors and denominator factors.

$$\frac{x^5}{y^3} \cdot \frac{(2)}{1} = \frac{2x^5}{y^3}$$

Objective 3 Convert units of measurement using dimensional analysis.

Review this example for Objective 3:

3. Convert 72 centimeters per second to yards per hour.

Solve for the desired units in the numerator.

$$\frac{72 \text{ cm}}{1 \text{ sec}} = \frac{1 \text{ in.}}{2.54 \text{ cm}} = \frac{1 \text{ ft}}{12 \text{ in.}} = \frac{1 \text{ yd}}{3 \text{ ft}} = \frac{0.787 \text{ yd}}{1 \text{ sec}}$$

Solve for the desired units in the denominator.

$$\frac{0.787 \text{ yd}}{1 \text{ sec}} = \frac{60 \text{ sec}}{1 \text{ min}} = \frac{60 \text{ min}}{1 \text{ hr}} = \frac{2825 \text{ yd}}{1 \text{ hr}}$$

Practice this exercise:

3. Convert 5 yards to inches.

ADDITIONAL EXERCISES

Objective 1 Multiply rational expressions.

For extra help, see Examples 1–2 on page 897 of your text and the Section 14.2 lecture video.

Multiply.

1. $\frac{12y}{7} \cdot \frac{y}{3} \cdot \frac{2}{y^2}$

2. $\frac{8-2z}{49} \cdot \frac{98}{7z-28}$

3. $\frac{9x}{x^2-4x+4} \cdot \frac{x^2-4}{18x^2}$

4. $\frac{x^2-7x+10}{x^2-4} \cdot \frac{x-2}{x^2-10x+25}$

5. $\frac{c^2+cf-cd-df}{c^2+5c+cf+5f} \cdot \frac{c^2-7c-35+5c}{c^2-7d+cd-7c}$

Objective 2 Divide rational expressions.

For extra help, see Examples 3–4 on pages 898–899 of your text and the Section 14.2 lecture video.

Divide.

6. $\frac{20x^6}{5y^5} \div \frac{16x^4}{10y^2}$

7. $\frac{w^3-7w^2+12w}{2x} \div \frac{2w-6}{4w+16}$

8. $\frac{w+4}{2} \div \frac{3w+12}{8}$

9. $\frac{v^2+8v+16}{v+3} \div \left(5v^2+17v-12\right)$

Name: Date:
Instructor: Section:

10. $\frac{k+4}{3} \div \frac{2k+8}{9}$

Objective 3 Convert units of measurement using dimensional analysis.
For extra help, see Example 5 on page 900 of your text and the Section 7.2 lecture video.
Convert.

11. Convert 2.5 tons to ounces.

12. Convert 90 miles per hour to feet per second.

Name: Date:
Instructor: Section:

Chapter 14 RATIONAL EXPRESSIONS AND EQUATIONS

14.3 Adding and Subtracting Rational Expressions with the Same Denominator

Learning Objectives
1 Add or subtract rational expressions with the same denominator.

GUIDED EXAMPLES AND PRACTICE

Objective 1 Add or subtract rational expressions with the same denominator.

Review these examples for Objective 1:

1. Add.

$$\frac{c}{c+4}+\frac{8}{c+4}$$

Add the numerators and keep the same denominator.

$$\frac{c}{c+4}+\frac{8}{c+4}=\frac{c+8}{c+4}$$

Because c and 8 are not like terms, we express their sum as a polynomial.

2. Subtract.

$$\frac{a^2}{a+2}-\frac{4}{a+2}$$

Subtract the numerators and keep the same denominator.

$$\frac{a^2}{a+2}-\frac{4}{a+2}=\frac{a^2-4}{a+2}$$

Factor the numerator.

$$\frac{a^2-4}{a+2}=\frac{(a+2)(a-2)}{a+2}$$

Divide out the common factor.

$$\frac{(a+2)(a-2)}{a+2}=\frac{a-2}{1}$$

Practice these exercises:

1. Add.

$$\frac{h^3-14}{h+7}+\frac{7h^2-2h}{h+7}$$

2. Subtract.

$$\frac{8}{n^2}-\frac{7}{n^2}$$

Name: Date:
Instructor: Section:

ADDITIONAL EXERCISES

Objective 1 Add or subtract rational expressions with the same denominator.

For extra help, see Examples 1–5 on page 906–909 of your text and the Section 14.3 lecture video.

Add or subtract. Simplify your answers to lowest terms.

1. $\frac{3x}{104}+\frac{5x}{104}$

2. $\frac{y^2}{y+2}+\frac{2y}{y+2}$

3. $\frac{m+5}{m^2-49}-\frac{12}{m^2-49}$

4. $\frac{c-2b}{c+b}+\frac{c+4b}{c+b}$

5. $\frac{w^2+12}{w+1}+\frac{9-w^2}{w+1}$

6. $\frac{z^2}{z-2}+\frac{z-6}{z-2}$

7. $\frac{11c+71}{10y}-\frac{c+1}{10y}$

8. $\frac{y^2}{y^2+2y}-\frac{4}{y^2+2y}$ 9^2

9. $\frac{z^2}{z+4}+\frac{4z}{z+4}$

10. $\frac{b-7d}{b+d}+\frac{b+9d}{b+d}$

11. $\frac{5z+21}{4c}-\frac{z+1}{4c}$

12. $\frac{x^2}{x-14}+\frac{x-210}{x-14}$

13. $\frac{y^2+8y-8}{3y^3+9y^2-30y}+\frac{2y^2+6y+3}{3y^3+9y^2-30y}$

Name: Date:
Instructor: Section:

Chapter 14 RATIONAL EXPRESSIONS AND EQUATIONS

14.4 Adding and Subtracting Rational Expressions with Different Denominators

Learning Objectives
1 Find the LCD of two or more rational expressions.
2 Given two rational expressions, write equivalent rational expressions with their LCD.
3 Add or subtract rational expressions with different denominators.

GUIDED EXAMPLES AND PRACTICE
Objective 1 Find the LCD of two or more rational expressions

Review these examples for Objective 1:

1. Find the LCD of the two expressions.

$\frac{5y}{9x^3}, \frac{7y}{24x^2}$

First find the prime factors of the denominators.

$\frac{5y}{9x^3} = \frac{5y}{3^2x^3}$

$\frac{7y}{24x^2} = \frac{7y}{2^3(3)x^2}$

Multiply all the unique terms that are raised to the highest power, $3^2, 2^3, x^3$.

LCD $= 2^3 \cdot x^3 \cdot 3^2 = 72x^3$

Practice this exercise:

1. Find the LCD of the two expressions.

$\frac{2}{3a^5}, \frac{3}{11a^6}$

Objective 2 Given two rational expressions, write equivalent rational expressions with their LCD.

Review this example for Objective 2:

2. Write equivalent rational expressions.

$\frac{y}{8x^3}, \frac{7}{18x^2}$

LCD $= 72x^3$

For each rational expression, multiply both the numerator and denominator by an appropriate factor so that the denominator becomes $72x^3$.

Determine the appropriate factor by dividing the LCD by the original denominator for each expression.

$\frac{72x^3}{8x^3} = 9 \qquad \frac{72x^3}{18x^2} = 4x$

Multiply each expression by their factor.

$\frac{y}{8x^3} \cdot \frac{[9]}{[9]} = \frac{y[9]}{72x^3} = \frac{9y}{72x^3}$

$\frac{7}{18x^2} \cdot \frac{[4x]}{[4x]} = \frac{7[4x]}{72x^3} = \frac{28x}{72x^3}$

Practice this exercise:

2. Write equivalent rational expressions.

$\frac{2x}{x+4}, \frac{7}{x-1}$

Name: Date:
Instructor: Section:

Objective 3 Add or subtract rational expressions with different denominators.

Review this example for Objective 3:

3. Subtract.

$$\frac{5y}{y^2-13y+36}-\frac{4y}{y^2-14y+45}$$

First, find the LCD by factoring the denominators.

$y^2-13y+36=(y-9)(y-4)$

$y^2-14y+45=(y-9)(y-5)$

LCD = $(y-9)(y-5)(y-4)$

Write equivalent rational expressions with the LCD.

$$\frac{(y-9)(y-5)(y-4)}{(y-9)(y-4)}=(y-5)$$

$$\frac{(y-9)(y-5)(y-4)}{(y-9)(y-5)}=(y-4)$$

Distribute in the numerators.

$$\frac{5y}{(y-9)(y-4)}\cdot\frac{(y-5)}{(y-5)}=\frac{5y(y-5)}{(y-9)(y-4)(y-5)}$$

$$=\frac{5y^2-25}{(y-9)(y-4)(y-5)}$$

$$\frac{4y}{(y-9)(y-5)}\cdot\frac{(y-4)}{(y-4)}=\frac{4y(y-4)}{(y-9)(y-5)(y-4)}$$

$$=\frac{4y^2-16}{(y-9)(y-5)(y-4)}$$

Combine and subtract numerators.

$$\frac{(5y^2-25)-(4y^2-16)}{(y-9)(y-4)(y-5)}=\frac{y^2-9}{(y-9)(y-4)(y-5)}$$

Factor the numerator and divide out the common factor.

$$\frac{(y-3)(y+3)}{(y-9)(y-4)(y-5)}$$

Practice this exercise:

3. Subtract.

$$\frac{9}{z+3}-\frac{5}{z-3}$$

Name: Date:
Instructor: Section:

ADDITIONAL EXERCISES

Objective 1 Find the LCD of two or more rational expressions.

For extra help, see Examples 1–2 on pages 912–913 of your text and the Section 14.4 lecture video.
Find the least common denominator for the rational expressions.

1. $\dfrac{8x}{3x-6}, \dfrac{5x}{4x-8}$

2. $\dfrac{2}{p^2-4}, \dfrac{7+p}{p+2}$

Objective 2 Given two rational expressions, write equivalent rational expressions with their LCD.

For extra help, see Examples 3–4 on pages 913–914 of your text and the Section 14.4 lecture video.
Write equivalent rational expressions with the LCD.

3. $\dfrac{4x+2}{x^2-9}, \dfrac{2x-1}{x^2-x-6}$

4. $\dfrac{5y}{(x-3)^2}, \dfrac{4}{5(x-3)}$

Objective 3 Adding or Subtracting Rational Expressions with Different Denominators.

For extra help, see Examples 5–7 on pages 914–917 of your text and the Section 14.4 lecture video.
Add or subtract as indicated.

5. $\dfrac{8}{z}+\dfrac{9}{5z}$

6. $\dfrac{y}{y-c}-\dfrac{c}{c-y}$

7. $\dfrac{9}{y-5}+\dfrac{7}{5-y}$

8. $\dfrac{y}{y^2-2y-35}+\dfrac{8}{y+5}$

9. $\dfrac{x-3}{x+5}+\dfrac{x+6}{x-8}$

Name: Date:
Instructor: Section:

Chapter 14 RATIONAL EXPRESSIONS AND EQUATIONS

14.5 Complex Rational Expressions

Learning Objectives
1 Simplify complex rational expressions.

Key Terms
Use the vocabulary term listed below to complete the statement in Exercise 1.

complex rational expression

1. A rational expression that contains rational expressions in the numerator or denominator is a ______________.

GUIDED EXAMPLES AND PRACTICE
Objective 1 Simplify complex rational expressions.

Review these examples for Objective 1:

1. Simplify using Method 1.

$$\frac{\frac{4}{2x+1}}{\frac{x}{5x-4}}$$

Rewrite as a horizontal division problem.

$$\frac{\frac{4}{2x+1}}{\frac{x}{5x-4}}=\frac{4}{2x+1}\div\frac{x}{5x-4}$$

Write an equivalent multiplication problem.

$$\frac{4}{2x+1}\div\frac{x}{5x-4}=\frac{4}{2x+1}\cdot\frac{5x-4}{x}$$

Multiply.

$$\frac{4}{2x+1}\cdot\frac{5x-4}{x}=\frac{20x-16}{2x^2+x}=\frac{4(5x-4)}{x(2x+1)}$$

Practice these exercises:

1. Simplify using Method 1.

$$\frac{-\frac{1}{5}}{\frac{1}{35}}$$

Name: Date:
Instructor: Section:

2. Simplify using Method 2.

$$\frac{\frac{4}{2x+1}}{\frac{x}{5x-4}}$$

Multiply the numerator and denominator by the LCD and divide out common factors, if necessary.

LCD $= (2x+1)(5x-4)$

$$\frac{\left(\frac{4}{2x+1}\right)\cdot(2x+1)(5x-4)}{\left(\frac{x}{5x-4}\right)\cdot(2x+1)(5x-4)} = \frac{4(5x-4)}{x(2x+1)}$$

2. Simplify using Method 2.

$$\frac{-\frac{1}{5}}{\frac{1}{35}}$$

ADDITIONAL EXERCISES

Objective 1 Simplify complex rational expressions.

For extra help, see Example 1 on pages 922–925 of your text and the Section 14.5 lecture video.

Simplify.

1. $\dfrac{\frac{t}{u^2}}{\frac{t^2}{u}}$

2. $\dfrac{\frac{8}{7}+\frac{3}{14}}{\frac{6}{5}-\frac{2}{10}}$

3. $\dfrac{\frac{1}{y}+3}{\frac{1}{y}-6}$

4. $\dfrac{1+\frac{9}{49}}{1-\frac{3}{7}}$

5. $\dfrac{t-\frac{1}{t}}{t+\frac{1}{t}}$

6. $\dfrac{1+\frac{7}{y}}{1-\frac{49}{y^2}}$

7. $\dfrac{\frac{9}{x-7}-\frac{9}{x+7}}{\frac{9}{x^2-49}}$

8. $\dfrac{\frac{1}{y+b}-\frac{1}{y}}{b}$

9. $\dfrac{1+\frac{2}{y}-\frac{3}{y^2}}{\frac{1}{y}+\frac{3}{y^2}}$

Name: Date:
Instructor: Section:

Chapter 14 RATIONAL EXPRESSIONS AND EQUATIONS

14.6 Solving Equations Containing Rational Expressions

Learning Objectives
1 Solve equations containing rational expressions.

Key Terms
Use the vocabulary term listed below to complete the statement in Exercise 1.

extraneous solution

1. If by solving an equation, you obtain a number that causes an expression in the original equation to be undefined, we say that the number is a(n) __________________ and we discard it.

GUIDED EXAMPLES AND PRACTICE

Objective 1 Solving Equations Containing Rational Expressions

Review these examples for Objective 1:

1. Check the given value to see if it is a solution to the equation.

$\frac{8}{x+2}+1=3; 4$

Substitute 4 for x in the equation.

$\frac{8}{4+2}+1\stackrel{?}{=}3$

Add the denominator.

$\frac{8}{6}+1\stackrel{?}{=}3$

Multiply both sides by the LCD, 6.

$\cancel{6}\cdot\frac{8}{\cancel{6}}+6\cdot 1\stackrel{?}{=}6\cdot 3$

$8+6\stackrel{?}{=}18$

Add the left side of the equation.

$13\neq 18$

Since, $13\neq 18$, 4 is not a solution to the equation.

Practice these exercises:

1. Check the given value to see if it is a solution to the equation.

$\frac{-5}{y-2}+\frac{7}{2}=\frac{27}{2}; \frac{3}{2}$

Name: Date:
Instructor: Section:

2. Solve.

$$\frac{1}{3x^2-3}+\frac{2}{x-1}=\frac{3}{x+1}$$

Inspect the denominators to check for values that make the rational expressions undefined.

$x+1=0 \qquad x=-1$

$x-1=0 \qquad x=+1$

$3x^2-3=0 \qquad x=\pm 1$

Those values are $x=+1$ and $x=-1$.

Factor the quadratic expression.

$3x^2-3=3(x-1)(x+1)$

$\text{LCD}=3(x-1)(x+1)$

Multiply both sides by the LCD.

$$\left(3(x-1)(x+1)\right)\left(\frac{1}{3x^2-3}+\frac{2}{x-1}\right)=\left(\frac{3}{x+1}\right)\left(3(x-1)(x+1)\right)$$

Distribute and divide out the common factors.

$$\left(3(x-1)(x+1)\right)\left(\frac{1}{3x^2-3}+\frac{2}{x-1}\right)=\left(\frac{3}{x+1}\right)\left(3(x-1)(x+1)\right)$$

$$\left(\frac{\left(3(x-1)(x+1)\right)}{3x^2-3}+\frac{2\left(3(x-1)(x+1)\right)}{x-1}\right)=\left(\frac{3\left(3(x-1)(x+1)\right)}{x+1}\right)$$

$$1+6(x+1)=9(x-1)$$

Distribute and combine like terms.

$1+6(x+1)=9(x-1)$

$6x+7=9x-9$

Get the variables on one side of the equation and the constants on the other side and solve for x.

$16=3x$

$x=\frac{16}{3}$

Check the solution in the original equation.

$$\frac{1}{3x^2-3}+\frac{2}{x-1}=\frac{3}{x+1}$$

$$\frac{1}{3(\frac{16}{3})^2-3}+\frac{2}{\frac{16}{3}-1}=\frac{3}{\frac{16}{3}+1}$$

$$\frac{3}{247}+\frac{6}{13}=\frac{9}{19}$$

$$\frac{117}{247}=\frac{117}{247}$$

2. Solve.

$$\frac{a}{4}-\frac{a-4}{8}=\frac{7}{8}$$

Name: Date:
Instructor: Section:

ADDITIONAL EXERCISES

Objective 1 Solving Equations Containing Rational Expressions.

For extra help, see Examples 1–5 on page 930–933 of your text and the Section 14.6 lecture video.

Check the given values to see if they are solutions to the equation.

1. $\frac{9}{x}+\frac{8}{9}=\frac{17}{18}$; 162

2. $\frac{3}{5x}+\frac{2}{3}=\frac{5}{6x}$; 18

Solve and check. Identify any extraneous solutions.

3. $\frac{m}{m-6}-7=\frac{6}{m-6}$

4. $\frac{8}{11y}+\frac{5}{y}=1$

5. $\frac{z+3}{z}=\frac{4}{3}$

6. $\frac{y-6}{y+2}=\frac{3}{5}$

7. $\frac{4}{y+7}=\frac{3}{y-6}$

8. $\frac{3}{z^2-6z+9}-\frac{1}{z-3}=\frac{1}{6z-18}$

9. $\frac{1}{2x^2-72}+\frac{8}{x-6}=\frac{5}{x+6}$

10. $\frac{6}{6c^2-c-24}=\frac{3}{3c^2-3c-8}$

11. $\frac{6}{y^2-14y+49}-\frac{1}{y-7}=\frac{1}{8y-56}$

Name: Date:
Instructor: Section:

Chapter 14 RATIONAL EXPRESSIONS AND EQUATIONS

14.7 Applications with Rational Expressions, Including Variation

Learning Objectives
1. Use tables to solve problems with two unknowns involving rational expressions.
2. Solve problems involving direct variation.
3. Solve problems involving inverse variation.
4. Solve problems involving joint variation.
5. Solve problems involving combined variation.

Key Terms
Use the vocabulary terms listed below to complete each statement in Exercises 1–3.

direct variation **inverse variation** **joint variation**

1. Two variables x and y have __________________ if $y = \frac{k}{x}$.

2. Two variables x and y have __________________ if $y = kx$.

3. y has __________________ with x and z if $y = kxz$, where k is the constant of variation.

GUIDED EXAMPLES AND PRACTICE

Objective 1 Use tables to solve problems with two unknowns involving rational expressions.

Review this example for Objective 1:

1. It takes Holly 2 hours to shovel the snow in her whole driveway by hand. Using a snowplow, Rob can clear the same driveway in $\frac{3}{4}$ hour. If they work together, how long does it take them to clear the driveway of now?

Holly shovels at a rate of 1 driveway in 2 hours, or $\frac{1}{2}$ of a driveway per hour. Rob snowplows at a rate of 1 driveway in $\frac{3}{4}$ hour, or $\frac{1}{\frac{3}{4}} = \frac{4}{3}$ driveways per hour. Let t represent the amount of time it takes Rob and Holly to work together.

Practice this exercise:

1. Carmella can paint a room in 6 hours while it takes Tony 3 hours to paint the same room. How long would it take them to paint the room if they worked together?

Name:
Instructor:

Date:
Section:

	Rate	Time	Amount Completed
Holly	$\frac{1}{2}$	t	$\frac{1}{2}t$
Rob	$\frac{3}{4}$	t	$\frac{3}{4}t$

The total job is 1 driveway, so we can write an equation that combines the individual expressions for the tasks completed and set this sum equal to 1 driveway.

Holly's amount completed + Rob's amount completed = 1 driveway

$$\frac{1}{2}t+\frac{4}{3}t=1$$

Solve for t.

$$\frac{1}{2}t+\frac{4}{3}t=1$$

$$t\left(\frac{1}{2}+\frac{4}{3}\right)=1$$

$$t\left(\frac{11}{6}\right)=1$$

$$t=\frac{6}{11}$$

Working together, it takes Rob and Holly $\frac{6}{11}$, or $1\frac{5}{6}$, hours to clear the driveway.

Objective 2 Solve problems involving direct variation.

Review this example for Objective 2:

2. If x varies directly as y, and $x=15$ when $y=5$, find x when $y=8$.

Translate "x varies directly as y"

$x=ky$

Use the given values and solve for k.

$x=ky$

$15=k5$

$3=k$

Replace k with the new value of 3 into the original equation.

$x=3y$

Now find x when $y=8$.

$x=3y$

$x=3(8)$

$x=24$

Practice this exercise:

2. The number of kilograms of water in a human body varies directly as the mass of the body. A 90-kilogram person contains 60 kilograms of water. How many kilograms of water are in a 60-kilogram person?

Name: Date:
Instructor: Section:

Objective 3 Solve problems involving inverse variation.

Review this example for Objective 3:

3. The current I in an electrical conductor varies inversely as the resistance R of the conductor. The current is $\frac{1}{2}$ ampere when the resistance is 96 ohms. What is the current when the resistance is 216 ohms?

Translate "I varies inversely as R"

$$I = \frac{k}{R}$$

Use the given values and solve for k.

$$I = \frac{k}{R}$$

$$\frac{1}{2} = \frac{k}{96}$$

$$48 = k$$

Replace k with the new value of 48 into the original equation.

$$I = \frac{48}{R}$$

Now find x when $R = 216$.

$$I = \frac{48}{216}$$

$$I = \frac{2}{9}$$

Practice this exercise:

3. Suppose m varies inversely as n. If n is 4 when m is 9.5, what is n when m is 8?

Objective 4 Solve problems involving joint variation.

Review this example for Objective 4:

4. For a given interest rate, simple interest varies jointly as the principal and time. If $2000 left in an account for 6 years earned interest of $720, then how much interest would be earned in 7 years?

Let p represent the principal and t represent time.
Translate "I varies jointly with p and t"

$$I = kpt$$

Use the given values and solve for k.

$$I = kpt$$

$$720 = k \cdot 2000 \cdot 6$$

$$0.06 = k$$

Practice this exercise:

4. Suppose j varies jointly with g and v, and $j = 3$ when $g = 4$ and $v = 5$. Find j when $g = 10$ and $v = 11$.

Name: Date:
Instructor: Section:

Replace k with the new value of .06 into the original equation.

$I = 0.06pt$

Now find I when $p = 2000$ and $t = 7$.

$I = 0.06 \cdot 2000 \cdot 7$

$I = 840$

Objective 5 Solve problems involving combined variation.

Review this example for Objective 5:

5. Suppose that y varies directly with w and the square of x and inversely with z, and $y = 4$ when $w = 1$, $x = 2$, and $z = 5$. Find y when $w = 4$, $x = 5$, and $z = 20$.

Translate "y varies directly with w and the square of x and inversely with z."

$$y = \frac{kwx^2}{z}$$

Use the given values and solve for k.

$$y = \frac{kwx^2}{z}$$

$$4 = \frac{k \cdot 1 \cdot 2^2}{5}$$

$$4 = \frac{k \cdot 4}{5}$$

$$5 = k$$

Replace k with the new value of 5 into the original equation.

$$y = \frac{5wx^2}{z}$$

Now find y when $w = 4$, $x = 5$, and $z = 20$.

$$y = \frac{5 \cdot 4 \cdot 5^2}{20}$$

$$y = 25$$

Practice this exercise:

5. Suppose that y varies jointly with w and the square of x and inversely with z, and $y = 1400$ when $w = 4$, $x = 6$, and $z = 10$. Find y when $w = 3$, $x = 5$, and $z = 12$.

ADDITIONAL EXERCISES

Objective 1 Use tables to solve problems with two unknowns involving rational expressions.

For extra help, see Examples 1–2 on pages 939–941 of your text and the Section 14.7 lecture video.

Solve.

1. Cheryl, an experienced shipping clerk, can fill a certain order in 4 hours. Max, a new clerk, needs 5 hours to do the same job. Working together, how long will it take them to fill the order?

2. A steamboat travels 12 kilometers per hour faster than a freighter. The steamboat travels 82 kilometers in the same time the freighter travels 50 kilometers. Find the speed of each boat.

3. The speed of a stream is 3 miles per hour. A boat travels 7 miles upstream in the same time it takes to travel 13 miles downstream. What is the speed of the boat in still water?

Objective 2 Solve problems involving direct variation.
For extra help, see Examples 3–4 on page 942 of your text and the Section 14.7 lecture video.
Solve.

4. If x varies directly as y, and $x = 18.4$ when $y = 80$, find x when $y = 30$.

5. If y varies directly as x, and $y = 12$ when $x = 18$, find y when $x = 48$.

Objective 3 Solve problems involving inverse variation.
For extra help, see Example 5 on page 943 of your text and the Section 14.7 lecture video.
Solve.

6. The wavelength W of a radio wave varies inversely as its frequency F. A wave with a frequency of 1400 kilohertz has a length of 150 meters. What is the length of a wave with a frequency of 500 kilohertz?

7. Suppose p varies inversely as q. If p is 16 when q is 6.5, what is q when p is 4?

Objective 4 Solve problems involving joint variation.
For extra help, see Example 6 on page 944 of your text and the Section 14.7 lecture video.
Solve.

8. If the width of a rectangular solid is held constant, the volume varies jointly with the length and the height. If the volume is 160 cubic inches when the length is 40 inches and the height is 10 inches, find the volume when the length is 30 inches and the height is 15 inches.

9. Suppose x varies jointly with y and the square of z. If $x = 80$ when $y = 2$ and $z = 4$, find x when $y = 4$ and $z = 2$.

Objective 5 Solve problems involving combined variation.
For extra help, see Example 7 on pages 944–945 of your text and the Section 14.7 lecture video.
Solve.

10. Suppose p varies jointly with q and r and inversely with the square of s. If $p = 70$ when $q = 16$, $r = 7$, and $s = 2$ find p when $q = 10$, $r = 18$, and $s = 5$.

11. Suppose x varies directly with y and inversely with z the square of w. If $x = 4$ when $y = 48$, $z = 9$, and $w = 2$ find x when $y = 54$, $z = 2$, and $w = 3$.

Name: Date:
Instructor: Section:

Chapter 15 RATIONAL EXPONENTS, RADICALS, AND COMPLEX NUMBERS

15.1 Radical Expressions and Functions

Learning Objectives

1. Find the *n*th root of a number.
2. Approximate roots using a calculator.
3. Simplify radical expressions.
4. Evaluate radical functions.
5. Find the domain of radical functions.
6. Solve applications involving radical functions.

Key Terms

Use the vocabulary terms listed below to complete each statement in Exercises 1–2.

***n*th root** **radical function**

1. The number b is a(n) ________________ of a number a if $b^n = a$.

2. A function containing a radical expression whose radicand has a variable is a(n) ________________.

GUIDED EXAMPLES AND PRACTICE

Objective 1 Find the *n*th root of a number.

Review this example for Objective 1:

1. Evaluate each root, if possible.

a) $\sqrt{169}$

b) $\sqrt{-25}$

c) $\sqrt[3]{-64}$

a) $\sqrt{169} = 13$ because $13^2 = 169$.

b) $\sqrt{-25}$ is not a real number because there is no real number whose square is negative.

c) $\sqrt[3]{64} = -4$ because $(-4)^3 = -64$.

Practice this exercise:

1. Evaluate each root, if possible.

a) $\sqrt[4]{-16}$

b) $\sqrt{36}$

c) $\sqrt[3]{-27}$

Objective 2 Approximate roots using a calculator.

Review this example for Objective 2:

2. Use a calculator to approximate $\sqrt{13}$ to the nearest thousandth.

$\sqrt{13} \approx 3.606$

Practice this exercise:

2. Use a calculator to approximate $\sqrt[3]{26}$ to the nearest thousandth.

Name: Date:
Instructor: Section:

Objective 3 Simplify radical expressions.

Review this example for Objective 3:

3. Find the root. Assume that variables represent nonnegative values.

$\sqrt{x^{16}}$

$\sqrt{x^{16}} = x^8$

Practice this exercise:

3. Find the root. Assume that variables represent nonnegative values.

$\sqrt{36x^{16}y^{34}}$

Objective 4 Evaluate radical functions.

Review this example for Objective 4:

4. Given $f(x) = \sqrt{12x+9}$, find $f(x)$.

To find $f(6)$, substitute 6 for x and simplify.

$$\begin{aligned} f(6) &= \sqrt{12(6)+9} \\ &= \sqrt{72+9} \\ &= \sqrt{81} \\ &= 9 \end{aligned}$$

So, $f(6) = 9$.

Practice this exercise:

4. Given $f(x) = \sqrt{3x+88}$, find $f(11)$.

Objective 5 Find the domain of radical functions.

Review this example for Objective 5:

5. Find the domain of $f(x) = \sqrt{3x-9}$.

Because $3x-9$ must be nonnegative, solve $3x-9 \geq 0$.

$$\begin{aligned} 3x-9 &\geq 0 \\ 3x &\geq 9 \\ x &\geq 3 \end{aligned}$$

Domain: $\{x \mid x \geq 3\}$, or $[3, \infty)$.

Practice this exercise:

5. Find the domain of $f(x) = \sqrt{5x+9}$.

Name: Date:
Instructor: Section:

Objective 6 Solve applications involving radical functions.

Review this example for Objective 6:

6. The velocity v of an object, in meters per second, that has fallen h meters is given by $v=-\sqrt{19.6h}$. Find the velocity of a stone that has fallen 45 meters. Round to the nearest tenth.

We must use the formula $v=-\sqrt{19.6h}$ with $h=45$.

$$v=-\sqrt{19.6(45)}$$
$$=-\sqrt{882}$$
$$\approx -29.7$$

The velocity of the stone is approximately -29.7 meters per second.

Practice this exercise:

6. Three pieces of lumber are to be connected to form a right triangle that will be part of a frame for a roof. If the horizontal piece is 24 feet and the vertical piece is 10 feet, how long must the connecting piece be?

ADDITIONAL EXERCISES

Objective 1 Find the *n*th root of a number.

For extra help, see Example 1 on page 964 of your text and the Section 15.1 lecture video.

Evaluate each root, if possible.

1. $\sqrt{81}$

2. $-\sqrt{4}$

3. $\sqrt[3]{-343}$

Objective 2 Approximate roots using a calculator.

For extra help, see Example 2 on page 965 of your text and the Section 15.1 lecture video.

Use a calculator to approximate each root to the nearest thousandth.

4. $\sqrt{62}$

5. $\sqrt[3]{57}$

Objective 3 Simplify radical expressions.

For extra help, see Examples 3–4 on pages 965–966 of your text and the Section 15.1 lecture video.

Find the root. Assume that variables represent nonnegative values.

6. $\sqrt[4]{256c^8}$

Name: Date:
Instructor: Section:

Find the root. Assume that variables represent any real number.

7. $\sqrt{(x-6)^2}$

8. $\sqrt{(x+6)^{12}}$

9. $\sqrt[3]{(y-2)^9}$

Objective 4 Evaluate radical functions.
For extra help, see Example 5 on page 966 of your text and the Section 15.1 lecture video.
Find the indicated value of the function.

10. $f(x)=\sqrt{3x+5}$; find $f(4)$.

11. $f(x)=\sqrt[3]{25x+12}$; find $f(20)$.

Objective 5 Find the domain of radical functions.
For extra help, see Example 6 on page 967 of your text and the Section 15.1 lecture video.
Find the domain.

12. $f(x)=\sqrt{-2x+5}$

13. $f(x)=\sqrt{\frac{4}{5}x+8}$

Objective 6 Solve applications involving radical functions.
For extra help, see Example 7 on page 967–968 of your text and the Section 15.1 lecture video.

The formula $S=\frac{7}{2}\sqrt{2D}$ *can be used to approximate the speed, S, in miles per hour, that a car was traveling prior to braking and skidding a distance, D, in feet, on asphalt.*

14. Find the speed of a car if the skid distance is 32 feet.

Name: Date:
Instructor: Section:

Chapter 15 RATIONAL EXPONENTS, RADICALS, AND COMPLEX NUMBERS

15.2 Rational Exponents

Learning Objectives
1 Evaluate rational exponents.
2 Write radicals as expressions raised to rational exponents.
3 Simplify expressions with rational number exponents using the rules of exponents.
4 Use rational exponents to simplify radical expressions.

Key Terms
Use the vocabulary term listed below to complete the statement in Exercise 1.

rational exponent

1. A(n) ____________________ is an exponent that is a rational number.

GUIDED EXAMPLES AND PRACTICE

Objective 1 Evaluate rational exponents.

Review these examples for Objective 1:

1. Rewrite the expression $16^{1/4}$ using radicals; then simplify, if possible.

$16^{1/4} = \sqrt[4]{16} = 2$

2. Rewrite the expression $(3yz)^{3/4}$ using radicals; then simplify, if possible. Assume that all variables represent nonnegative values.

Use the general rule for rational exponents.
$(3yz)^{3/4} = \sqrt[4]{(3yz)^3} = \sqrt[4]{27y^3z^3}$

Practice these exercises:

1. Rewrite the expression $-25^{1/2}$ using radicals; then simplify, if possible.

2. Rewrite the expression $(3y+x)^{5/6}$ using radicals; then simplify, if possible. Assume that all variables represent nonnegative values.

Objective 2 Write radicals as expressions raised to rational exponents.

Review this example for Objective 2:

3. Write $\sqrt[3]{121}$ in exponential form.

$\sqrt[3]{121} = \sqrt[3]{11^2} = 11^{2/3}$

Practice this exercise:

3. Write $\sqrt[3]{19}$ in exponential form.

Name: Date:
Instructor: Section:

Objective 3 Simplify expressions with rational number exponents using the rules of exponents.

Review this example for Objective 3:

4. Use the rules of exponents to simplify. Write the answer with positive exponents. Assume that all variables represent positive values.

$\left(25x^4y^6\right)^{3/2}$

Use the raising a product to a power rule, $(ab)^n = a^n b^n$.

$\left(25x^4y^6\right)^{3/2} = 25^{3/2}\left(x^4\right)^{3/2}\left(y^6\right)^{3/2}$

Now use the raising a power to a power rule, $\left(a^m\right)^n = a^{mn}$.

$25^{3/2}\left(x^4\right)^{3/2}\left(y^6\right)^{3/2} = 125\left(x^{4(3/2)}\right)y^{6(3/2)}$

Now multiply the exponents.

$125\left(x^{4(3/2)}\right)y^{6(3/2)} = 125x^6y^9$

Practice this exercise:

4. Use the rules of exponents to simplify. Write the answer with positive exponents. Assume that all variables represent positive values.

$\left(8x^6y^{18}\right)^{4/3}$

Objective 4 Use rational exponents to simplify radical expressions.

Review these examples for Objective 4:

5. Represent $\sqrt[10]{b^6}$ as a radical with a smaller root index. Assume that the variable represents a nonnegative value.

$$\sqrt[10]{b^6} = b^{6/10} = b^{3/5} = \sqrt[5]{b^3}$$

6. Perform the indicated operations. Write the result using a radical. Assume that the variable represents a positive value.

$\sqrt[4]{z}\cdot\sqrt[3]{z^2}$

$$\sqrt[4]{z}\cdot\sqrt[3]{z^2} = z^{1/4}\cdot z^{2/3} = z^{1/4+2/3} = z^{3/12+8/12} = z^{11/12} = \sqrt[12]{z^{11}}$$

Practice these exercises:

5. Represent $\sqrt[6]{a^4}$ as a radical with a smaller root index. Assume that the variable represents a nonnegative value.

6. Perform the indicated operations. Write the result using a radical. Assume that the variable represents a positive value. $\sqrt[7]{y}\cdot\sqrt{y}$

Name: Date:
Instructor: Section:

7. Write as a single radical. Assume that all variables represent nonnegative values.

$\sqrt[4]{\sqrt[2]{x}}$

$$\sqrt[4]{\sqrt[2]{x}} = \left(x^{1/2}\right)^{1/4}$$
$$= x^{1/8}$$
$$= \sqrt[8]{x}$$

7. Write as a single radical. Assume that all variables represent nonnegative values.

$\sqrt[5]{\sqrt[3]{z}}$

ADDITIONAL EXERCISES

Objective 1 Evaluate rational exponents.

For extra help, see Examples 1–3 on pages 973–976 of your text and the Section 15.2 lecture video.

Rewrite each of the following using radicals; then simplify, if possible. Assume that all variables represent nonnegative values.

1. $x^{1/5}$

2. $216^{4/3}$

3. $256^{-5/4}$

Objective 2 Write radicals as expressions raised to rational exponents.

For extra help, see Example 4 on page 976 of your text and the Section 15.2 lecture video.

Write each of the following in exponential form. Assume that all variables represent positive values.

4. $\sqrt[6]{n^5}$

5. $\dfrac{5x}{\sqrt[5]{z^3}}$

Objective 3 Simplify expressions with rational number exponents using the rules of exponents.

For extra help, see Example 5 on pages 977–978 of your text and the Section 15.2 lecture video.

Use the rules of exponents to simplify. Write the answers with positive exponents. Assume that all variables represent positive values.

6. $a^{7/3} \cdot a^{1/6}$

7. $\dfrac{6^{9/4}}{6^{7/4}}$

8. $\left(-6y^{5/6}\right)\left(7y^{-2/3}\right)$

Name: Date:
Instructor: Section:

Objective 4 Use rational exponents to simplify radical expressions.

For extra help, see Examples 6–8 on page 978–980 of your text and the Section 15.2 lecture video.

Represent each of the following as a radical with a smaller root index. Assume that all variables represent nonnegative values.

9. $\sqrt[3]{x^9 y^{12} z^3}$

10. $\sqrt[8]{m^4 n^2}$

Perform the indicated operations. Write the result using a radical. Assume that all variables represent positive values.

11. $\dfrac{\sqrt[3]{y^5}}{\sqrt[6]{y^5}}$

12. $\dfrac{\sqrt[2]{n^5}}{\sqrt[4]{n^3}}$

Name: Date:
Instructor: Section:

Chapter 15 RATIONAL EXPONENTS, RADICALS, AND COMPLEX NUMBERS

15.3 Multiplying, Dividing, and Simplifying Radicals

Learning Objectives
1 Multiply radical expressions.
2 Divide radical expressions.
3 Use the product rule to simplify radical expressions.

GUIDED EXAMPLES AND PRACTICE

Objective 1 Multiply radical expressions.

Review this example for Objective 1:

1. Find the product and simplify. Assume that variables represent positive values.

$\sqrt{7a^5b}\cdot\sqrt{7a^3b}$

$$\sqrt{7a^5b}\cdot\sqrt{7a^3b} = \sqrt{7a^5b\cdot 7a^3b}$$
$$= \sqrt{49a^8b^2}$$
$$= 7a^4b$$

Practice this exercise:

1. Find the product and simplify. Assume that variables represent positive values.

$\sqrt{10pn^7}\cdot\sqrt{10pn}$

Objective 2 Divide radical expressions.

Review this example for Objective 2:

2. Simplify the expression.

$\sqrt{\dfrac{25}{121}}$

$$\sqrt{\frac{25}{121}} = \frac{\sqrt{25}}{\sqrt{121}}$$
$$= \frac{5}{11}$$

Practice this exercise:

2. Simplify the expression.

$\sqrt{\dfrac{16}{49}}$

Objective 3 Use the product rule to simplify radical expressions.

Review these examples for Objective 3:

3. Simplify $\sqrt{200}$.

$$\sqrt{200} = \sqrt{2\cdot 2\cdot 2\cdot 5\cdot 5}$$
$$= 2\cdot 5\sqrt{2}$$
$$= 10\sqrt{2}$$

Practice these exercises:

3. Simplify $\sqrt{40}$.

Name: Date:
Instructor: Section:

4. Find the product and write the answer in simplest form. Assume that variables represent nonnegative values.

$-6x^4y\sqrt{448x^7y^{19}}$

$$\begin{aligned}-6x^4y\sqrt{448x^7y^{19}} &= -6x^4y\sqrt{64\cdot 7\cdot x^6\cdot x\cdot y^{18}\cdot y}\\ &= -6x^4y(64x^3y^9)\sqrt{7xy}\\ &= -384x^7y^9\sqrt{7xy}\end{aligned}$$

4. Find the product and write the answer in simplest form. Assume that variables represent nonnegative values.

$3\sqrt{2x^5}\cdot 5\sqrt{6x^3}$

ADDITIONAL EXERCISES

Objective 1 Multiply radical expressions.

For extra help, see Example 1 on page 983 of your text and the Section 15.3 lecture video.

Find the product and simplify. Assume that variables represent positive values.

1. $\sqrt[3]{5a}\cdot\sqrt[3]{17b}$

2. $\sqrt{\dfrac{x}{5}}\cdot\sqrt{\dfrac{7}{y}}$

Objective 2 Divide radical expressions.

For extra help, see Example 2 on page 984 of your text and the Section 15.3 lecture video.

Simplify.

3. $\dfrac{\sqrt{490}}{\sqrt{10}}$

4. $\dfrac{\sqrt[3]{189}}{\sqrt[3]{7}}$

Objective 3 Use the product rule to simplify radical expressions.

For extra help, see Examples 3–6 on pages 985–988 of your text and the Section 15.3 lecture video.

Simplify. Assume that variables represent nonnegative values.

5. $\sqrt{x^{16}y^{10}}$

6. $\sqrt[3]{243x^8}$

Find the product and write the answer in simplest form. Assume that variables represent nonnegative values.

7. $\sqrt{10}\cdot\sqrt{14}$

8. $a^2\sqrt{a^2b}\cdot b\sqrt{a^3b^4}$

Name: Date:
Instructor: Section:

Find the quotient and write the answer in simplest form. Assume that variables represent positive values.

9. $\dfrac{\sqrt{525}}{\sqrt{3}}$

10. $\dfrac{\sqrt{x^8 y^4}}{\sqrt{x^2 y^2}}$

11. $\dfrac{72\sqrt{56x^9 y^{14}}}{8\sqrt{7x^6 y^5}}$

Name: Date:
Instructor: Section:

Chapter 15 RATIONAL EXPONENTS, RADICALS, AND COMPLEX NUMBERS

15.4 Adding, Subtracting, and Multiplying Radical Expressions.

Learning Objectives
1 Add or subtract like radicals
2 Use the distributive property in expressions containing radicals.
3 Simplify radical expressions that contain mixed operations.

Key Terms
Use the vocabulary term listed below to complete the statement in Exercise 1.

like radicals

1. Radical expressions with identical radicands and identical root indices are called ___________________.

GUIDED EXAMPLES AND PRACTICE

Objective 1 Add or subtract like radicals

Review this example for Objective 1:

1. Add.

$9\sqrt{32}+5\sqrt{50}$

$$\begin{aligned}9\sqrt{32}+5\sqrt{50} &= 9\sqrt{16\cdot 2}+5\sqrt{25\cdot 2}\\ &= 9\cdot 4\sqrt{2}+5\cdot 5\sqrt{2}\\ &= 36\sqrt{2}+25\sqrt{2}\\ &= 61\sqrt{2}\end{aligned}$$

Practice this exercise:

1. Subtract.

$8\sqrt{5}-3\sqrt{5}$

Objective 2 Use the distributive property in expressions containing radicals.

Review these examples for Objective 2:

2. Use the distributive property to simplify.

$\sqrt{5}\left(\sqrt{4}+\sqrt{5}\right)$

$$\begin{aligned}\sqrt{5}\left(\sqrt{6}+\sqrt{5}\right) &= \sqrt{5}\cdot\sqrt{6}+\sqrt{5}\cdot\sqrt{5}\\ &= \sqrt{30}+5\\ &= 5+\sqrt{30}\end{aligned}$$

Practice these exercises:

2. Use the distributive property to simplify.

$\sqrt{7}\left(\sqrt{7}+\sqrt{10}\right)$

Name: Date:
Instructor: Section:

3. Multiply. (Use FOIL.)
$\left(4\sqrt{3}+3\sqrt{7}\right)\left(5\sqrt{11}-11\sqrt{5}\right)$

$\left(4\sqrt{3}+3\sqrt{7}\right)\left(5\sqrt{11}-11\sqrt{5}\right)=$
$20\sqrt{33}-44\sqrt{15}+15\sqrt{77}-15\sqrt{35}$

3. Multiply. (Use FOIL.)
$\left(7\sqrt{7}-4\sqrt{5}\right)\left(8\sqrt{3}+3\sqrt{2}\right)$

Objective 3 Simplify radical expressions that contain mixed operations.

Review this example for Objective 3:

4. Simplify.
$\sqrt{14}\cdot\sqrt{2}+11\sqrt{7}$

$$\begin{aligned}\sqrt{14}\cdot\sqrt{2}+11\sqrt{7}&=\sqrt{2\cdot 7}\cdot\sqrt{2}+11\sqrt{7}\\&=\sqrt{2}\cdot\sqrt{7}\cdot\sqrt{2}+11\sqrt{7}\\&=2\sqrt{7}+11\sqrt{7}\\&=13\sqrt{7}\end{aligned}$$

Practice this exercise:

4. Simplify.
$\sqrt{15}\cdot\sqrt{5}+15\sqrt{3}$

ADDITIONAL EXERCISES

Objective 1 Add or subtract like radicals

For extra help, see Examples 1–2 on pages 992–993 of your text and the Section 15.4 lecture video.
Add or subtract. Assume that variables represent nonnegative values.

1. $6b\sqrt[3]{3}-2b\sqrt[3]{3}$

2. $\sqrt{80}-\sqrt{45}$

3. $6\sqrt{125}-9\sqrt{180}$

4. $\sqrt{25x^5}-2\sqrt{36x^5}$

5. $\sqrt[3]{54x}-\sqrt[3]{2x^4}$

Objective 2 Use the distributive property in expressions containing radicals.

For extra help, see Examples 3–4 on pages 993–994 of your text and the Section 15.4 lecture video.
Use the distributive property. Assume that variables represent nonnegative values.

6. $2\sqrt{5x}\left(3\sqrt{5x}-7\sqrt{10x}\right)$

Name: Date:
Instructor: Section:

Multiply. (Use FOIL.)

7. $\left(\sqrt{3}+5\right)\left(\sqrt{15}-5\right)$

8. $\left(5+2\sqrt{3}\right)\left(5-2\sqrt{3}\right)$

9. $\left(2\sqrt[3]{7}+\sqrt[3]{2}\right)\left(\sqrt[3]{7}-2\sqrt[3]{2}\right)$

10. $\left(6+\sqrt{10}\right)^2$

Multiply the conjugates. Assume that variables represent nonnegative values.

11. $\left(\sqrt{2}+4\sqrt{12}\right)\left(\sqrt{2}-4\sqrt{12}\right)$

12. $\left(\sqrt{3c}+\sqrt{d}\right)\left(\sqrt{3c}-\sqrt{d}\right)$

Objective 3 Simplify radical expressions that contain mixed operations.
For extra help, see Example 5 on page 994 of your text and the Section 15.4 lecture video.
Simplify.

13. $6\sqrt{3}\cdot 2\sqrt{12}-8\sqrt{18}\cdot\sqrt{6}$

Name: Date:
Instructor: Section:

Chapter 15 RATIONAL EXPONENTS, RADICALS, AND COMPLEX NUMBERS

15.5 Rationalizing Numerators and Denominators of Radical Expressions

Learning Objectives
1 Rationalize denominators.
2 Rationalize denominators that have a sum or difference with a square root term.
3 Rationalize numerators.

GUIDED EXAMPLES AND PRACTICE

Objective 1 Rationalize denominators.

Review this example for Objective 1:

1. Rationalize the denominator. Assume that the variable represents a positive value.

$$\frac{28}{\sqrt{x^3}}$$

$$\frac{28}{\sqrt{x^3}} = \frac{28}{\sqrt{x^3}} \cdot \frac{\sqrt{x}}{\sqrt{x}}$$
$$= \frac{28\sqrt{x}}{\sqrt{x^4}}$$
$$= \frac{28\sqrt{x}}{x^2}$$

Practice this exercise:

1. Rationalize the denominator. Assume that the variable represents a positive value.

$$\frac{7}{\sqrt{6y}}$$

Objective 2 Rationalize denominators that have a sum or difference with a square root term.

Review this example for Objective 2:

2. Rationalize the denominator and simplify.

$$\frac{4}{3-\sqrt{7}}$$

$$\frac{4}{3-\sqrt{7}} = \frac{4}{3-\sqrt{7}} \cdot \frac{3+\sqrt{7}}{3+\sqrt{7}}$$
$$= \frac{12+4\sqrt{7}}{3^2-\left(\sqrt{7}\right)^2}$$
$$= \frac{12+4\sqrt{7}}{9-7}$$
$$= 6+2\sqrt{7}$$

Practice this exercise:

2. Rationalize the denominator and simplify.

$$\frac{9}{2+\sqrt{5}}$$

Name: Date:
Instructor: Section:

Objective 3 Rationalize numerators.

Review this example for Objective 3:

3. Rationalize the numerator. Assume that variables represent positive values.

$$\frac{\sqrt{5y}}{3}$$

$$\frac{\sqrt{5y}}{3} = \frac{\sqrt{5y}}{3} \cdot \frac{\sqrt{5y}}{\sqrt{5}}$$
$$= \frac{5y}{3\sqrt{5}}$$

Practice this exercise:

3. Rationalize the numerator. Assume that variables represent positive values.

$$\frac{\sqrt{7x}}{6}$$

ADDITIONAL EXERCISES

Objective 1 Rationalize denominators.

For extra help, see Examples 1–2 on pages 999–1001 of your text and the Section 15.5 lecture video.

Rationalize the denominator. Assume that variables represent positive values.

1. $\frac{5}{\sqrt{28}}$

2. $\sqrt{\frac{4}{125}}$

3. $\frac{\sqrt{3x^2}}{\sqrt{45}}$

4. $\frac{5}{\sqrt[3]{2}}$

5. $\sqrt[3]{\frac{2}{3}}$

6. $\frac{2}{\sqrt[3]{3x^2}}$

7. $\sqrt[3]{\frac{3}{25z^2}}$

8. $\frac{2\sqrt{5}}{\sqrt{5}-6}$

Objective 2 Rationalize denominators that have a sum or difference with a square root term.

For extra help, see Example 3 on page 1002 of your text and the Section 15.5 lecture video.

Rationalize the denominator and simplify. Assume that variables represent positive values.

9. $\frac{\sqrt{7}}{\sqrt{7}-6}$

10. $\frac{4\sqrt{y}}{\sqrt{y}+6}$

11. $\frac{\sqrt{x}}{\sqrt{x}+\sqrt{y}}$

Objective 3 Rationalize numerators.

For extra help, see Example 4 on page 1003 of your text and the Section 15.5 lecture video.

Rationalize the numerator. Assume that variables represent positive values.

12. $\frac{\sqrt{11}+2}{7}$

Name: Date:
Instructor: Section:

Chapter 15 RATIONAL EXPONENTS, RADICALS, AND COMPLEX NUMBERS

15.6 Radical Equations and Problem Solving

Learning Objectives
1 Use the power rule to solve radical equations.

Key Terms
Use the vocabulary terms listed below to complete each statement in Exercises 1–7.

radical equation

1. A(n) ________________ is an equation containing at least one radical expression whose radicand has a variable.

GUIDED EXAMPLES AND PRACTICE

Objective 1 Use the power rule to solve radical equations.

Review this example for Objective 1:

1. Solve $\sqrt{y+8}-2=\sqrt{y}$.

Solve for y.

$$\sqrt{y+8}-2=\sqrt{y}$$
$$\sqrt{y+8}=\sqrt{y}+2$$
$$\left(\sqrt{y+8}\right)^2=\left(\sqrt{y}+2\right)^2$$
$$y+8=y+4\sqrt{y}+4$$
$$4=4\sqrt{y}$$
$$1=\sqrt{y}$$
$$1=y$$

Check that $y=1$ is a valid solution.

$$\sqrt{1+8}-2\stackrel{?}{=}\sqrt{1}$$
$$\sqrt{9}-2\stackrel{?}{=}1$$
$$3-2\stackrel{?}{=}1$$
$$1=1 \quad \text{True}$$

Practice this exercise:

1. Solve $8+\sqrt{x}=\sqrt{2x+64}$.

Name: Date:
Instructor: Section:

ADDITIONAL EXERCISES

Objective 1 Use the power rule to solve radical equations.

For extra help, see Examples 1–6 on pages 1009–1013 of your text and the Section 15.6 lecture video.

Solve.

1. $\sqrt{x} = 7$

2. $\sqrt[3]{x} = -4$

3. $\sqrt{x+9} = 11$

4. $\sqrt{x-3} = 21$

5. $\sqrt[3]{x+1} = 4$

6. $\sqrt{y+5} - 5 = 3$

7. $7 + \sqrt{x-7} = 9$

8. $\sqrt[4]{y+3} + 2 = -4$

Solve. Identify any extraneous solutions.

9. $\sqrt{8y-3} = \sqrt{7y+6}$

10. $\sqrt{2y+4} = \sqrt{2y-6}$

11. $\sqrt[4]{x+4} = \sqrt[4]{2x}$

12. $\sqrt{x+31} = x+1$

13. $\sqrt{y-1} = y-1$

14. $\sqrt[3]{5x+8} + 4 = 3$

Name: Date:
Instructor: Section:

Chapter 15 RATIONAL EXPONENTS, RADICALS, AND COMPLEX NUMBERS

15.7 Complex Numbers

Learning Objectives
1. Write imaginary numbers using i.
2. Perform arithmetic operations with complex numbers.
3. Raise i to powers.

Key Terms
Use the vocabulary terms listed below to complete each statement in Exercises 1–4.

imaginary unit **imaginary number** **complex number** **complex conjugate**

1. The ___________________ of a number $a+bi$ is the number $a-bi$.

2. A number that can be expressed as the product of a real number and the imaginary unit is call a(n) ___________________.

3. The ___________________ is the number represented by i where $i=\sqrt{-1}$.

4. A(n) ___________________ can be expressed in the form $a+bi$ where a and b are real numbers and i is the imaginary unit.

GUIDED EXAMPLES AND PRACTICE

Objective 1 Write imaginary numbers using *i*.

Review this example for Objective 1:

1. Write the imaginary number as a product of a real number and i.
$\sqrt{-294}$

$$\begin{aligned}\sqrt{-294} &= \sqrt{294\cdot(-1)}\\ &= \sqrt{294}\cdot\sqrt{-1}\\ &= \sqrt{49\cdot 6}i\\ &= 7i\sqrt{6}\end{aligned}$$

Practice this exercise:

1. $\sqrt{-45}$

Name: Date:
Instructor: Section:

Objective 2 Perform arithmetic operations with complex numbers.

Review this example for Objective 2:

2. Multiply.

$(8+i)(2-i)$

$$\begin{aligned}(8+i)(2-i) &= 16-8i+2i-i^2 \\ &= 16-6i-(-1) \\ &= 17+6i\end{aligned}$$

Practice this exercise:

2. $(-2+4i)^2$

Objective 3 Raise i to powers.

Review this example for Objective 3:

3. Find the power of i.

i^{33}

$$\begin{aligned}i^{33} &= i^{32}\cdot i \\ &= \left(i^4\right)^8 \cdot i \\ &= 1^8 \cdot i \\ &= i\end{aligned}$$

Practice this exercise:

3. Find the power of i.

i^9

ADDITIONAL EXERCISES

Objective 1 Write imaginary numbers using i.

For extra help, see Example 1 on page 1017 of your text and the Section 15.7 lecture video.

Write the imaginary number using i.

1. $\sqrt{-121}$

2. $\sqrt{-6}$

Objective 2 Perform arithmetic operations with complex numbers.

For extra help, see Examples 2–4 on pages 1019–1021 of your text and the Section 15.7 lecture video.

Add or subtract.

3. $(6+8i)+(3-7i)$

4. $(3+4i)-(-5i)$

5. $(-4+15i)+(-3-9i)+(13+9i)$

Multiply.

6. $5i(-7+8i)$

7. $(5+9i)(7-6i)$

Name: Date:
Instructor: Section:

Divide and write in standard form.

8. $\frac{3}{i}$

9. $\frac{8}{5+i}$

10. $\frac{7-4i}{4-i}$

Objective 3 Raise i to powers.

For extra help, see Example 5 on page 1021 of your text and the Section 15.7 lecture video.

Find the powers of i.

11. i^{36}

12. i^{14}

Name: Date:
Instructor: Section:

Chapter 16 QUADRATIC EQUATIONS AND FUNCTIONS

16.1 The Square Root Principle and Completing the Square

Learning Objectives
1 Use the square root principle to solve quadratic equations.
2 Solve quadratic equations by completing the square.

GUIDED EXAMPLES AND PRACTICE

Objective 1 Use the square root principle to solve quadratic equations.

Review this example for Objective 1:

1. Solve using the square root principle.

$(2x+5)^2 = 81$

Use the square root principle and simplify.

$2x+5 = \pm 9$

Isolate the x.

$2x = -5 \pm 9$

$x = \dfrac{-5 \pm 9}{2}$

Simplify by separating the two solutions.

$x = \dfrac{-5+9}{2}$ or $x = \dfrac{-5-9}{2}$

$x = 2$ $x = -7$

Practice this exercise:

1. Solve using the square root principle.

$(0.2x+5.8)^2 = 0.49$

Objective 2 Solve quadratic equations by completing the square.

Review this example for Objective 2:

2. Solve the equation by completing the square.

$6x^2 - 7 = -11x$

Rewrite in the form $ax^2 + bx = c$.

$6x^2 + 11x = 7$

Divide both sides by 6.

$x^2 + \dfrac{11}{6}x = \dfrac{7}{6}$

Add $\left(\dfrac{11}{12}\right)^2$ to both sides to complete the square.

$x^2 + \dfrac{11}{6}x + \dfrac{121}{144} = \dfrac{7}{6} + \dfrac{121}{144}$

Factor the left side and simplify the right side.

$\left(x + \dfrac{11}{12}\right)^2 = \dfrac{289}{144}$

Practice this exercise:

2. Solve the equation by completing the square.

$3x^2 + 20x - 3 = 0$

Name: Date:
Instructor: Section:

Isolate the variable on the left-hand side and simplify the square root.

$$\left(x+\frac{11}{12}\right)^2=\frac{289}{144}$$

$$x+\frac{11}{12}=\pm\frac{17}{12}$$

$$x=\frac{-11\pm17}{12}$$

$$x=\frac{1}{2}\text{ or }-\frac{7}{3}$$

ADDITIONAL EXERCISES

Objective 1 Use the square root principle to solve quadratic equations.

For extra help, see Examples 1–3 on pages 1038–1040 of your text and the Section 16.1 lecture video.

Solve and check.

1. $z^2=1.96$

2. $t^2=27$

3. $x^2=-64$

Solve and check. Begin by using the addition or multiplication principles of equality to isolate the squared term.

4. $x^2-43=6$

5. $x^2+2=38$

6. $4n^2=576$

Solve and check. Use the square root principle to eliminate the square.

7. $(3k+2)^2=100$

8. $(x-8)^2=-81$

Name: Date:
Instructor: Section:

Objective 2 Solve quadratic equations by completing the square.

For extra help, see Examples 4–5 on pages 1041–1043 of your text and the Section 16.1 lecture video.

Solve the equation by completing the square.

9. $z^2 + 2z + 17 = 0$

10. $x^2 = -2x + 3$

11. $x^2 + 3x - 7 = 21$

Solve the equation by completing the square. Begin by writing the equation in the form $ax^2 + bx = c$.

12. $12x^2 - 5 = -17x$

13. $8x^2 - 15 = 37x$

Name: Date:
Instructor: Section:

Chapter 16 QUADRATIC EQUATIONS AND FUNCTIONS

16.2 Solving Quadratic Equations Using the Quadratic Formula

Learning Objectives
1 Solve quadratic equations using the quadratic formula.
2 Use the discriminant to determine the number of real solutions that a quadratic equation has.
3 Find the x- and y-intercepts of a quadratic function.
4 Solve applications using the quadratic formula.

Key Terms
Use the vocabulary term listed below to complete the statement in Exercise 1.

discriminant

1. The radicand, $b^2 - 4(a)(c)$, is the __________________ in the quadratic equation.

GUIDED EXAMPLES AND PRACTICE

Objective 1 Solve quadratic equations using the quadratic formula.

Review this example for Objective 1:

1. Use the quadratic formula to solve $2x^2 - 7x = 1$ for x.

$2x^2 - 7x = 1$

$2x^2 - 7x - 1 = 0$

So, $a = 2$, $b = -7$, and $c = -1$.

Substitute these values into the quadratic formula.

$$x = \frac{-b \pm \sqrt{b^2 - 4ac}}{2a}$$

$$x = \frac{-(-7) \pm \sqrt{(-7)^2 - 4(2)(-1)}}{2(2)}$$

$$= \frac{7 \pm \sqrt{49 + 8}}{4}$$

$$= \frac{7 \pm \sqrt{57}}{4}$$

Practice this exercise:

1. Use the quadratic formula to solve $\frac{3}{5}x^2 - x - \frac{10}{3} = 0$.

Name: Date:
Instructor: Section:

Objective 2 **Use the discriminant to determine the number of real solutions that a quadratic equation has.**

Review this example for Objective 2:

2. Use the discriminant to determine the number and type of solutions for the equation. If solution(s) are real, state whether they are rational or irrational.

$2x^2 - 9x + 7 = 0$

The discriminant of the polynomial $ax^2 + bx + c$, is $b^2 - 4ac$. Substitute the appropriate values for $2x^2 - 9x + 7$.

$$\begin{aligned} b^2 - 4ac &= (-9)^2 - 4(2)(7) \\ &= 81 - 56 \\ &= 25 \end{aligned}$$

Since the discriminant is greater than zero, there are two real roots. Since the discriminant is a perfect square, the roots are rational.

Practice this exercise:

2. Use the discriminant to determine the number and type of solutions for the equation. If solution(s) are real, state whether they are rational or irrational.

$x^2 - 4x + 9 = 0$

Objective 3 **Find the x- and y-intercepts of a quadratic function.**

Review this example for Objective 3:

3. Find the x- and y-intercepts of $y = -x^2 - 2x + 15$.

To find the x-intercepts, solve $-x^2 - 2x + 15 = 0$.

$$\begin{aligned} 0 &= -x^2 - 2x + 15 \\ &= x^2 + 2x - 15 \\ &= (x - 3)(x + 5) \end{aligned}$$

$x - 3 = 0$ or $x + 5 = 0$

$x = 3$ $\quad$ $x = -5$

Therefore, the x-intercepts are $(-5, 0)$ and $(3, 0)$. The y-intercept is $(0, 15)$.

Practice this exercise:

3. Find the x- and y-intercepts of $y = -x^2 - 6x + 16$.

Name: Date:
Instructor: Section:

Objective 4 Solve applications using the quadratic formula.

Review this example for Objective 4:

4. The height of an object after t seconds is given by $h(t) = 16.1t^2 + 37t + 6$. At what time does the object hit the ground? Round to the nearest thousandth.

The height of the object is zero when it hits the ground, so solve the equation $h(t) = 0$ for t.

$0 = -16.1t^2 + 37t + 6$

$$t = \frac{-37 \pm \sqrt{37^2 - 4(-16.1)(6)}}{2(-16.1)}$$

$$t = \frac{-37 \pm \sqrt{1755.4}}{-32.2}$$

$t \approx -0.152$ or 2.450

Since time cannot be negative, the object hits the ground at approximately 2.450 seconds.

Practice this exercise:

4. If an object is thrown downward with an initial velocity of v_0, then the distance it travels is given by $s = 4.9t^2 + v_0 t$. An object is thrown downward from an airplane 200 meters from the ground, with an initial velocity of 8 meters per second. How long does it take for the object to reach the ground? Round to the nearest tenth.

ADDITIONAL EXERCISES

Objective 1 Solve quadratic equations using the quadratic formula.

For extra help, see Example 1 on pages 1049-1050 of your text and the Section 16.2 lecture video.

Solve using the quadratic formula.

1. $x^2 - 3x - 18 = 0$

2. $x^2 - 4x = -4$

3. $x^2 - x = -7$

Objective 2 Use the discriminant to determine the number of real solutions that a quadratic equation has.

For extra help, see Example 2 on page 1051 of your text and the Section 16.2 lecture video.

Use the discriminant to determine the number and type of solutions for the equation. If the solutions(s) are real, state whether they are rational or irrational.

4. $x^2 - 5x - 5 = 0$

5. $\frac{1}{4}x^2 - 2x + 4 = 0$

6. $x^2 - 8x + 16 = 0$

Name: Date:
Instructor: Section:

Objective 3 Find the *x*- and *y*-intercepts of a quadratic function.
For extra help, see Example 3 on pages 1052-1053 of your text and the Section 16.2 lecture video.
Find the x- and y-intercepts.

7. $y = x^2 - x - 6$

Objective 4 Solve applications using the quadratic formula.
For extra help, see Example 4 on page 1053 of your text and the Section 16.2 lecture video.
Translate to a quadratic equation; then solve using the quadratic formula.

8. A positive integer squared plus 4 times its consecutive integer is equal to 81. Find the integers.

9. The length of a rectangular air filter is 3 inches less than twice the width. Find the length and width of the filter if the area is 527 square inches.

Name: Date:
Instructor: Section:

Chapter 16 QUADRATIC EQUATIONS AND FUNCTIONS

16.3 Solving Equations That Are Quadratic in Form

Learning Objectives
1 Solve equations by rewriting them in quadratic form.
2 Solve equations that are quadratic in form by using substitution.
3 Solve application problems using equations that are quadratic in form.

Key Terms
Use the vocabulary term listed below to complete the statement in Exercise 1.

quadratic in form

1. An equation is _______________ if it can be written as a quadratic equation au^2+bu+c, where $a \neq 0$ and u is a variable of an expression.

GUIDED EXAMPLES AND PRACTICE

Objective 1 Solve equations by rewriting them in quadratic form.

Review this example for Objective 1:

1. Solve the equation. Identify any extraneous solutions.

$$\frac{1}{3x-1}-\frac{1}{3x+1}=1$$

Multiply both sides by the LCD, $(3x+1)(3x-1)$.

$$(3x+1)(3x-1)\left(\frac{1}{3x-1}-\frac{1}{3x+1}\right)=1\cdot(3x+1)(3x-1)$$

Distribute $(3x+1)(3x-1)$ and simplify both sides.

$$3x+1-(3x-1)=9x^2-1$$

$$3=9x^2$$

$$\pm\frac{\sqrt{3}}{3}=x$$

Practice this exercise:

1. Solve the equation. Identify any extraneous solutions.

$$6-t^{-1}-5t^{-2}=0$$

Name: Date:
Instructor: Section:

Check for extraneous solutions.

Check $\frac{\sqrt{3}}{3}$.

$$\frac{1}{3\left(\frac{\sqrt{3}}{3}\right)-1}-\frac{1}{3\left(\frac{\sqrt{3}}{3}\right)+1}\stackrel{?}{=}1$$

$$\frac{1}{\sqrt{3}-1}-\frac{1}{\sqrt{3}+1}\stackrel{?}{=}1$$

$$\frac{\sqrt{3}+1}{\sqrt{3}+1}\cdot\frac{1}{\sqrt{3}-1}-\frac{1}{\sqrt{3}+1}\cdot\frac{\sqrt{3}-1}{\sqrt{3}-1}\stackrel{?}{=}1$$

$$\frac{\sqrt{3}+1}{2}-\frac{\sqrt{3}-1}{2}\stackrel{?}{=}1$$

$$\frac{2}{2}\stackrel{?}{=}1$$

$$1=1 \qquad \text{True.}$$

$\frac{\sqrt{3}}{3}$ is a solution.

Check $-\frac{\sqrt{3}}{3}$.

$$\frac{1}{3\left(\frac{-\sqrt{3}}{3}\right)-1}-\frac{1}{3\left(\frac{-\sqrt{3}}{3}\right)+1}\stackrel{?}{=}1$$

$$\frac{1}{-\sqrt{3}-1}-\frac{1}{-\sqrt{3}+1}\stackrel{?}{=}1$$

$$\frac{-\sqrt{3}+1}{-\sqrt{3}+1}\cdot\frac{1}{-\sqrt{3}-1}-\frac{1}{-\sqrt{3}+1}\cdot\frac{-\sqrt{3}-1}{-\sqrt{3}-1}\stackrel{?}{=}1$$

$$\frac{-\sqrt{3}+1}{2}-\frac{-\sqrt{3}-1}{2}\stackrel{?}{=}1$$

$$\frac{2}{2}\stackrel{?}{=}1$$

$$1=1 \qquad \text{True.}$$

$-\frac{\sqrt{3}}{3}$ is a solution.

Name: Date:
Instructor: Section:

Objective 2 Solve equations that are quadratic in form by using substitution.

Review this example for Objective 2:

2. Use substitution to solve.

$y^{1/3} - y^{1/6} - 12 = 0.$

Substitute $u = y^{1/6}$ into the equation.

$$\left(y^{1/6}\right)^2 - y^{1/6} - 12 = 0$$
$$u^2 - u - 12 = 0$$
$$(u-4)(u+3) = 0$$
$$u - 4 = 0 \quad \text{or} \quad u + 3 = 0$$
$$u = 4 \qquad\qquad u = -3$$

Since $y^{1/6}$ can never be negative, $y^{1/6}$ is an extraneous solution. So, $y^{1/6} = 4$ and $y = 4^6 = 4096.$

Practice this exercise:

2. $3x^{1/2} + x^{1/4} = 2$

Objective 3 Solve application problems using equations that are quadratic in form.

Review this example for Objective 3:

3. The average speed of a motorcycle is 25 miles per hour more than the average speed of a car. The car takes 1 hour longer to travel 150 miles. Find how long it takes the motorcycle to travel 150 miles.

First, organize the information given.

Vehicle	d	t	r
motorcycle	150	x	$\frac{150}{x}$
car	150	$x+1$	$\frac{150}{x+1}$

Write an equation, for the average speed of the vehicles, and solve for x.

$$\frac{150}{x} = \frac{150}{x+1} + 25$$
$$x(x+1)\left(\frac{150}{x}\right) = \left(\frac{150}{x+1} + 25\right)x(x+1)$$
$$150x + 150 = 150x + 25x^2 + 25x$$
$$0 = 25x^2 + 25 - 150$$
$$0 = x^2 + x - 6$$
$$= (x+3)(x-2)$$
$$x + 3 = 0 \quad \text{or} \quad x - 2 = 0$$
$$x = -3 \qquad\qquad x = 2$$

Since time cannot be negative, It takes the motorcycle 2 hours to travel 150 miles.

Practice this exercise:

3. Georgia and Benny commute to work daily. Benny drives 60 miles and averages 5 miles per hour more than Georgia. Georgia drives 70 miles, and she is on the road one-half hour longer than Benny. How fast does each person drive?

Name: Date:
Instructor: Section:

ADDITIONAL EXERCISES

Objective 1 Solve equations by rewriting them in quadratic form.
For extra help, see Examples 1 and 2 on pages 1059–1060 of your text and the Section 16.3 lecture video.
Solve the equations with rational expressions. Identify any extraneous solutions.

1. $\frac{110}{x}-\frac{110}{x-5}=-\frac{1}{5}$

2. $\frac{1}{x}+\frac{1}{x+3}=\frac{1}{2}$

3. $\frac{1}{2x-1}-\frac{1}{2x+1}=\frac{1}{4}$

Solve the equations with radical expressions. Identify any extraneous solutions.

4. $x-6\sqrt{x}+5=0$

5. $x-7=\sqrt{x-1}$

6. $\sqrt{9x-2}-x-2=0$

7. $\sqrt{3x-2}-\sqrt{4x+1}=-5$

Objective 2 Solve equations that are quadratic in form by using substitution.
For extra help, see Example 3 on pages 1061-1062 of your text and the Section 16.3 lecture video.
Solve using substitution. Identify any extraneous solutions.

8. $x^4-41x^2+400=0$

9. $10x^4-9x^2+2=0$

10. $(x+3)^2-13(x+3)+42=0$

11. $x^{2/3}-2x^{1/3}-3=0$

Objective 3 Solve application problems using equations that are quadratic in form.
For extra help, see Example 4 on pages 1062-1064 of your text and the Section 16.3 lecture video.
Solve.

12. Jane and Mike are professional painters. Working alone, it takes Mike 3 hours longer to paint one floor of a home than it takes Jane working alone. Together they can paint one floor in $2\frac{1}{2}$ hours. How long does it take each person working alone?

Name: Date:
Instructor: Section:

Chapter 16 QUADRATIC EQUATIONS AND FUNCTIONS

16.4 Graphing Quadratic Functions

Learning Objectives

1 Graph quadratic functions of the form $f(x) = ax^2$.
2 Graph quadratic functions of the form $f(x) = ax^2 + k$.
3 Graph quadratic functions of the form $f(x) = a(x-h)^2$.
4 Graph quadratic functions of the form $f(x) = a(x-h)^2 + k$.
5 Graph quadratic functions of the form $f(x) = ax^2 + bx + c$.
6 Solve applications involving parabolas.

GUIDED EXAMPLES AND PRACTICE

Objective 1 Graph quadratic functions of the form $f(x) = ax^2$.

Review this example for Objective 1:

1. For the function $f(x) = \frac{2}{3}x^2$,

a) State whether the graph opens upward or downward.
b) Find the coordinates of the vertex.
c) Write the equation of the axis of symmetry.
d) Graph.

a) The graph of a quadratic function is a parabola. Since $a = \frac{2}{3}$ is positive, we know that the parabola opens upward and $f(x) \geq 0$.

b) The graph of a function of the form $f(x) = a(x-h)^2 + k$ has a vertex at (h, k), so the vertex of $f(x) = \frac{2}{3}x^2$ is $(0, 0)$.

c) For a function of the form $f(x) = a(x-h)^2 + k$, the axis of symmetry is $x = h$. The axis of symmetry of $f(x) = \frac{2}{3}x^2$ is $x = 0$.

Practice this exercise:

1. For the function $f(x) = -x^2$,

a) State whether the graph opens upward or downward.
b) Find the coordinates of the vertex.
c) Write the equation of the axis of symmetry.
d) Graph.

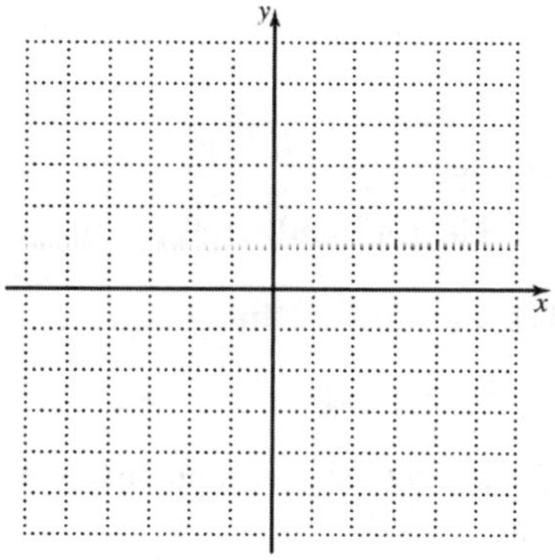

Name: Date:
Instructor: Section:

d) To complete the graph of $f(x)=\frac{2}{3}x^2$ find a few points on either side of the axis of symmetry.

x	$f(x)=\frac{2}{3}x^2$
1	$\frac{2}{3}$
−1	$\frac{2}{3}$
2	$\frac{8}{3}$
−2	$\frac{8}{3}$

Graph the parabola.

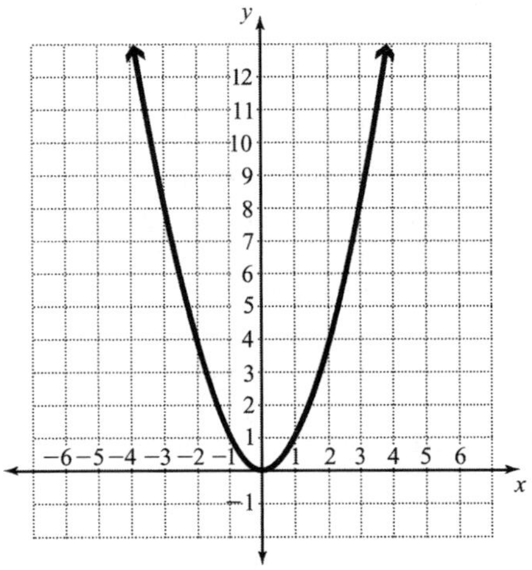

$f(x)=\frac{2}{3}x^2$

Objective 2 Graph quadratic functions of the form $f(x)=ax^2+k$.

Review this example for Objective 2:

2. For the function $f(x)=\frac{2}{3}x^2+2$,

a) State whether the graph opens upward or downward.
b) Find the coordinates of the vertex.
c) Write the equation of the axis of symmetry.
d) Graph.

a) The graph of a quadratic function is a parabola. Since $a=\frac{2}{3}$ is positive, we know that the parabola opens upward and $f(x)\geq 0$.

Practice this exercise:

2. For the function $f(x)=-x^2+3$,

a) State whether the graph opens upward or downward.
b) Find the coordinates of the vertex.
c) Write the equation of the axis of symmetry.
d) Graph.

Name:
Instructor:

Date:
Section:

b) The graph of a function of the form $f(x) = a(x-h)^2 + k$ has a vertex at (h, k), so the vertex of $f(x) = \frac{2}{3}x^2 + 2$ is $(0, 2)$.

c) For a function of the form $f(x) = a(x-h)^2 + k$, the axis of symmetry is $x = h$. The axis of symmetry of $f(x) = \frac{2}{3}x^2 + 2$ is $x = 0$.

d) To complete the graph of $f(x) = \frac{2}{3}x^2 + 2$ find a few points on either side of the axis of symmetry.

x	$f(x) = \frac{2}{3}x^2 + 2$
1	$2\frac{2}{3}$
−1	$2\frac{2}{3}$
2	$4\frac{2}{3}$
−2	$4\frac{2}{3}$

Graph the parabola.

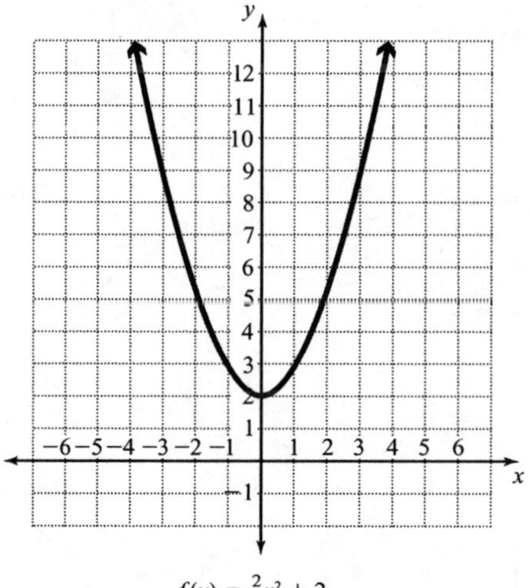

$f(x) = \frac{2}{3}x^2 + 2$

Notice that the axis of symmetry is located at $x = 0$ (the y-axis) and adding a constant $k = 2$ to $f(x) = \frac{2}{3}x^2$ shifts the graph up from the origin 2.

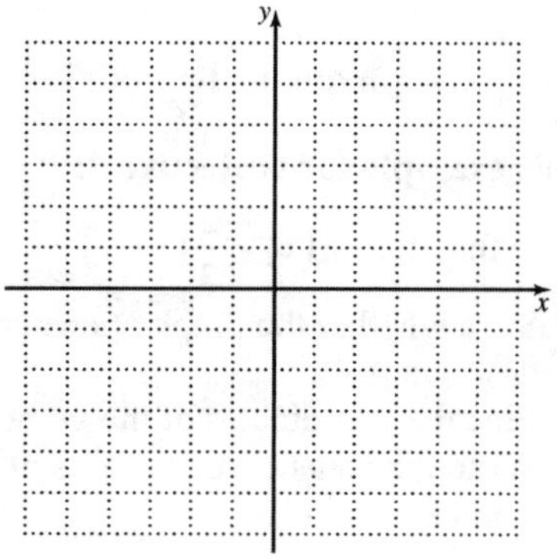

Name:
Instructor:

Date:
Section:

Objective 3 Graph quadratic functions of the form $f(x) = a(x-h)^2$.

Review this example for Objective 3:

3. For the function $f(x) = \frac{2}{3}(x+2)^2$,

a) State whether the graph opens upward or downward.
b) Find the coordinates of the vertex.
c) Write the equation of the axis of symmetry.
d) Graph.

a) The graph of a quadratic function is a parabola. Since $a = \frac{2}{3}$ is positive, we know that the parabola opens upward and $f(x) \geq 0$.

b) The graph of a function of the form $f(x) = a(x-h)^2 + k$ has a vertex at (h, k), so the vertex of $f(x) = \frac{2}{3}(x+2)^2$ is $(-2, 0)$.

c) For a function of the form $f(x) = a(x-h)^2 + k$, the axis of symmetry is $x = h$. The axis of symmetry of $f(x) = \frac{2}{3}(x+2)^2$ is $x = -2$.

d) To complete the graph of $f(x) = \frac{2}{3}(x+2)^2$ find a few points on either side of the axis of symmetry.

x	$y = \frac{2}{3}(x+2)^2$
1	6
−1	$\frac{2}{3}$
−3	$\frac{2}{3}$
−5	6

Practice this exercise:

3. For the function $f(x) = -(x+3)^2$,

a) State whether the graph opens upward or downward.
b) Find the coordinates of the vertex.
c) Write the equation of the axis of symmetry.
d) Graph.

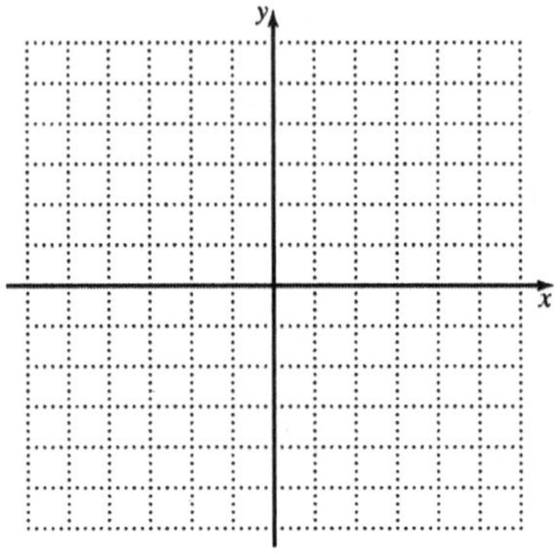

Name: Date:
Instructor: Section:

Graph the parabola.

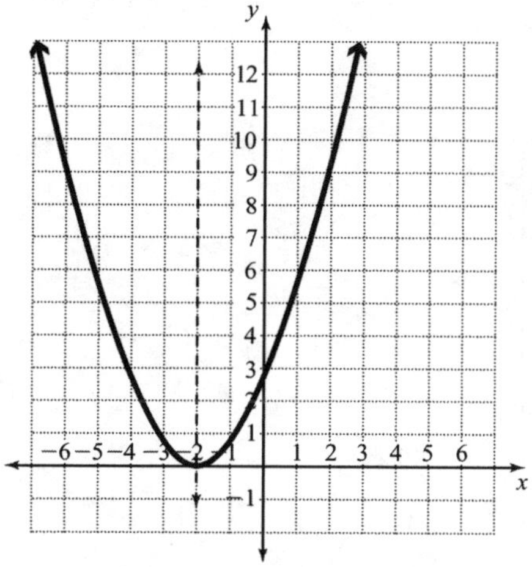

$f(x) = \frac{2}{3}(x + 2)^2$

Objective 4 Graph quadratic functions of the form $f(x) = a(x - h)^2 + k$.

Review this example for Objective 4:

4. For the function $f(x) = -(x-3)^2 + 4$,

a) State whether the graph opens upward or downward.
b) Find the coordinates of the vertex.
c) Write the equation of the axis of symmetry.
d) Graph.
e) Find the domain and range.

a) The graph of a quadratic function is a parabola. Since $a = -1$ is negative, we know that the parabola opens downward and $f(x) \le 0$.

b) The graph of a function of the form $f(x) = a(x-h)^2 + k$ has a vertex at (h, k), so the vertex of $f(x) = -(x-3)^2 + 4$ is $(3, 4)$.

c) For a function of the form $f(x) = a(x-h)^2 + k$, the axis of symmetry is $x = h$. The axis of symmetry of $f(x) = -(x-3)^2 + 4$ is $x = 3$.

d) To complete the graph of $f(x) = -(x-3)^2 + 4$ find a few points on either side of the axis of symmetry.

Practice this exercise:

4. For the function $h(x) = -6(x-7)^2 + 3$,

a) State whether the graph opens upward or downward.
b) Find the coordinates of the vertex.
c) Write the equation of the axis of symmetry.
d) Graph.
e) Find the domain and range.

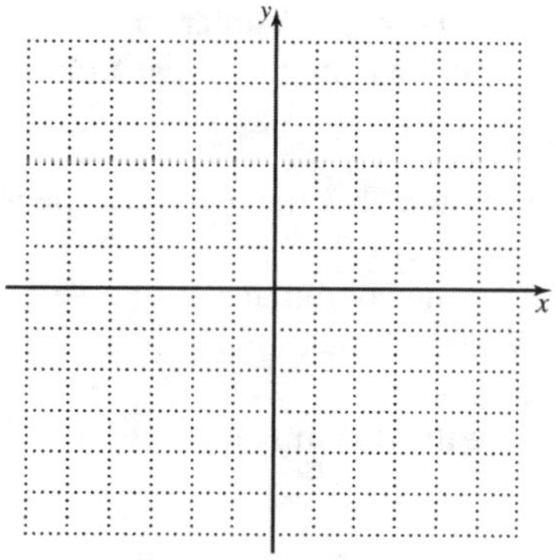

Name: Date:
Instructor: Section:

x	$f(x)=-(x-3)^2+4$
0	5
1	0
5	0
6	−5

Graph the parabola.

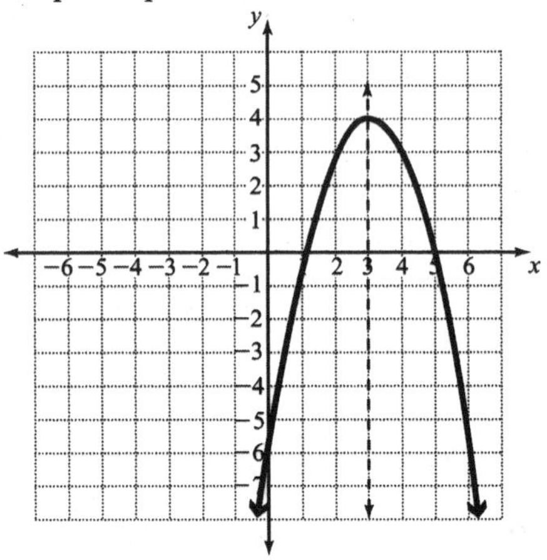

$f(x) = -(x + 3)^2 + 4$

e) Domain: $\{x|x \text{ is a real number}\}$, or $(-\infty, \infty)$

Range: $\{y|y \leq 4\}$, or $[4, -\infty)$

Objective 5 Graph quadratic functions of the form $f(x) = ax^2 + bx + c$.

Review this example for Objective 5:

5. For the function $g(x) = x^2 + 6x + 3$,

- **a)** Find the x- and y-intercepts.
- **b)** Write the equation in the form $g(x) = a(x-h)^2 + k$.
- **c)** State whether the parabola opens upward or downward.
- **d)** Find the coordinates of the vertex.
- **e)** Write the equation of the axis of symmetry.
- **f)** Graph.
- **g)** Find the domain and range.

Practice this exercise:

5. For the function $k(x) = x^2 + 4x + 7$,

- **a)** Find the x- and y-intercepts.
- **b)** Write the equation in the form $g(x) = a(x-h)^2 + k$.
- **c)** State whether the parabola opens upward or downward.
- **d)** Find the coordinates of the vertex.
- **e)** Write the equation of the axis of symmetry.
- **f)** Graph.
- **g)** Find the domain and range.

Name: Date:
Instructor: Section:

a) To find the x-intercepts, let $y = 0$.

$0 = x^2 + 6x + 3$

$0 = \frac{-6 \pm \sqrt{6^2 - 4(1)(3)}}{2(1)}$

$0 = \frac{-6 \pm \sqrt{36-12}}{2}$

$0 = \frac{-6 \pm \sqrt{24}}{2} = \frac{-6 \pm \sqrt{24}}{2}$

$0 = -3 \pm \sqrt{6} \approx -0.55 \text{ or } -5.45$

The x-intercepts are $(-0.55, 0)$ and $(-5.45, 0)$.

To find the y-intercept, let $x = 0$.

$y = x^2 + 6x + 3$

$y = (0)^2 + 6(0) + 3$

$y = 3$

The y-intercept is $(0, 3)$.

b) Transform $g(x) = x^2 + 6x + 3$ in to the form $g(x) = a(x-h)^2 + k$ by completing the square.

$y = x^2 + 6x + 3$

$y - 3 = x^2 + 6x$

$y - 3 + \left(\frac{6}{2}\right)^2 = x^2 + 6x + \left(\frac{6}{2}\right)^2$

$y + 6 = (x+3)^2$

$y = (x+3)^2 - 6$

c) $a = 1$ so the parabola opens upward.

d) The vertex is at (h, k). Thus, the vertex is at $(-3, -6)$.

e) The axis of symmetry is $x = h$ or $x = -3$.

f) To complete the graph of $g(x) = x^2 + 6x + 3$ find a few points on either side of the axis of symmetry.

x	$g(x) = x^2 + 6x + 3$
0	3
−1	−2
−5	−2
−6	3

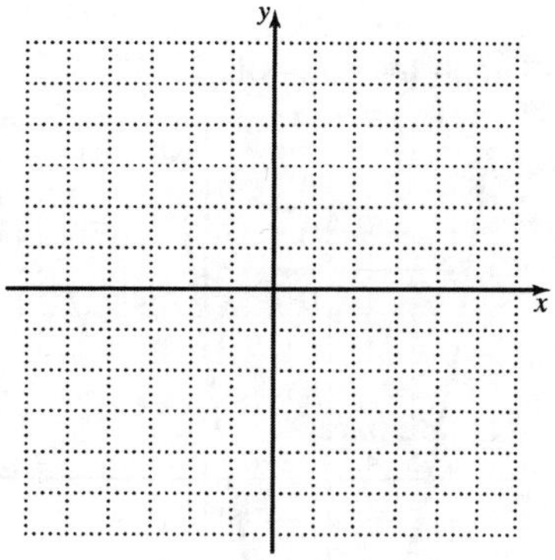

Name: Date:
Instructor: Section:

Graph the parabola.

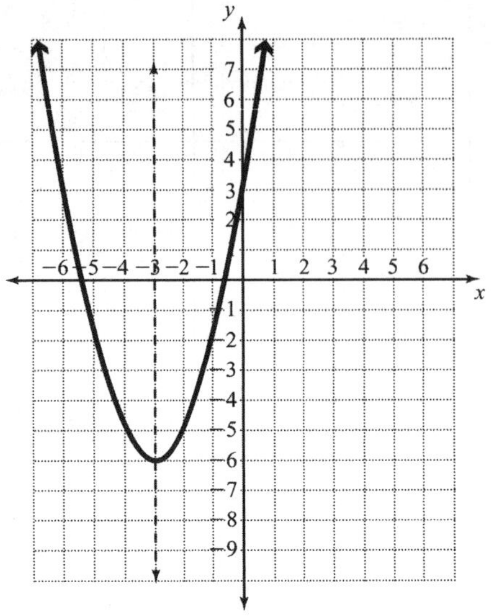

$g(x) = x^2 + 6x + 3$

g) Domain: $\{x | x \text{ is a real number}\}$, or $(-\infty, \infty)$

Range: $\{y | y \geq -6\}$, or $[-6, \infty)$

Objective 6 Solve applications involving parabolas.

Review this example for Objective 6:

6. An archer shoots an arrow straight up in the air from ground level with an initial velocity of 35 feet per second. The equation $h = -14t^2 + 35t$ describes the height, h, of the arrow t seconds after it was shot. Use the equation to find the maximum height that the arrow reaches.

Because the graph of $h = -14t^2 + 35t$ is a parabola that opens downward $(a = -14)$, the maximum height occurs at the vertex.
The t coordinate of the vertex:

$$t = \frac{-b}{2a} = \frac{-35}{2(-14)} = \frac{5}{4} = 1.25 \text{ seconds}$$

To find the h-coordinate of the vertex, replace t in $h = -14t^2 + 35t$ with 1.25.

$h = -14t^2 + 35t$

$h = -14(1.25)^2 + 35(1.25)$

$h = 21.875$

The arrow reaches the maximum height of 21.875 feet at 1.25 seconds.

Practice this exercise:

6. The amount of nitrogen dioxide, A, in parts per million that was present in the air in the city of Ovid on a certain day in June is modeled by the equation $A = -2t^2 + 40t + 12$, where t is the number of hours after 6:00 A.M.Use this equation to find the time at which the nitrogen dioxide level was at its maximum.

Name: Date:
Instructor: Section:

ADDITIONAL EXERCISES

Objective 1 Graph quadratic functions of the form $f(x) = ax^2$.

For extra help, see Example 1 on pages 1068-1069 of your text and the Section 16.4 lecture video.

a) *State whether the parabola opens upward or downward.*
b) *Find the coordinates of the vertex.*
c) *Write the equation of the axis of symmetry.*
d) *Graph.*

1. $f(x) = -\frac{1}{3}x^2$

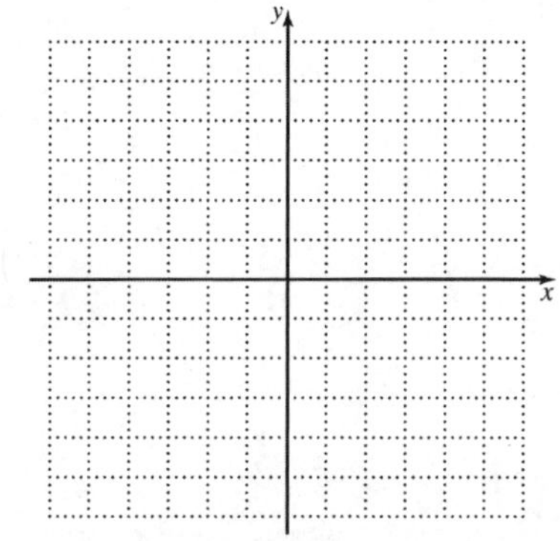

Objective 2 Graph quadratic functions of the form $f(x) = ax^2 + k$.

For extra help, see Example 2 on page 1070 of your text and the Section 16.4 lecture video.

a) *State whether the parabola opens upward or downward.*
b) *Find the coordinates of the vertex.*
c) *Write the equation of the axis of symmetry.*
d) *Graph.*

2. $f(x) = \frac{1}{3}x^2 + 5$

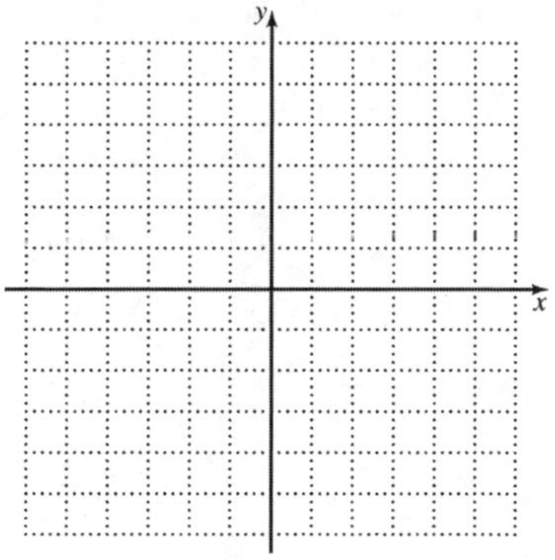

3. $g(x) = -2x^2 - 2$

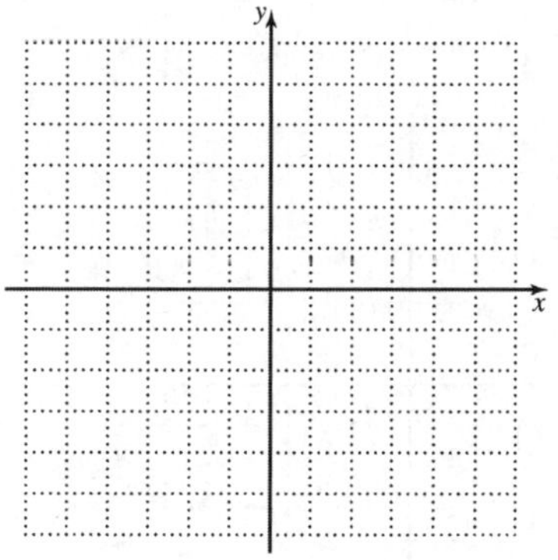

Name: Date:
Instructor: Section:

Objective 3 Graph quadratic functions of the form $f(x) = a(x-h)^2$.

For extra help, see Example 3 on pages 1070-1071 of your text and the Section 16.4 lecture video.

a) *State whether the parabola opens upward or downward.*
b) *Find the coordinates of the vertex.*
c) *Write the equation of the axis of symmetry.*
d) *Graph.*

4. $f(x) = -\frac{1}{2}(x+3)^2$

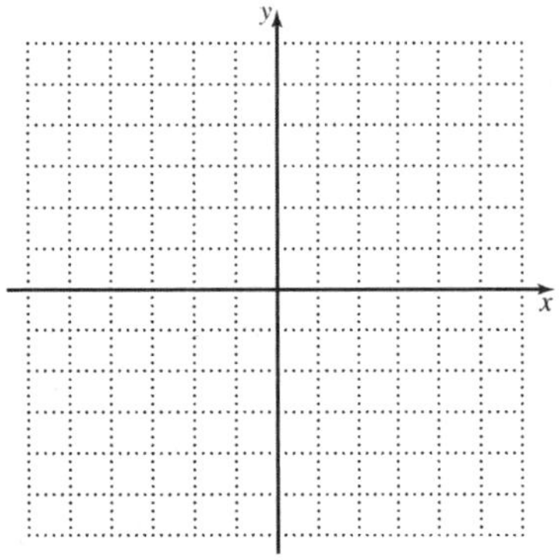

Objective 4 Graph quadratic functions of the form $f(x) = a(x-h)^2 + k$.

For extra help, see Example 4 on page 1071 of your text and the Section 16.4 lecture video.

a) *State whether the parabola opens upward or downward.*
b) *Find the coordinates of the vertex.*
c) *Write the equation of the axis of symmetry.*
d) *Graph.*
e) *Find the domain and range.*

5. $f(x) = \frac{1}{4}(x+2)^2 + 1$

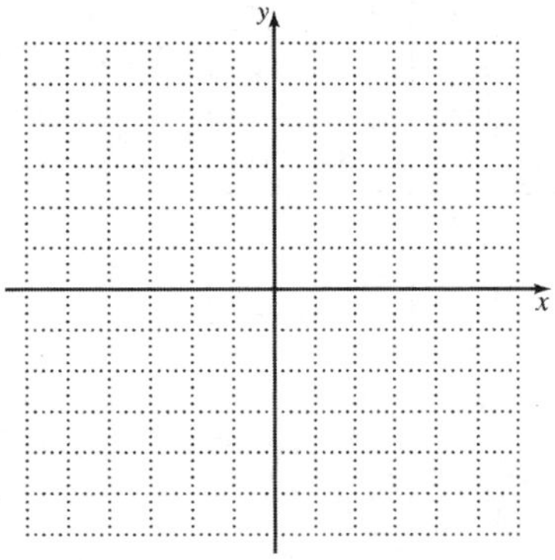

Name: Date:
Instructor: Section:

Objective 5 Graph quadratic functions of the form $f(x) = ax^2 + bx + c$.

For extra help, see Examples 5–6 on pages 1072–1074 of your text and the Section 16.4 lecture video.

a) *Find the x- and y-intercepts.*

b) *Write the equation in the form* $f(x) = a(x-h)^2 + k$.

c) *State whether the parabola opens upward or downward.*

d) *Find the coordinates of the vertex.*

e) *Write the equation of the axis of symmetry.*

f) *Graph.*

g) *Find the domain and range.*

6. $h(x) = x^2 + 12x + 36$

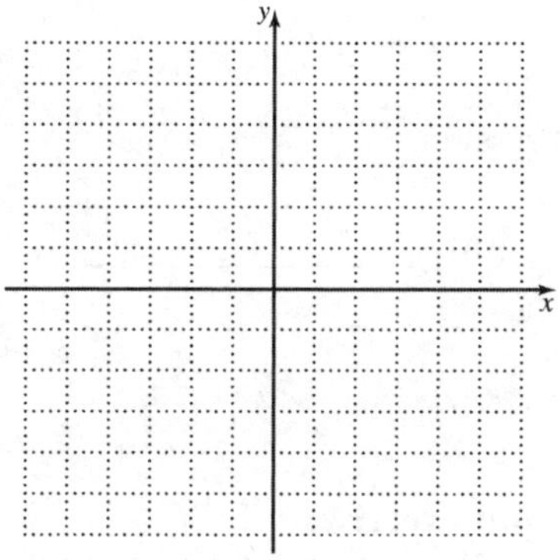

7. $g(x) = -x^2 - 2x + 3$

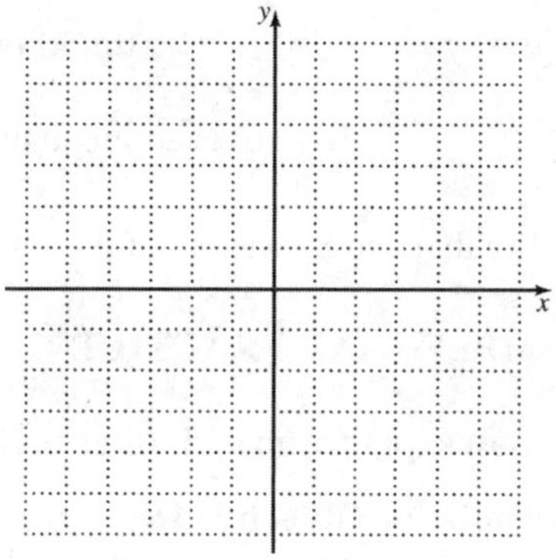

Objective 6 Solve applications involving parabolas.

For extra help, see Examples 7–8 on pages 1074–1076 of your text and the Section 16.4 lecture video.

Solve.

8. A toy rocket is launched upward from ground level with an initial velocity of 48 meters per second. The equation $h = -4.9t^2 + 48t$ describes the height of the rocket, where h represents the height in meters and t represents the flight time in seconds.

a) Graph the equation.

b) What is the maximum height the rocket reaches? Round to the nearest thousandth of a meter.

c) How long is the rocket in flight? Round to the nearest thousandth of a second.

9. A carpenter is building a rectangular room with a fixed perimeter of 336 feet.

a) What dimensions would yield the maximum area?

b) What is the maximum area?

Name:
Instructor:

Date:
Section:

Chapter 16 QUADRATIC EQUATIONS AND FUNCTIONS

16.5 Solving Nonlinear Inequalities

Learning Objectives
1 Solve quadratic and other inequalities.
2 Solve rational inequalities.

Key Terms
Use the vocabulary terms listed below to complete each statement in Exercises 1 and 2.

quadratic inequality **rational inequality**

1. A(n) __________________ is an inequality containing a rational expression.

2. An inequality that can be written in the form $ax^2 + bx + c > 0$, where is a(n) __________________.

GUIDED EXAMPLES AND PRACTICE

Objective 1 Solve quadratic and other inequalities.

Review this example for Objective 1:

1. Solve. Write the solution set using interval notation; then graph the solution set on a number line.
$x^2 - 15x + 50 < 0$

First find the solution of the related equation.

$x^2 - 15x + 50 = 0$

$(x-5)(x-10) = 0$

$x - 5 = 0$ or $x - 10 = 0$

$x = 5$ $x = 10$

Plot the 5 and 10 on a number line, which divides the number line into three intervals.

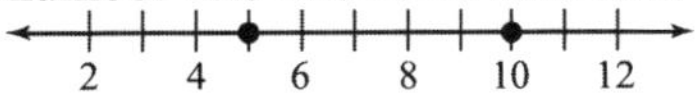

Choose a test number from each interval and substitute that value into $x^2 - 15x + 50 < 0$.

Interval	$(-\infty, 5)$	$(5,10)$	$(10, \infty)$
Test Number	0	8	1000
Test Results	$50 < 0$	$-6 < 0$	$985050 < 0$
True or False	False	True	False

The solution set is $(5,10)$.

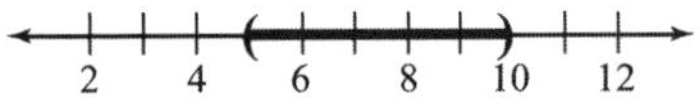

Practice this exercise:

1. Solve. Write the solution set using interval notation; then graph the solution set on a number line.
$(x+6)(x+5)(x-2) < 0$

Name: Date:
Instructor: Section:

Objective 2 Solve rational inequalities.

Review this example for Objective 2:

2. Solve. Write the solution set using interval notation; then graph the solution set on a number line.

$$\frac{(x+2)^2}{x+5} \geq 0$$

First find the values that make the denominator equal to 0.

$x+5=0$

$x=-5$

Now solve the related equation.

$$\frac{(x+2)^2}{x+5} = 0$$

$$(x+5)\frac{(x+2)^2}{x+5} = (x+5)0$$

$$(x+2)^2 = 0$$

$x=-2$

Plot -5 and -2 on a number line and label the intervals.

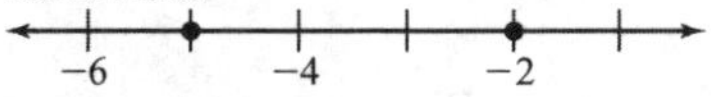

Use a table.

Interval	$(-\infty,-5)$	$(-5,-2)$	$(-2,\infty)$
Test Number	-6	-4	0
Test Results	$-16 \geq 0$	$4 \geq 0$	$\frac{4}{5} \geq 0$
True or False	False	True	True

The solution set is $(-5,\infty)$.

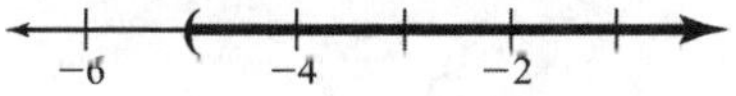

Practice this exercise:

2. Solve. Write the solution set using interval notation; then graph the solution set on a number line.

$$\frac{(3x-10)^2}{x+3} > 0$$

ADDITIONAL EXERCISES

Objective 1 Solve quadratic and other inequalities.

For extra help, see Examples 1 and 2 on pages 1082–1083 of your text and the Section 16.5 lecture video.

Solve. Write the solution set using interval notation; then graph the solution set on a number line.

1. $(x+3)(x-4)<0$

2. $m^2-13m+40>0$

3. $r^2-15r+44<0$

4. $x^2-2x+1 \geq 0$

5. $x^2-7x+21<0$

6. $(x+5)(3x+12)(2x-10)>0$

Name: Date:
Instructor: Section:

Objective 2 Solve rational inequalities.

For extra help, see Example 3 on pages 1084–1085 of your text and the Section 16.5 lecture video.

Solve the rational inequalities. Write the solution set using interval notation; then graph the solution set on a number line.

7. $\dfrac{x-6}{x+5} \le 0$

8. $\dfrac{8}{x-2} \ge 2$

9. $\dfrac{b}{b+4} < 2$

10. $\dfrac{x-7}{x-8} > 3$

Name: Date:
Instructor: Section:

Chapter 16 QUADRATIC EQUATIONS AND FUNCTIONS

16.6 Function Operations

Learning Objectives
1 Add or subtract functions.
2 Multiply functions.
3 Divide functions.

GUIDED EXAMPLES AND PRACTICE

Objective 1 Add or subtract functions.

Review this example for Objective 1:

1. Given $f(x)=x^2-4$ and $g(x)=x+2$, find $f+g$ and $f-g$.

For $f+g$, write a sum of the two functions and combine like terms.

$$(f+g)(x)=f(x)+g(x)$$
$$=(x^2-4)+(x+2)$$
$$=x^2+x-2$$

For $f-g$, write a difference of the two functions and combine like terms.

$$(f-g)(x)=f(x)-g(x)$$
$$=(x^2-4)-(x+2)$$
$$=x^2-x-6$$

Practice this exercise:

1. Given $f(x)=x^3-2x+4$ and $g(x)=2x^3-x^2+7x$, find $f+g$ and $f-g$.

Objective 2 Multiply functions.

Review this example for Objective 2:

2. Given $f(x)=x^2-4$ and $g(x)=x+2$, find $f\cdot g$.

Write a product of the two functions, then use FOIL to multiply the binomials.

$$(f\cdot g)(x)=f(x)\cdot g(x)$$
$$=(x^2-4)\cdot(x+2)$$
$$=x^3+2x^2-4x-8$$

Practice this exercise:

2. Given $f(x)=x^2-3x+2$, and $g(x)=x^2+2x-9$ find $f\cdot g$.

Name: Date:
Instructor: Section:

Objective 3 Divide functions.

Review these examples for Objective 3:

3. Given $f(x)=x^2-4$ and $g(x)=x+2$, find f/g.

$$(f/g)(x)=\frac{f(x)}{g(x)}$$
$$=\frac{x^2-4}{x+2}$$
$$=\frac{(x+2)(x-2)}{x+2}$$
$$=x-2$$

So, $(f/g)(x)=x-2$ for all x not equal to -2.

Practice these exercises:

3. Given $f(x)=5x^3-10x^2$ and $g(x)=5x$, find f/g.

ADDITIONAL EXERCISES

Objective 1 Add or subtract functions.

For extra help, see Example 1 on page 1091 of your text and the Section 16.6 lecture video.

Find $f+g$ and $f-g$.

1. $f(x)=4x^2,\ g(x)=8x-7$

2. $f(x)=-5x+3,\ g(x)=x^2+2$

3. $f(x)=x+2,\ g(x)=x-2$

4. $f(x)=x-1,\ g(x)=x^2-x$

5. $f(x)=2x^2+3,\ g(x)=x^3+6x$

6. $f(x)=-4x^2+3x+2,\ g(x)=7x^2-x+5$

7. $f(x)=2x^2+6x-1,\ g(x)=2x^2+3x-4$

Objective 2 Multiply functions.

For extra help, see Example 2 on page 1092 of your text and the Section 16.6 lecture video.

Find $f\cdot g$.

8. $f(x)=2x-9,\ g(x)=3x+1$

9. $f(x)=x^2-18,\ g(x)=14-x$

Objective 3 Divide functions.

For extra help, see Example 3 on page 1092 of your text and the Section 16.6 lecture video.

Find f/g.

10. $f(x)=x^2-7x+10,\ g(x)=x-2$

11. $f(x)=x^2-1,\ g(x)=x+1$

Name: Date:
Instructor: Section:

Chapter 17 EXPONENTIAL AND LOGARITHMIC FUNCTIONS

17.1 Composite and Inverse Functions

Learning Objectives

1. Find the composition of two functions.
2. Show that two functions are inverses.
3. Show that a function is one-to-one.
4. Find the inverse of a function.
5. Graph a given function's inverse function.

Key Terms

Use the vocabulary terms listed below to complete each statement in Exercises 1–3.

composition of functions **inverse functions** **one-to-one function**

1. Given a function's graph, the function is a(n) _______________ if every horizontal line that can intersect the graph does so at one and only one point.

2. Two functions *f* and *g* are _______________ if and only if $(f \circ g)(x) = x$ for all *x* in the domain of *g* and $(f \circ g)(x) = x$ for all *x* in the domain of *f*.

3. If *f* and *g* are functions, the _______________ *f* and *g* is defined as $(f \circ g)(x) = f[g(x)]$ for all *x* in the domain *g* for which *g*(*x*) is in the domain of *f*.

GUIDED EXAMPLES AND PRACTICE

Objective 1 Find the composition of two functions.

Review this example for Objective 1:

1. If $f(x) = 5x + 2$ and $g(x) = x^2 + 3$, find $(f \circ g)(x)$.

The definition of a composition is below.

$(f \circ g)(x) = f[g(x)]$

Replace $g(x)$ with $x^2 + 3$.

$f[g(x)] = f(x^2 + 3)$

In $f(x)$, replace x with $x^2 + 3$.

$f[g(x)] = 5(x^2 + 3) + 2$

Multiply and add.

$= 5x^2 + 15 + 2$

$= 5x^2 + 17$

Practice this exercise:

1. Find $(f \circ g)(x)$ and $(g \circ f)(x)$.

$f(x) = 2x^2 + 4, \ g(x) = 5x - 3$

Name: Date:
Instructor: Section:

Objective 2 Show that two functions are inverses.

Review this example for Objective 2:

2. Verify that *f* and *g* are inverses.

$f(x)=5x+3,\ g(x)=\dfrac{x-3}{5}$

Substitute $\dfrac{x-3}{5}$ for $g(x)$.

$f[g(x)]=f\left(\dfrac{x-3}{5}\right)$

Substitute $\dfrac{x-3}{5}$ for x in $f(x)$.

$f[g(x)]=5\left(\dfrac{x-3}{5}\right)+3$

Multiply and simplify.

$=x-3+3$

$=x$

So, $f[g(x)]=x$.

Substitute $5x+3$ for $f(x)$.

$g[f(x)]=g(5x+3)$

Substitute $5x+3$ for x in $g(x)$.

$g[f(x)]=\dfrac{(5x+3)-3}{5}$

Add and simplify

$=\dfrac{5x}{5}$

$=x$

So, $g[f(x)]=x$.

Because $f[g(x)]=x$ and $g[f(x)]=x$, *f* and *g* are inverses.

Practice this exercise:

2. Verify that *f* and *g* are inverses.

$f(x)=x^2+8,\ x\geq 0;\ \ g(x)=\sqrt{x-8},\ x\geq 8$

Name: Date:
Instructor: Section:

Objective 3 Show that a function is one to one.

Review this example for Objective 3:

3. Determine whether the following graph is of a one-to-one function.

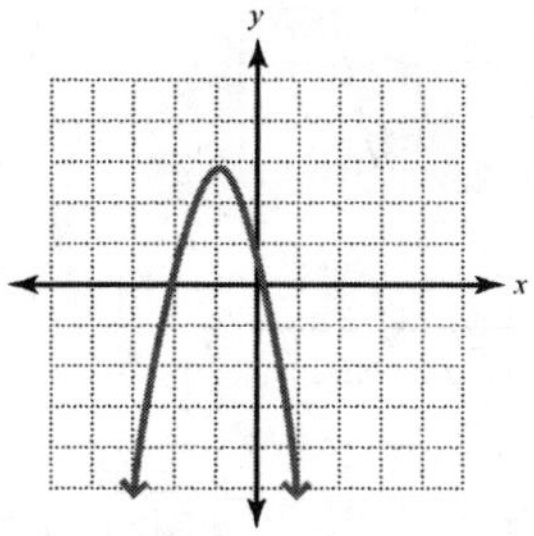

A horizontal line can intersect this graph in more than one point, so the function is not one to one.

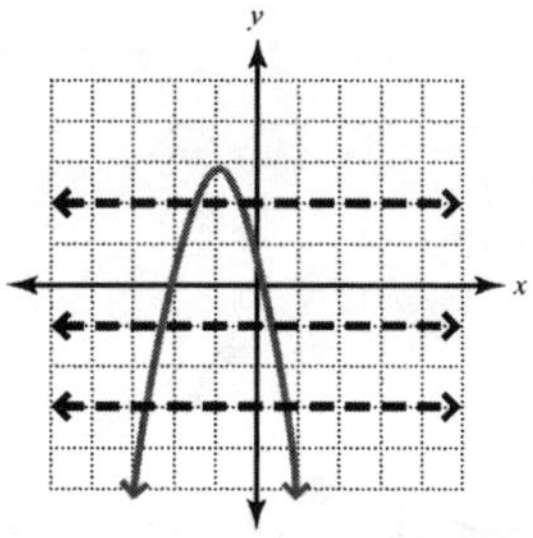

Practice this exercise:

3. Determine whether the following graph is of a one-to-one function.

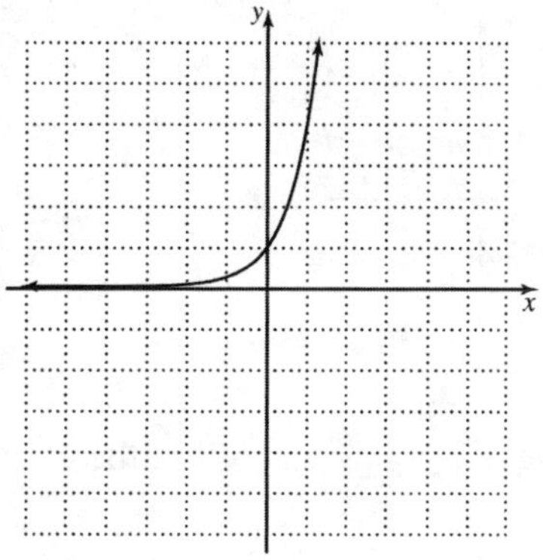

Objective 4 Find the inverse of a function.

Review this example for Objective 4:

4. If $f(x)=x^3-9$, find $f^{-1}(x)$.

Replace $f(x)$ with y.

$y=x^3-9$

Interchange x and y.

$x=y^3-9$

Solve for y.

$x+9=y^3$

$\sqrt[3]{x+9}=y$

Replace y with $f^{-1}(x)$.

$f(x)=\sqrt[3]{x+9}$

Verify by showing that $f\left[f^{-1}(x)\right]=x$ and $f^{-1}\left[f(x)\right]=x$.

Practice these exercises:

4. If $f(x)=4x+1$, find $f^{-1}(x)$.

Name: Date:
Instructor: Section:

Objective 5 Graph a given function's inverse function.

Review this example for Objective 5:

5. Sketch the graph of the inverse of the following function.

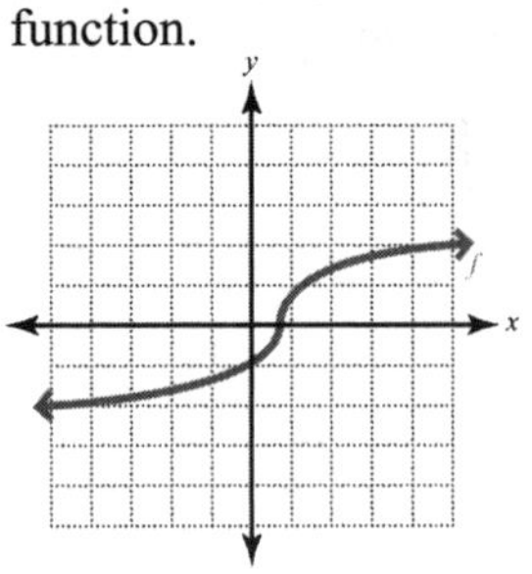

Draw the line $y = x$ and reflect the graph in the line.

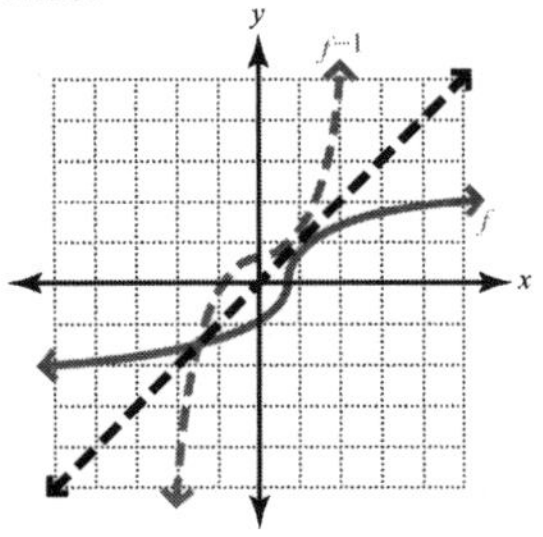

Practice this exercise:

5. Sketch the graph of the inverse of the following function.

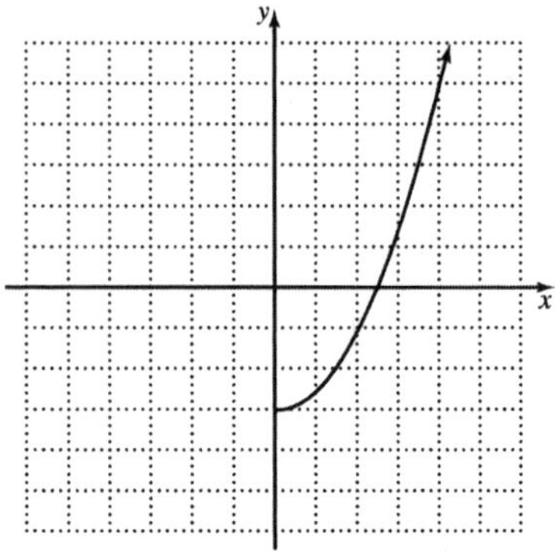

ADDITIONAL EXERCISES

Objective 1 Find the composition of two functions.

For extra help, see Example 1 on pages 1110–1111 of your text and the Section 17.1 lecture video.

Find $(f \circ g)(x)$ *and* $(g \circ f)(x)$.

1. $f(x) = 7x + 5,\ g(x) = \sqrt{x+8}$

2. $f(x) = x^2 + 2,\ g(x) = \sqrt{6-x}$

Objective 2 Show that two functions are inverses.

For extra help, see Example 2 on page 1112 of your text and the Section 17.1 lecture video.

Verify that f and g are inverses.

3. $f(x) = 10x - 9,\ g(x) = 4 - 2x$

Name: Date:
Instructor: Section:

Objective 3 Show that a function is one to one.
For extra help, see Example 3 on pages 1113–1114 of your text and the Section 17.1 lecture video.
Determine whether the following graphs are of one to one functions.

4.

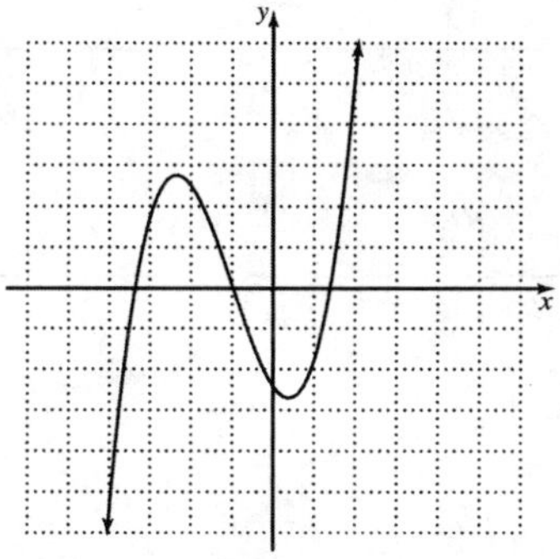

5.

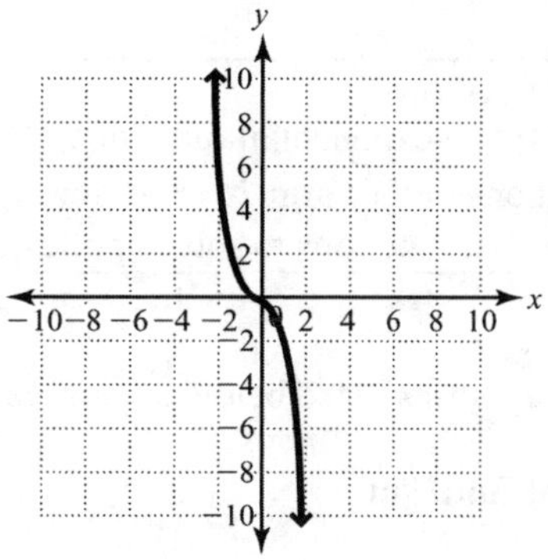

Objective 4 Find the inverse of a function.
For extra help, see Example 4 on page 1115 of your text and the Section 17.1 lecture video.
Find $f^{-1}(x)$ for each of the following one-to-one functions.

6. $f(x)=\dfrac{x-8}{x+9}$

7. $f(x)=5x^3-6$

Objective 5 Graph a given function's inverse function.
For extra help, see Example 5 on pages 1116–1117 of your text and the Section 17.1 lecture video.
Sketch the inverse of the functions whose graphs are shown.

8.

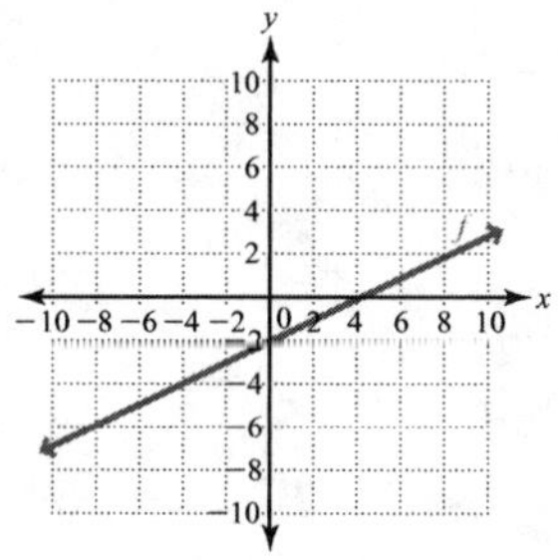

9.

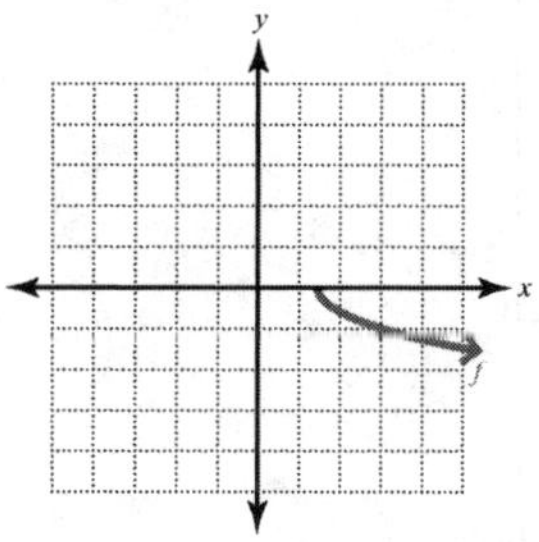

Name: Date:
Instructor: Section:

Chapter 17 EXPONENTIAL AND LOGARITHMIC FUNCTIONS

17.2 Exponential Functions

Learning Objectives
1 Define and graph exponential functions.
2 Solve equations of the form $b^x = b^y$ for x.
3 Use exponential functions to solve application problems.

Key Terms
Use the vocabulary terms listed below to complete each statement in Exercise 1.

exponential function

1. If $b > 0$, $b \neq 1$, and x is any real number, then the equation $f(x) = b^x$ is an example of a(n) ________________.

GUIDED EXAMPLES AND PRACTICE

Objective 1 Define and graph exponential functions.

Review this example for Objective 1:

1. Graph the function.
$f(x) = 3^{3x-1}$

Find some ordered pairs, plot them, and draw the graph.

x	$y = f(x)$
-1	$3^{-4} = \frac{1}{81}$
0	$3^{-1} = \frac{1}{3}$
1	$3^2 = 9$
2	$3^5 = 243$

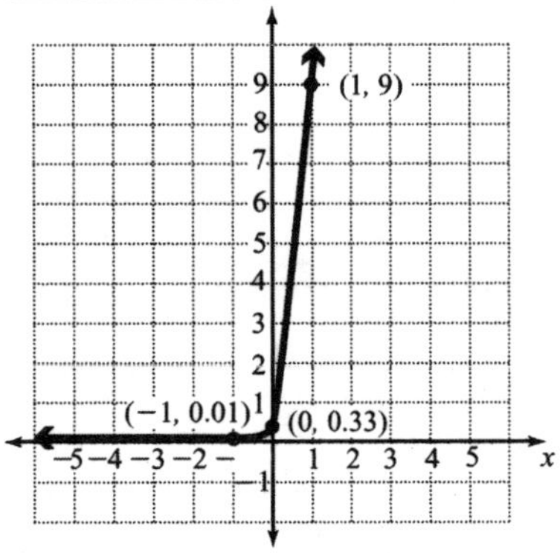

Practice this exercise:

1. Graph the function.
$f(x) = 2^{x-3}$

Name: Date:
Instructor: Section:

Objective 2 **Solve equations of the form $b^x = b^y$ for x.**

Review this example for Objective 2:

2. Solve the equation.

$7^x = 2401$

Write 2401 as 7^4, so that both sides have the same base. The equation then becomes;

$7^x = 7^4$

Now set the exponents equal to each other.

$x = 4$

Practice this exercise:

2. Solve the equation.

$4^x = 8$

Objective 3 **Use exponential functions to solve application problems.**

Review these examples for Objective 3:

3. Find the accumulated amount in an account if $6000 is deposited at 6% compounded quarterly for 12 years. Round the answer to the nearest cent.

If *P* dollars are invested at an annual interest rate of *r* (written as a decimal) compounded *n* times per year for *t* years, the accumulated amount *A* in the account is given by the formula;

$$A = P\left(1 + \frac{r}{n}\right)^{nt}$$

Substitute 6000 for *P*, 0.06 for *r*, 4 for *n*, and 12 for *t*, then solve for *A*.

$$A = 6000\left(1 + \frac{0.06}{4}\right)^{4(12)}$$

$$A = 6000(1 + 0.015)^{48}$$

$$A = 6000(1.015)^{48}$$

$$A = 12{,}260.87$$

After 12 years, the accumulated amount in the account would be $12,260.87.

Practice these exercises:

3. If $19,000 is deposited into an account paying 6% interest compounded quarterly, how much will be in the account after 7 years? Use the formula $A = P\left(1 + \frac{r}{n}\right)^{nt}$.

ADDITIONAL EXERCISES

Objective 1 **Define and graph exponential functions.**

For extra help, see Examples 1–2 on pages 1122–1123 of your text and the Section 17.2 lecture video.

Graph.

1. $f(x) = 2^x + 3$

2. $f(x) = \left(\frac{1}{6}\right)^x$

Name: Date:
Instructor: Section:

Objective 2 Solve equations of the form $b^x = b^y$ for x.
For extra help, see Example 3 on page 1124 of your text and the Section 17.2 lecture video.
Solve each equation.

3. $5^x = \dfrac{1}{625}$

4. $\left(\dfrac{3}{4}\right)^x = \dfrac{64}{27}$

Objective 3 Use exponential functions to solve application problems.
For extra help, see Examples 4–6 on pages 1125–1126 of your text and the Section 17.2 lecture video.
A culture of a bacteria doubles in size every 40 minutes. If A_0 is the initial amount and t is the number of minutes passed, the amount present is A, where $A = A_0(2)^{t/40}$.

5. If a culture of the bacteria began with 2000 cells, how many cells will be present after 20 minutes? Round to the nearest whole number.

Name: Date:
Instructor: Section:

Chapter 17 EXPONENTIAL AND LOGARITHMIC FUNCTIONS

17.3 Logarithmic Functions

Learning Objectives
1 Convert between exponential and logarithmic forms.
2 Solve logarithmic equations by changing to exponential form.
3 Graph logarithmic functions.
4 Solve applications involving logarithms.

Key Terms
Use the vocabulary terms listed below to complete each statement in Exercises 1–9.

logarithm

1. If $b > 0$ and $b \neq 1$, then $y = \log_b x$ is equivalent to $x = b^y$. In the equation $y = \log_b x$, the variable y is called a(n) _______________.

GUIDED EXAMPLES AND PRACTICE

Objective 1 Convert between exponential and logarithmic forms.

Review this example for Objective 1:

1. Write in logarithmic form.
$5^3 = 125$

$\log_3 125 = 5$

2. Write in exponential form.
$\log_6 216 = 3$

$6^3 = 216$

Practice this exercise:

1. Write in logarithmic form.
$3^5 = 243$

2. Write in exponential form.
$\log_5 625 = 4$

Objective 2 Solve logarithmic equations by changing to exponential form.

Review this example for Objective 2:

3. Solve.
$\log_b 8 = 3$

$b^3 = 8$ Write in exponential form.
$b = 2$ Find the cube root of 8.

Practice this exercise:

3. Solve.
$\log_2 x = -1$

Name: Date:
Instructor: Section:

Objective 3 Graph logarithmic functions.

Review this example for Objective 3:

4. Graph.

$f(x) = \log_2 2x$

$y = \log_2 2x$ Replace $f(x)$ with y.

$2^y = 2x$ or $2^{y-1} = x$ Write in exponential form.

Choose values for y and x.

y	0	1	2	3	-1	-2
x	$2^{-1} = \frac{1}{2}$	$2^0 = 1$	$2^1 = 2$	$2^2 = 4$	$2^{-2} = \frac{1}{4}$	$2^{-3} = \frac{1}{8}$

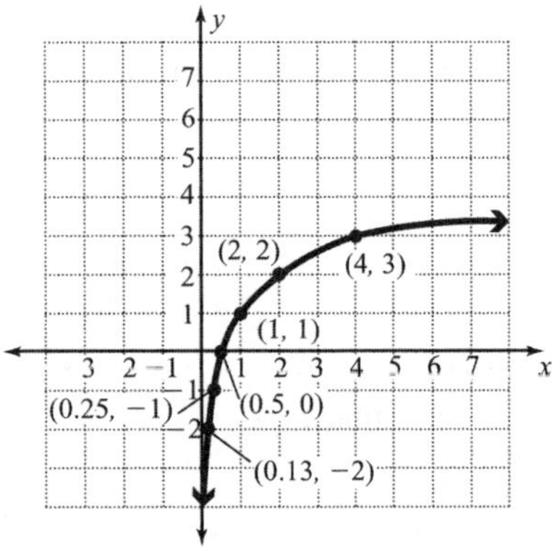

Practice this exercise:

4. Graph.

$f(x) = \log_6 x$

Objective 4 Solve applications involving logarithms.

Review this example for Objective 4:

5. The function $P = 95 - 30\log_2 x$ models the percent, P, of students that recall the important features of a classroom lecture over time, where x is the number of days that have elapsed since the lecture was given. What percent of the students recall the important features of a lecture 8 days after it was given?

$P = 95 - 30\log_2 8$ Substitute $x = 8$.

$P = 95 - 30(3)$ $\log_2 8 = 3$, since $2^3 = 8$.

$P = 95 - 90$ Multiply.

$P = 5$ Simplify.

Practice this exercise:

5. The percent of adult length attained by a 2- to 18-month old reptile of a certain species can be approximated by $f(x) = 38 + 30\log_3(x - 11)$, where x is the reptile's age in months and $f(x)$ is the percent. At an age of 14 months, what percent of its adult length has the reptile reached?

Name: Date:
Instructor: Section:

ADDITIONAL EXERCISES

Objective 1 Convert between exponential and logarithmic forms.
For extra help, see Examples 1–2 on page 1132 of your text and the Section 17.3 lecture video.
Write in logarithmic form.

1. $64^{1/3} = 4$

2. $\left(\frac{1}{5}\right)^{-3} = 125$

Write in exponential form.

3. $\log_{10} \frac{1}{100{,}000} = -5$

4. $\log_2 \frac{1}{4} = -2$

Objective 2 Solve logarithmic equations by changing to exponential form.
For extra help, see Examples 3–5 on pages 1133–1134 of your text and the Section 17.3 lecture video.
Solve.

5. $\log_4 256 = x$

6. $\log_2 \frac{1}{8} = x$

7. $\log_x \frac{1}{27} = -3$

8. $\log_4 x + 5 = 8$

9. $3\log_b 64 = 9$

Objective 3 Graph logarithmic functions.
For extra help, see Example 6 on pages 1134–1135 of your text and the Section 17.3 lecture video.
Graph.

10. $f(x) = \log_{1/2} x$

Objective 4 Solve applications involving logarithms.
For extra help, see Example 7 on page 1135 of your text and the Section 17.3 lecture video.
Solve.

11. The percent of adult length attained by a 2- to 18-month old reptile of a certain species can be approximated by the equation $f(x) = 28 + 46\log_5(x-8)$, where x is the reptile's age in months and $f(x)$ is the percent. At an age of 13 months, what percent of its adult length has the reptile reached?

Name: Date:
Instructor: Section:

Chapter 17 EXPONENTIAL AND LOGARITHMIC FUNCTIONS

17.4 Properties of Logarithms

Learning Objectives
1 Apply the inverse properties of logarithms.
2 Apply the product, quotient, and power properties of logarithms.

GUIDED EXAMPLES AND PRACTICE

Objective 1 Apply the inverse properties of logarithms.

Review these examples for Objective 1:

1. Find the value.
$4^{\log_4 9}$

$4^{\log_4 9} = 9$ Use $b^{\log_b x} = x$.

2. Find the value.
$\log_2 2^5$

$\log_2 2^5 = 5$ Use $\log_b b^x = x$.

Practice these exercises:

1. Find the value.
$12^{\log_{12} f}$

2. Find the value.
$\log_q q^{18}$

Objective 2 Apply the product, quotient, and power properties of logarithms.

Review these examples for Objective 2:

3. Use the product rule to write the expression as a sum of logarithms.
$\log_{20}(xyz)$

Use the product rule $\log_b xy = \log_b x + \log_b y$;
$\log_{20}(xyz) = \log_{20} xy + \log_{20} z$
Use the product rule again;
$\log_{20} xy + \log_{20} z = (\log_{20} x + \log_{20} y) + \log_{20} z$
Thus, $\log_{20}(xyz) = \log_{20} x + \log_{20} y + \log_{20} z$.

4. Use the product rule to write the expression as a single logarithm.
$\log_4 6 + \log_4 3$

Use the product rule $\log_b x + \log_b y = \log_b xy$;
$\log_4 6 + \log_4 3 = \log_4(6\cdot 3) = \log_4 18$

Practice these exercises:

3. Use the product rule to write the expression as a sum of logarithms.
$\log_d(aqr)$

4. Use the product rule to write the expression as a single logarithm.
$\log_3 H + \log_3 Z$

Name: Date:
Instructor: Section:

5. Use the quotient rule to write the expression as a difference of logarithms. Leave your answer in simplest form.

$\log_7 \frac{7}{10}$

Use the quotient rule $\log_b \frac{x}{y} = \log_b x - \log_b y$;

$\log_7 \frac{7}{10} = \log_7 7 - \log_7 10$

$= 1 - \log_7 10$ Since $\log_7 7 = 1$

6. Use the quotient rule to write the expression as a single logarithm. Leave your answers in simplest form.

$\log_6 9 - \log_6 x$

Use the quotient rule $\log_b x - \log_b y = \log_b \frac{x}{y}$;

$\log_6 9 - \log_6 x = \log_6 \frac{9}{x}$

7. Use the power rule to write the expression as a multiple of a logarithm.

$\log_3 a^5$

Use the power rule $\log_b x^r = r \log_b x$;

$\log_3 a^5 = 5 \log_3 a$

8. Use the power rule to write the expression as a logarithm of a quantity to a power. Leave your answers in simplest form without negative or fractional exponents.

$6 \log_7 x$

Use the power rule $r \log_b x = \log_b x^r$;

$6 \log_7 x = \log_7 x^6$

5. Use the quotient rule to write the expression as a difference of logarithms. Leave your answer in simplest form.

$\log_6 \frac{13}{7}$

6. Use the quotient rule to write the expression as a single logarithm. Leave your answers in simplest form.

$\log_c 35 - \log_c 5$

7. Use the power rule to write the expression as a multiple of a logarithm.

$\log_N t^{15}$

8. Use the power rule to write the expression as a logarithm of a quantity to a power. Leave your answers in simplest form without negative or fractional exponents.

$2 \log_6 7$

ADDITIONAL EXERCISES

Objective 1 Apply the inverse properties of logarithms.

For extra help, see Example 1 on page 1139 of your text and the Section 17.4 lecture video.

Find the value.

1. $\log_7 7^z$

2. $b^{\log_b 22}$

Name: Date:
Instructor: Section:

Objective 2 Apply the product, quotient, and power properties of logarithms.
For extra help, see Examples 2–9 on pages 1140–1143 of your text and the Section 17.4 lecture video.
Use the product rule to write the expression as a sum of logarithms.

3. $\log_b y(y-4)$

Use the product rule to write the expression as a single logarithm.

4. $\log_3(x+6)+\log_3(x+10)$

Use the quotient rule to write the expression as a difference of logarithms. Leave your answers in simplest form.

5. $\log_d \frac{d}{k}$

Use the quotient rule to write the expression as a single logarithm. Leave your answers in simplest form.

6. $\log_a(x^2-16)-\log_a(x-4)$

Use the power rule to write the expression as a multiple of a logarithm.

7. $\log_7 \sqrt[3]{b}$

Use the power rule to write the expression as a logarithm of a quantity to a power. Leave your answers in simplest form without negative or fractional exponents.

8. $-3\log_2 a$

Write the expression as the sum or difference of multiples of logarithms.

9. $\log_b \frac{xy^7}{z^6}$

Write the expression as a single logarithm. Leave your answer in simplest form without negative or fractional exponents.

10. $4\log_a x+7\log_a z$

Name: Date:
Instructor: Section:

Chapter 17 EXPONENTIAL AND LOGARITHMIC FUNCTIONS

17.5 Common and Natural Logarithms

Learning Objectives
1 Define common logarithms and evaluate them using a calculator.
2 Define natural logarithms and evaluate them using a calculator.
3 Solve applications using common logarithms.
4 Solve applications using natural logarithms.

Key Terms
Use the vocabulary terms listed below to complete each statement in Exercises 1–4.

common logarithms **natural logarithms**

1. Logarithms with a base of 10 are called _______________.

2. _______________ are logarithms with a base of *e*.

GUIDED EXAMPLES AND PRACTICE

Objective 1 Define common logarithms and evaluate them using a calculator.

Review this example for Objective 1:

1. Use a calculator to approximate the logarithm to four decimal places.
$\log 15$

$\log 15 \approx 1.1761$

Practice this exercise:

1. Use a calculator to approximate the logarithm to four decimal places.
$\log 93$

Objective 2 Define natural logarithms and evaluate them using a calculator.

Review this example for Objective 2:

2. Use a calculator to approximate the logarithm to four decimal places.
$\ln 95$

$\ln 95 \approx 4.5539$

Practice this exercise:

2. Use a calculator to approximate the logarithm to four decimal places.
$\ln 65.3$

Name: Date:
Instructor: Section:

Objective 3 Solve applications using common logarithms.

Review this example for Objective 3:

3. Sound intensity can be measured in watts per unit of area or, more commonly, in decibels. The function $d = 10\log\frac{I}{I_0}$ is used to calculate sound intensity, where d represents the intensity in decibels, I represents the intensity in watts per unit of area, and I_0 represents the faintest audible sound to the average human ear (which is about 10^{-12} watts per square meter). A loud radio has a sound intensity of about 10^{-4} watts per square meter. Find the decibel reading for the loud radio.

$d = 10\log\frac{10^{-4}}{10^{-12}}$ Substitute values.

$d = 10\log 10^8$ Subtract exponents.

$d = 10(8)$ Use $\log_b b^x = x$.

$d = 80$ Simplify.

Practice this exercise:

3. *Solve.*
The intensity I of a particular sound has a decibel reading of $d = 10\log\frac{I}{I_0}$. What is the decibel reading of this sound if its intensity is 10^{-7} and $I_0 = 10^{-12}$ watts/m^2?

Objective 4 Solve applications using natural logarithms.

Review this example for Objective 4:

4. An amount of \$3000 is deposited into an account earning 5% annual interest compounded continuously. How many years will it take until the account has reached \$6000? Round your answer to the nearest tenth of a year.

$t = \frac{1}{r}\ln\frac{A}{P}$ Interest formula.

$t = \frac{1}{0.05}\ln\frac{6000}{3000}$ Substitute values.

$t = 20\ln 2$ Divide.

$t = 13.9$ Multiply.

Practice this exercise:

4. An amount of \$1000 is deposited into an account earning 10% annual interest compounded continuously. How many years will it take until the account has reached \$5000? Round your answer to the nearest tenth of a year. Use the formula $t = \frac{1}{r}\ln\frac{A}{P}$.

ADDITIONAL EXERCISES

Objective 1 Define common logarithms and evaluate them using a calculator.

For extra help, see Example 1 on page 1147 of your text and the Section 17.5 lecture video.

Use a calculator to approximate each logarithm to four decimal places.

1. $\log(9.15\times10^{-3})$

2. $\log(5.51\times10^{4})$

Name: Date:
Instructor: Section:

Objective 2 Define natural logarithms and evaluate them using a calculator.
For extra help, see Example 2 on page 1148 of your text and the Section 17.5 lecture video.
Use a calculator to approximate each logarithm to four decimal places.

3. $\ln\left(4.54\times e^{7}\right)$

Objective 3 Solve applications using common logarithms.
For extra help, see Example 3 on pages 1148–1149 of your text and the Section 17.5 lecture video.
The pH of a substance determines whether it is a base (pH > 7) or an acid (pH < 7). To find the pH of a solution, we use the formula $pH = -\log\left[H_3O^+\right]$, *where* $\left[H_3O^+\right]$ *is the hydronium ion concentration in moles per liter.*

4. Find the pH of a certain agricultural product with a hydronium ion concentration of 3.8×10^{-4} moles per liter. Round to the nearest tenth.

5. The pH of a fruit juice is 3.6. Find the hydronium ion concentration of the juice.

Objective 4 Solve applications using natural logarithms.
For extra help, see Example 4 on page 1149 of your text and the Section 17.5 lecture video.
Solve.

6. The average number of miles per gallon for a certain SUV can be approximated by $y = 11.59 + 1.48\ln x$, where y is the average number of miles per gallon and x is the number of years beginning with 1985. ($x = 1$ corresponds with 1985.) Find the approximate average number of miles per gallon in 2008.

Name: Date:
Instructor: Section:

Chapter 17 EXPONENTIAL AND LOGARITHMIC FUNCTIONS

17.6 Exponential and Logarithmic Equations with Applications

Learning Objectives
1 Solve equations that have variables as exponents.
2 Solve equations containing logarithms.
3 Solve applications involving exponential and logarithmic functions.
4 Use the change-of-base formula.

GUIDED EXAMPLES AND PRACTICE

Objective 1 Solve equations that have variables as exponents.

Review this example for Objective 1:

1. Solve. Round your answer to four decimal places.
$7^x = 11$

$\log 7^x = \log 11$	If $x = y$, then $\log_b x = \log_b y$.
$x \log 7 = \log 11$	Use $\log_b x^r = r \log_b x$.
$x = \dfrac{\log 11}{\log 7}$	Divide both sides by $\log 7$.
$x \approx 1.2323$	Use a calculator.

Practice this exercise:

1. Solve. Round your answer to four decimal places.
$2^x = 6$

Objective 2 Solve equations containing logarithms.

Review this example for Objective 2:

2. Solve. Give an exact answer.
$\log_2(5x+3) = 2$

The equation is in the form $\log_b x = y$ with $b = 2$, $y = 2$, and $x = 5x+3$; so write it in exponential form, $b^y = x$.

$2^2 = 5x+3$	Exponential form.
$4 = 5x+3$	Evaluate 2^2.
$1 = 5x$	Subtract 3 from both sides.
$\dfrac{1}{5} = x$	Solve for x.

Practice this exercise:

2. Solve. Give an exact answer.
$\log_3(3x-6) = 3$

Name: Date:
Instructor: Section:

Objective 3 Solve applications involving exponential and logarithmic functions.

Review this example for Objective 3:

3. How long will it take $7500 invested at 5% interest compounded quarterly to grow to $15,000? Round the answer to the nearest tenth of a year.

The interest formula is below.

$$A = P\left(1+\frac{r}{n}\right)^{nt}$$

Replace A with 15,000, P with 7500, r with 0.05, and n with 4.

$$15{,}000 = 7500\left(1+\frac{0.05}{4}\right)^{4t}$$

Divide both sides by 7500 and simplify inside the parentheses.

$$2 = (1.0125)^{4t}$$

Use $\log_b x^r = r\log_b x$.

$$\log 2 = 4t\log 1.0125$$

Divide both sides by $4\log 1.0125$.

$$\frac{\log 2}{4\log 1.0125} = t$$

Evaluate using a calculator and round.

$t \approx 13.9$

Practice this exercise:

3. How long will it take for $28,000 to grow to $30,616,41 at 6% interest compounded quarterly? Round to the nearest tenth of a year. Use the formula $A = P(1+\frac{r}{n})^{nt}$.

Objective 4 Use the change-of-base formula.

Review this example for Objective 4:

4. Use the change-of-base formula to find the logarithm. Round the answer to four decimal places.

$\log_3 15$

$\log_3 15 = \dfrac{\log 15}{\log 3}$ Use $\log_a x = \dfrac{\log_b x}{\log_b a}$.

≈ 2.4650 Evaluate using a calculator.

Practice this exercise:

4. Use the change-of-base formula to find the logarithm. Round the answer to four decimal places.

$\log_7 10$

Name: Date:
Instructor: Section:

ADDITIONAL EXERCISES

Objective 1 Solve equations that have variables as exponents.

For extra help, see Examples 1–2 on pages 1153–1154 of your text and the Section 17.6 lecture video.

Solve. Round your answers to four decimal places.

1. $4^{-x+4} = 85$

2. $e^{0.006x} = 29$

3. $e^{-0.596x} = 5$

Objective 2 Solve equations containing logarithms.

For extra help, see Example 3 on pages 1154–1155 of your text and the Section 17.6 lecture video.

Solve. Give exact answers.

4. $\log x + \log(x+15) = 2$

5. $\log_3(x) + \log_3(x-1) = \log_3 2$

6. $\log 8x - \log(2x-3) = \log 9$

7. $\log_6(4t+7) - \log_6 t = \log_6 5$

Objective 3 Solve applications involving exponential and logarithmic functions.

For extra help, see Examples 4–7 on pages 1155–1157 of your text and the Section 17.6 lecture video.

Solve.

8. Suppose that \$70,000 is invested at 5% interest.

a) Find the amount of money in the account after 7 years if the interest is compounded annually.

b) Find the amount of money in the account after 7 years if the interest is compounded continuously.

9. The population of a city was 98 thousand in 1992. The exponential growth rate was 1.8% per year. The population t years after 1992 is given by $P = 98{,}000e^{0.018t}$.

a) Estimate the population in 2006 to the nearest thousand.

b) Find the time at which the population reached 110 thousand. Round to the nearest year.

Objective 4 Use the change-of-base formula.

For extra help, see Examples 8 and 9 on page 1158 of your text and the Section 17.6 lecture video.

Use $P = 95 - 30\log_2 x$ and the change-of-base formula to find the percentage, P, of the lecture retained after the given number of days. Round your answer to the nearest whole percent.

10. 6 days

Name: Date:
Instructor: Section:

Chapter 18 CONIC SECTIONS

18.1 Parabolas and Circles

Learning Objectives

1. Graph parabolas in the form $x = a(y-k)^2 + h$.
2. Find the distance and midpoint between two points.
3. Graph circles of the form $(x-h)^2 + (y-k)^2 = r^2$.
4. Find the equation of a circle with a given center and radius.
5. Graph circles of the form $x^2 + y^2 + dx + ey + f = 0$.

Key Terms

Use the vocabulary terms listed below to complete each statement in Exercises 1–3.

conic section **circle** **radius**

1. A set of points that are equally distant from a central point, where the central point is the center, is a(n) ________________.

2. A(n) ________________ is a curve in a plane that is the result of intersecting the plane with a cone, or more specifically, a circle, an ellipse, a parabola, or a hyperbola.

3. The distance from the center of a circle to any point on the circle is a(n) ________________.

GUIDED EXAMPLES AND PRACTICE

Objective 1 Graph parabolas in the form $x = a(y-k)^2 + h$.

Review this example for Objective 1:

1. Determine whether the graph opens left or right, find the vertex and axis of symmetry, and draw the graph.

$x - -y^2 + 2y + 1$

First, write the equation in the form $x = a(y-k)^2 + h$, where (h, k) is the vertex and $y = k$ is the axis of symmetry.

Start by subtracting 1 from both sides.

$x - 1 = -y^2 + 2y$

Factor out the –1 from the right side of the equation.

$x - 1 = -(y^2 - 2y)$

Complete the square by adding –1 to both sides of the equation.

$x - 1 - 1 = -(y^2 - 2y + 1)$

Practice this exercise:

1. Determine whether the graph opens left or right, find the vertex and axis of symmetry, and draw the graph.

$x = y^2 + 6y + 3$

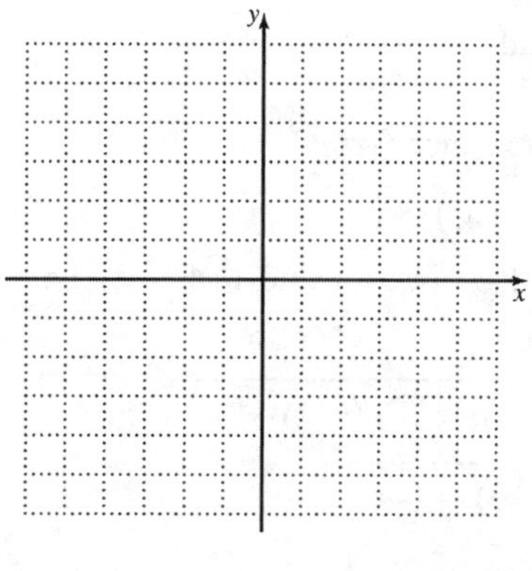

Name:
Instructor:

Date:
Section:

Simplify the left side and factor the right side.

$x-2=-(y-1)^2$

Isolate the variable, x.

$x=-(y-1)^2+2$

So, the vertex is (2, 1) and the axis of symmetry is $y=1$.

To complete the graph, let y equal values on each side of the axis of symmetry and find x.

x	y
−2	3
1	2
2	1
1	0
−2	−1

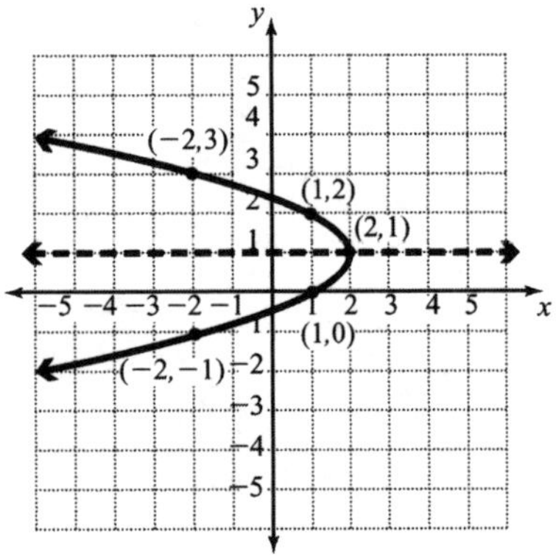

Objective 2 Find the distance and midpoint between two points.

Review these examples for Objective 2:

2. Find the distance and midpoint between the two points.

(6, 2) and (−4, −2)

Use the distance formula:

$d=\sqrt{(x_2-x_1)^2+(y_2-y_1)^2}$

Let (6, 2) be (x_1, y_1) and (−4, −2) be (x_2, y_2).

$d=\sqrt{((-4)-6)^2+((-2)-2)^2}$

$d=\sqrt{(-10)^2+(-4)^2}$

$d=\sqrt{100+16}$

$d=\sqrt{116}$

$d\approx 10.77$

Practice these exercises:

2. Find the distance and midpoint between the two points.

$(-10, -2)$ and $(10, -17)$

Name: Date:
Instructor: Section:

Use the midpoint formula.

$$\text{Midpoint} = \left(\frac{x_1 + x_2}{2}, \frac{y_1 + y_2}{2}\right)$$

Let $(6, 2)$ be (x_1, y_1) and $(-4, -2)$ be (x_2, y_2).

$$\text{Midpoint} = \left(\frac{6+(-4)}{2}, \frac{2+(-2)}{2}\right)$$

$$\text{Midpoint} = \left(\frac{2}{2}, \frac{0}{2}\right)$$

$$\text{Midpoint} = (1, 0)$$

Objective 3 Graph circles of the form $(x-h)^2 + (y-k)^2 = r^2$.

Review this example for Objective 3:

3. Find the center and radius of each circle and draw the graph.

$(x-1)^2 + (x+4)^2 = 4$

First, write the equation in the form $(x-h)^2 + (y-k)^2 = r^2$.

$(x-1)^2 + (x-(-4))^2 = 2^2$

$h = 1$, $k = -4$, and $r = 2$, so the center is $(1, -4)$ and the radius is 2.

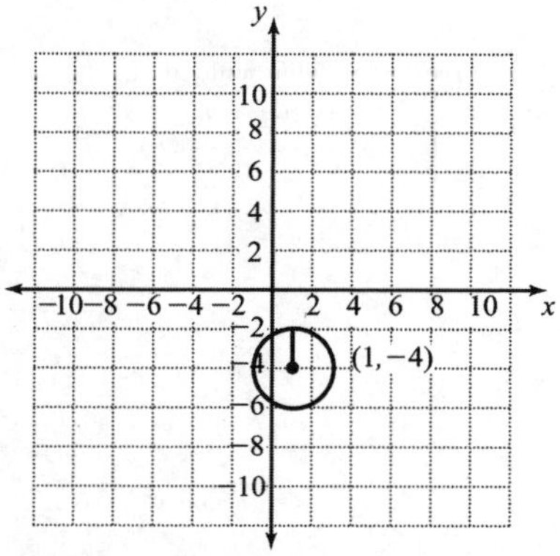

Practice this exercise:

3. Find the center and radius of each circle and draw the graph.

$(x-3)^2 + (y-2)^2 = 36$

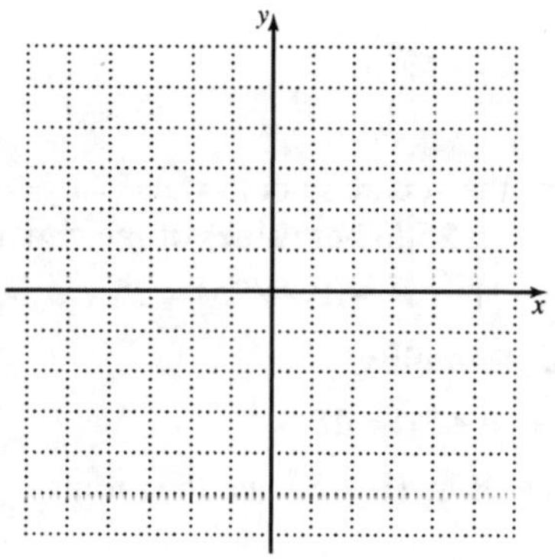

Name: Date:
Instructor: Section:

Objective 4 Find the equation of a circle with a given center and radius.

Review this example for Objective 4:

4. The center and radius of a circle are given. Write the equation of the circle in standard form.

Center: $(1, 7)$; radius: $\sqrt{7}$

Because the center is at $(1, 7)$, $h = 1$ and $k = 7$. Also $r = \sqrt{7}$. Plug these into the standard form of a circle.

$(x-h)^2 + (y-k)^2 = r^2$

$(x-1)^2 + (y-7)^2 = \sqrt{7}^2$

$(x-1)^2 + (y-7)^2 = 7$

Practice this exercise:

4. The center and radius of a circle are given. Write the equation of the circle in standard form.

Center: $(9, 4)$; radius: 2

Objective 5 Graph circles of the form $x^2 + y^2 + dx + ey + f = 0$.

Review this example for Objective 5:

5. Find the center and radius and draw the graph.

$x^2 + y^2 - 2x + 10y = 0$

Group x and y terms.

$(x^2 - 2x) + (y^2 + 10y) = 0$

Complete the square in both x and y by adding 1 and 25 to both sides of the equation.

$(x^2 - 2x + 1) + (y^2 + 10y + 25) = 25$

Factor and simplify.

$(x-1)^2 + (y+5)^2 = 25$

The center is at $(1, -5)$ and the radius is at 5.

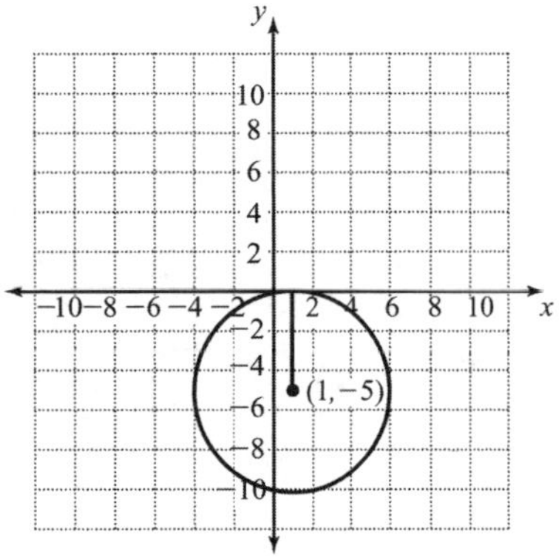

Practice this exercise:

5. Find the center and radius and draw the graph.

$x^2 + y^2 - 6x + 4y = 23$

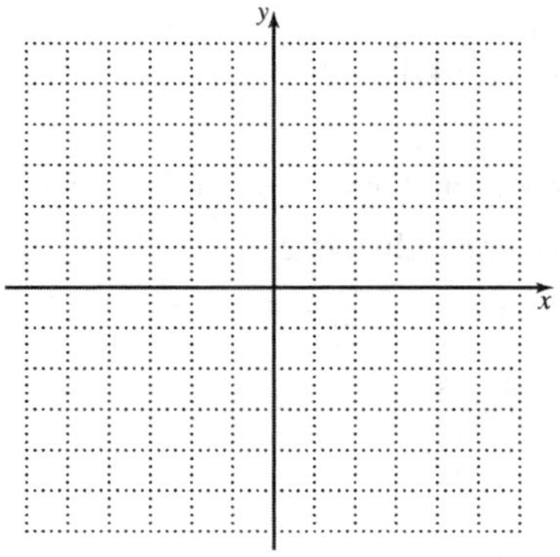

Name: Date:
Instructor: Section:

ADDITIONAL EXERCISES

Objective 1 Graph parabolas in the form $x = a(y-k)^2 + h$.

For extra help, see Example 2 on pages 1179-1180 of your text and the Section 18.1 lecture video.
Find the direction the parabola opens, the coordinates of the vertex, and the equation of the axis of symmetry and draw the graph.

1. $y = -x^2 - 4x - 3$

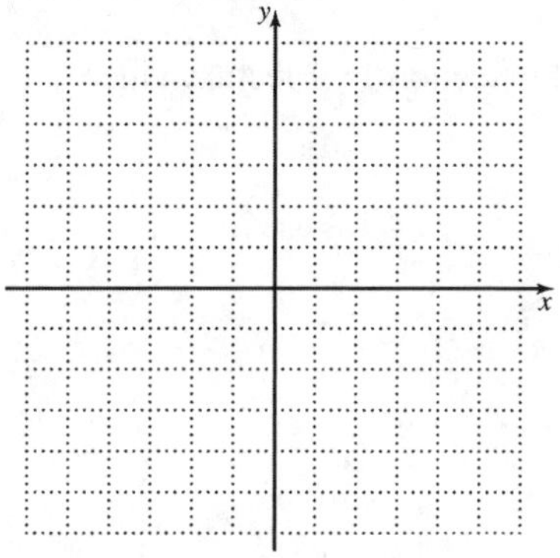

Objective 2 Find the distance and midpoint between two points.

For extra help, see Examples 3 and 4 on page 1181 of your text and the Section 18.1 lecture video.
Find the distance and midpoint between the two points.

2. $(5, 1)$ and $(2, -6)$

3. $(3, 2)$ and $(-3, -3)$

Objective 3 Graph circles of the form $(x-h)^2 + (y-k)^2 = r^2$.

For extra help, see Example 5 on page 1182 of your text and the Section 18.1 lecture video.
Find the center and radius and draw the graph.

4. $(x+2)^2 + (y+1)^2 = 9$

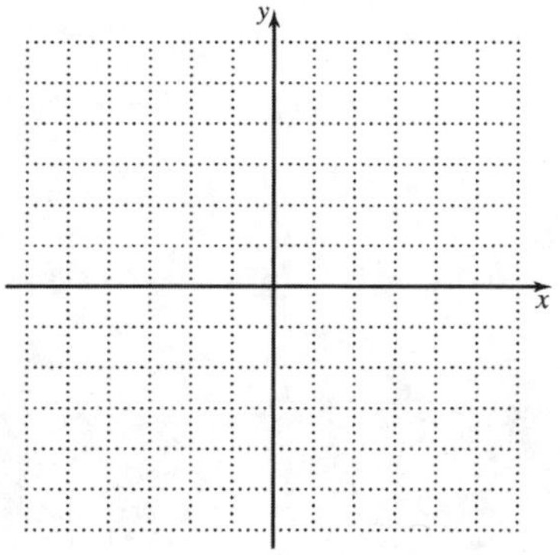

Name: Date:
Instructor: Section:

Objective 4 Find the equation of a circle with a given center and radius.
For extra help, see Example 6 on page 1183 of your text and the Section 18.1 lecture video.
The center and radius of a circle are given. Write the equation of the circle in standard form.

5. Center: $(-1, 3)$; radius: $\sqrt{6}$

6. Center: $(2, 2)$; radius: $\sqrt{51}$

Objective 5 Graph circles of the form $x^2 + y^2 + dx + ey + f = 0$.
For extra help, see Example 7–8 on pages 1183–1184 of your text and the Section 18.1 lecture video.
Find the center and radius and draw the graph.

7. $x^2 - 8x + y^2 + 4y + 16 = 0$

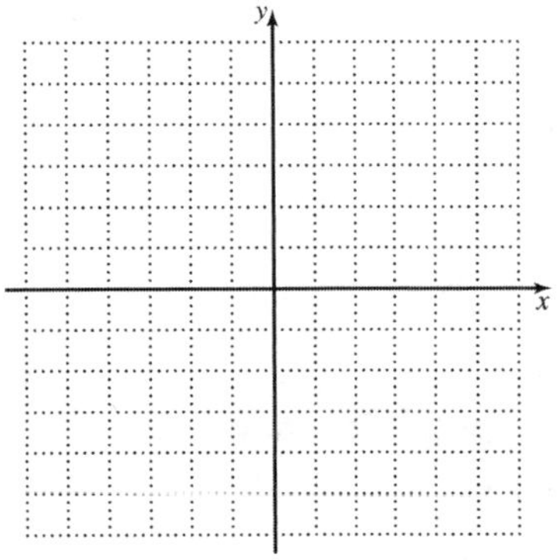

Name: Date:
Instructor: Section:

Chapter 18 CONIC SECTIONS

18.2 Ellipses and Hyperbolas

Learning Objectives
1 Graph ellipses.
2 Graph hyperbolas.

Key Terms
Use the vocabulary terms listed below to complete each statement in Exercises 1 and 2.

ellipse **hyperbola**

1. A(n) ___________________ is defined as the set of all points the difference of whose distances from two fixed points remains constant.

2. The set of all points the sum of whose distance from two fixed points remains constant is a(n) ___________________.

GUIDED EXAMPLES AND PRACTICE

Objective 1 Graph ellipses.

Review these examples for Objective 1:

1. Graph the ellipse. Label the x- and y-intercepts.
$x^2 + 9y^2 = 81$

First, write the equation in standard form.
$\dfrac{x^2}{a^2} + \dfrac{y^2}{b^2} = L$

Divide both sides by 81.
$\dfrac{x^2}{81} + \dfrac{9y^2}{81} = \dfrac{81}{81}$

$\dfrac{x^2}{81} + \dfrac{y^2}{9} - 1$

We can now see that this is an ellipse with $a = 9$ and $b = 3$, so the x-intercepts are $(9, 0)$ and $(-9, 0)$ and the y-intercepts are $(0, 3)$ and $(0, -3)$.

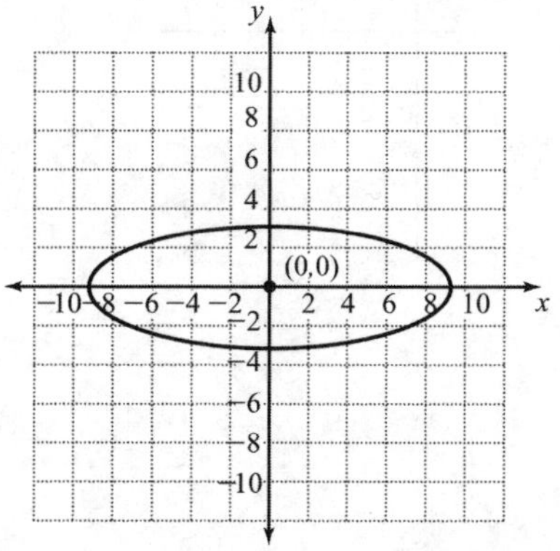

Practice these exercises:

1. Graph the ellipse. Label the x- and y-intercepts.
$\dfrac{x^2}{9} + \dfrac{y^2}{4} = 1$

Name: Date:
Instructor: Section:

2. Graph the ellipse. Label the center and the points directly above, below, to the left of, and to the right of the center.

$$\frac{(x-1)^2}{25}+\frac{(y+5)^2}{4}=1$$

The center of the ellipse is $(h, k)=(1, -5)$. Also, $a=5$ and $b=2$ so the ellipse passes through $(6, -5)$, $(-4, -5)$, $(1, -3)$, and $(1, -7)$.

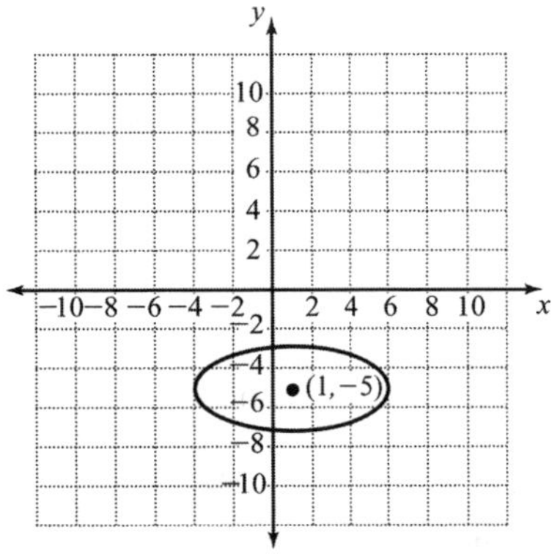

2. Graph the ellipse. Label the center and the points directly above, below, to the left of, and to the right of the center.

$$\frac{(x-3)^2}{16}+\frac{(y+2)^2}{9}=1$$

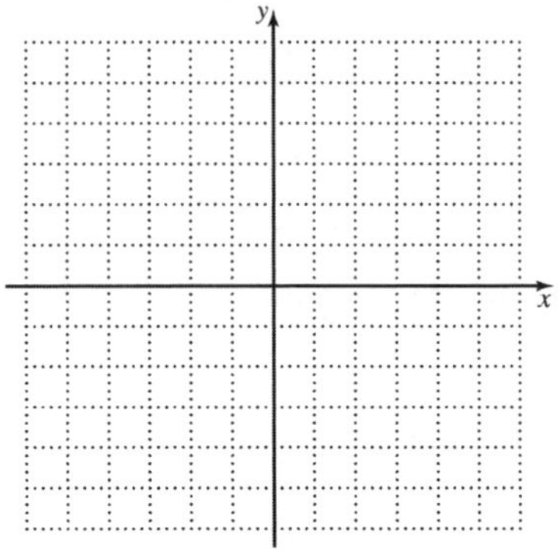

Objective 2 Graph hyperbolas.

Review this example for Objective 2:

3. Graph the hyperbola. Also show the fundamental rectangle with its corner points labeled, the asymptotes, and the intercepts.

$$\frac{x^2}{36}-\frac{y^2}{25}=1$$

The equation can be written as $\frac{x^2}{6^2}-\frac{y^2}{5^2}=1$, so we can see that $a=6$ and $b=5$. Because the x^2 term is positive, the graph has the x-intercepts at $(6, 0)$ and $(-6, 0)$. Therefore, the fundamental rectangle has vertices at $(6, 5)$, $(6, -5)$, $(-6, 5)$, and $(-6, -5)$.

Practice this exercise:

3. Graph the hyperbola. Also show the fundamental rectangle with its corner points labeled, the asymptotes, and the intercepts.

$$\frac{x^2}{4}-\frac{y^2}{16}=1$$

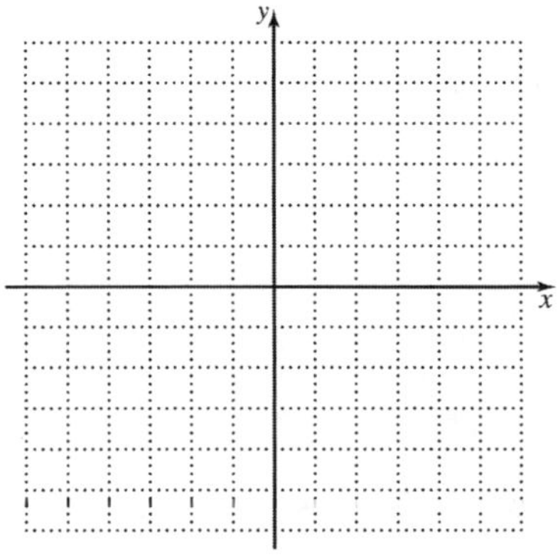

Name: Date:
Instructor: Section:

Now, we are able to draw the asymptotes and sketch the graph so it passes through the x-intercepts and approaches the asymptotes.

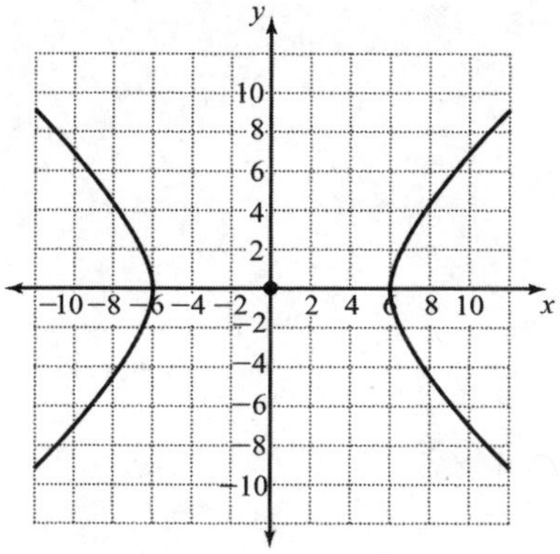

ADDITIONAL EXERCISES

Objective 1 Graph ellipses.

For extra help, see Examples 1–3 on pages 1192–1193 of your text and the Section 18.2 lecture video.

Graph each ellipse. Label the x- and y-intercepts.

1. $16x^2 + 4y^2 = 64$

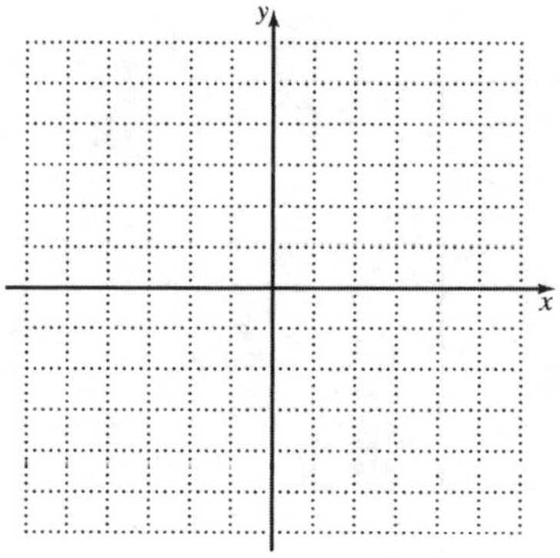

Name:
Instructor:

Date:
Section:

Graph the ellipse. Label the center and the points directly above, below, to the left, and to the right of the center.

2. $\dfrac{(x-3)^2}{4}+\dfrac{(y+3)^2}{9}=1$

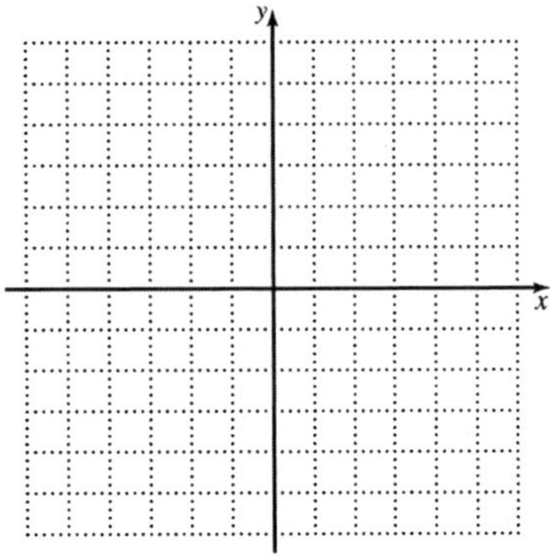

Objective 2 Graph hyperbolas.

For extra help, see Example 4 on page 1195 of your text and the Section 18.2 lecture video.

Graph each hyperbola. Also show the fundamental rectangle with its corner points labeled, the asymptotes, and the intercepts.

3. $16x^2-y^2=64$

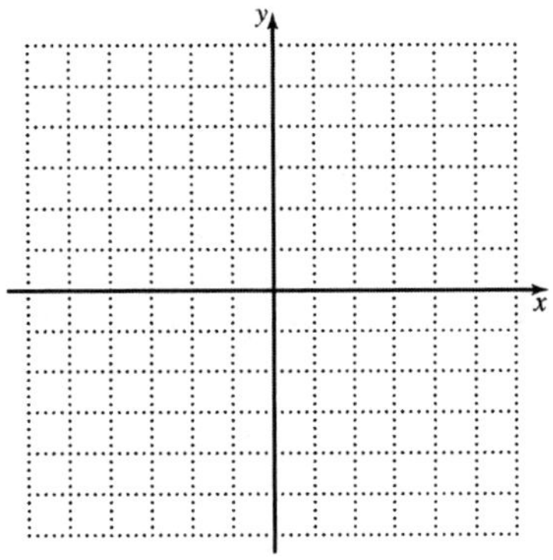

4. $25y^2-9x^2=225$

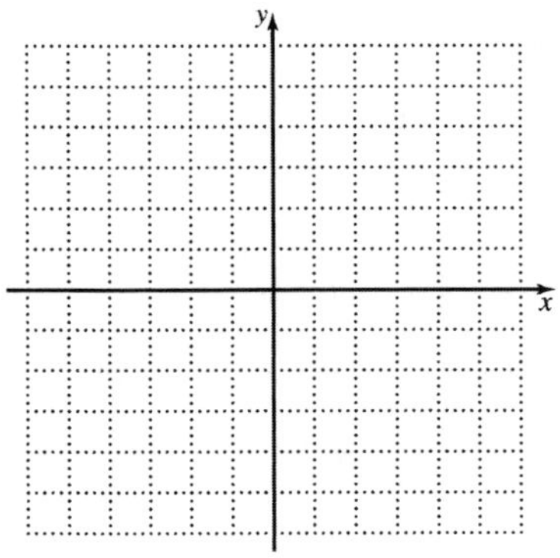

Name: Date:
Instructor: Section:

Chapter 18 CONIC SECTIONS

18.3 Nonlinear Systems of Equations

Learning Objectives
1 Solve nonlinear systems of equations using substitution.
2 Solve nonlinear systems of equations using elimination.

Key Terms
Use the vocabulary term listed below to complete the statement in Exercise 1.

nonlinear systems of equations

1. Some systems have equations that contain at least one nonlinear equation. These are called ____________________.

GUIDED EXAMPLES AND PRACTICE

Objective 1 Solve nonlinear systems of equations using substitution.

Review this example for Objective 1:

1. Solve the system by substitution.

$$\begin{cases} y = x^2 + 7x + 10 \\ x + y = 3 \end{cases}$$

First, solve $x + y = 3$ for *y*.

$y = 3 - x$

Next, substitute this expression for *y* in the first equation.

$y = x^2 + 7x + 10$

$3 - x = x^2 + 7x + 10$

Write the equation in the form $ax^2 + bx + c = 0$.

$-x^2 - 8x - 7 = 0$

$-(x^2 + 8x + 7) = 0$

$x^2 + 8x + 7 = 0$

Factor the left-hand side.

$(x+7)(x+1) = 0$

Use the zero-factor theorem.

$x + 7 = 0 \qquad x + 1 = 0$

Solve each equation.

$x + 7 = 0 \rightarrow x = -7$

$x + 1 = 0 \rightarrow x = -1$

To find *y*, substitute these solutions for *x* in either of the original equations.

When $x = -7$, $(-7) + y = 3 \rightarrow y = 10$

When $x = -1$, $(-1) + y = 3 \rightarrow y = 4$

Therefore, the solutions are $(-7, 10)$ and $(-1, 4)$.

Practice this exercise:

1. Solve the system by substitution.

$$\begin{cases} x^2 + y = 27 \\ 3x - y = 1 \end{cases}$$

Name: Date:
Instructor: Section:

Objective 2 Solve nonlinear systems of equations using elimination.

Review this example for Objective 2:

2. Solve the system by elimination.

$$\begin{cases} x^2 + y^2 = 106 \\ x^2 - y^2 = 56 \end{cases}$$

The graphs of these equations are a circle and hyperbola. To eliminate y^2, add the equations.

$$\begin{array}{l} x^2 + y^2 = 106 \\ \underline{x^2 - y^2 = \;\;56} \\ 2x^2 \qquad = 162 \end{array}$$

Divide both sides by 2.

$$2x^2 = 162$$
$$x^2 = 81$$

Find the square roots.

$$x = \pm\sqrt{81}$$
$$x = \pm\, 9$$

To find y, substitute the values for x in one of the original equations.

$$x^2 + y^2 = 106$$
$$(-9)^2 + y^2 = 106$$
$$81 + y^2 = 106$$

Subtract 81 from both sides.

$$y^2 = 25$$
$$y = \pm\sqrt{25}$$
$$y = \pm\, 5$$

Therefore, the solutions are $(9, 5)$, $(-9, 5)$, $(9, -5)$, and $(-9, -5)$.

Practice this exercise:

2. Solve the system by elimination.

$$\begin{cases} x^2 + y^2 = 100 \\ x^2 - y^2 = -28 \end{cases}$$

ADDITIONAL EXERCISES

Objective 1 Solve nonlinear systems of equations using substitution.

For extra help, see Example 1 on pages 1203-1204 of your text and the Section 18.3 lecture video.

Solve.

1. $\begin{cases} y = x^2 + 3x + 9 \\ x + y = 6 \end{cases}$

2. $\begin{cases} x^2 + y^2 = 100 \\ y - x = 2 \end{cases}$

3. $\begin{cases} y = x^2 - 24 \\ y = 2x \end{cases}$

4. $\begin{cases} x^2 + 225y^2 = 25 \\ x = y^2 + 5 \end{cases}$

Name: Date:
Instructor: Section:

Objective 2 Solve nonlinear systems of equations using elimination.

For extra help, see Example 2 on pages 1204-1205 of your text and the Section 18.3 lecture video.

Solve.

5. $\begin{cases} x^2 + y^2 = 20 \\ 3x^2 - y^2 = 44 \end{cases}$

6. $\begin{cases} xy = 100 \\ 16x^2 + y^2 = 800 \end{cases}$

7. $\begin{cases} y = x^2 + 2x - 3 \\ y = -x^2 + 4x + 1 \end{cases}$

8. $\begin{cases} -2x^2 + 5y^2 = 75 \\ 4x^2 + 3y^2 = 175 \end{cases}$

Name:
Instructor:

Date:
Section:

Chapter 18 CONIC SECTIONS

18.4 Nonlinear Inequalities and Systems of Inequalities

Learning Objectives	
1	Graph nonlinear inequalities.
2	Graph the solution set of a system of nonlinear inequalities.

GUIDED EXAMPLES AND PRACTICE

Objective 1 Graph nonlinear inequalities.

Review these examples for Objective 1:

1. Graph the inequality.

$$\frac{x^2}{16}+\frac{y^2}{4}\geq 1$$

The related equation is an ellipse with center $(0,0)$, $a=4$ and $b=2$ because the inequality is $\geq$, all ordered pairs on the ellipse are in the solution set, and we draw it with a solid curve. We need an ordered pair from each region to determine which region to shade.
Let's choose $(2, 0)$, which is in the ellipse, and $(5, 1)$, which is outside the ellipse.
Plug in the ordered pairs to the inequality determine the shaded region. Let's start with $(2,0)$.

$$\frac{x^2}{16}+\frac{y^2}{4}\geq 1$$
$$\frac{2^2}{16}+\frac{0^2}{4}\geq 1$$
$$\frac{1}{4}\geq 1 \qquad \text{False}$$

Now, plug in $(5, 1)$.

$$\frac{x^2}{16}+\frac{y^2}{4}\geq 1$$
$$\frac{5^2}{16}+\frac{1^2}{4}\geq 1$$
$$\frac{25}{16}+\frac{1}{4}\geq 1$$
$$\frac{29}{16}\geq 1 \qquad \text{True}$$

Practice this exercise:

1. Graph the inequality.

$x^2+y^2<25$

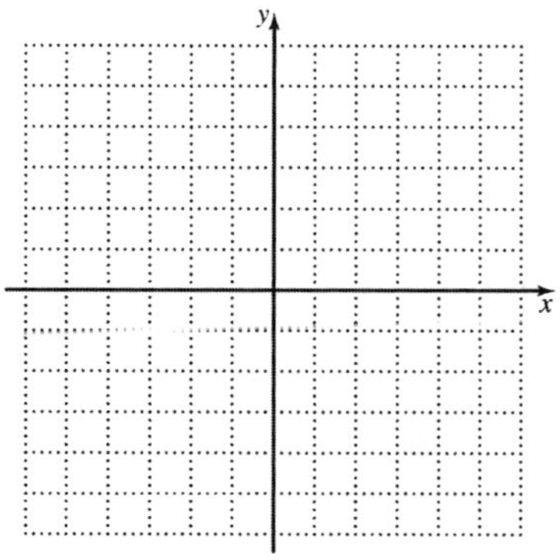

Name: Date:
Instructor: Section:

Since (5, 1) is true, the region outside the ellipse is shaded.

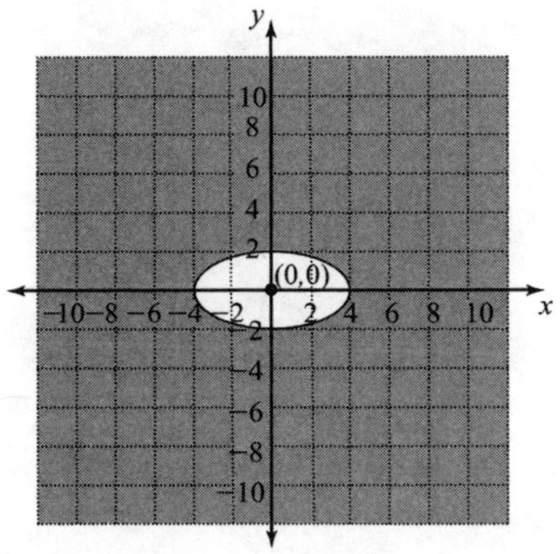

Objective 2 Graph the solution set of a system of nonlinear inequalities.

Review this example for Objective 2:

2. Graph the solution set of the system of inequalities.

$$\begin{cases} \frac{x^2}{9} - \frac{y^2}{9} > 1 \\ y > 1 \end{cases}$$

Graph each inequality separately. For each inequality, begin by graphing the related equation, and determine whether the graph should be solid or dashed. Then, use a test point to determine which region to shade.

First, lets look at $\frac{x^2}{9} - \frac{y^2}{9} > 1$.

$\frac{x^2}{3^2} - \frac{y^2}{3^2} > 1$, so we can see that $a = 3$ and $b = 3$. Because the x^2 term is positive, the graph has the x-intercepts at (3, 0) and (−3, 0). Since the inequality is a $>$, the curves will be dashed.

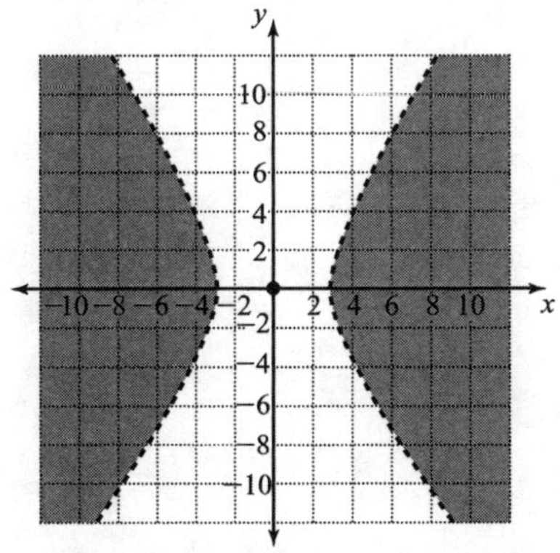

Practice this exercise:

2. Graph the solution set of the system of inequalities.

$$\begin{cases} y \le x^2 \\ 3x - 4y < 12 \end{cases}$$

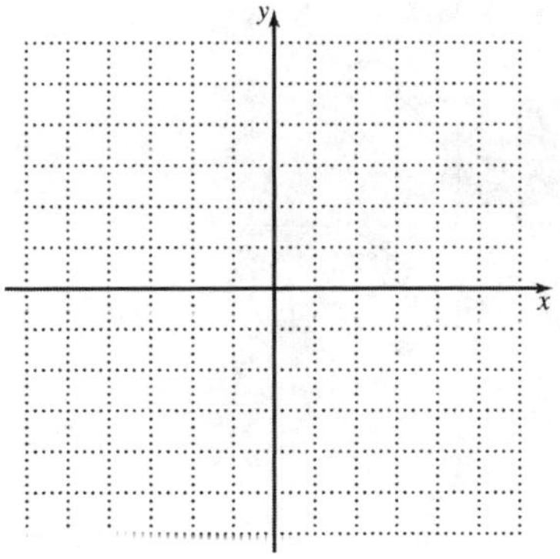

Now, graph $y > 1$. This graph is simply a straight line, with all values for y being true. This line will be dashed since the inequality is a $>$.

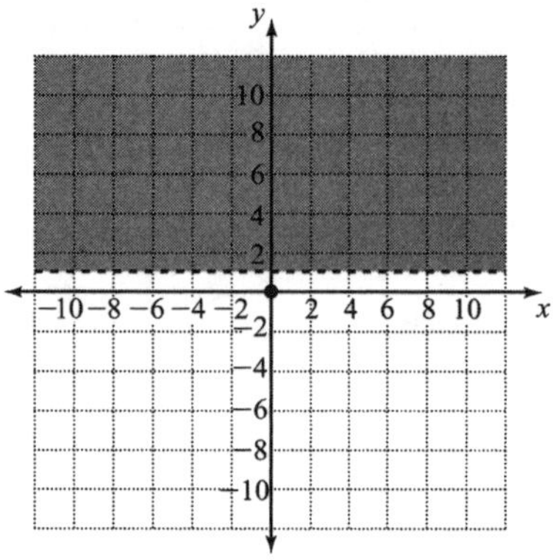

Finally, graph both inequalities on the same graph. The solution set for the system contains all ordered pairs in the region where the two graphs overlap.

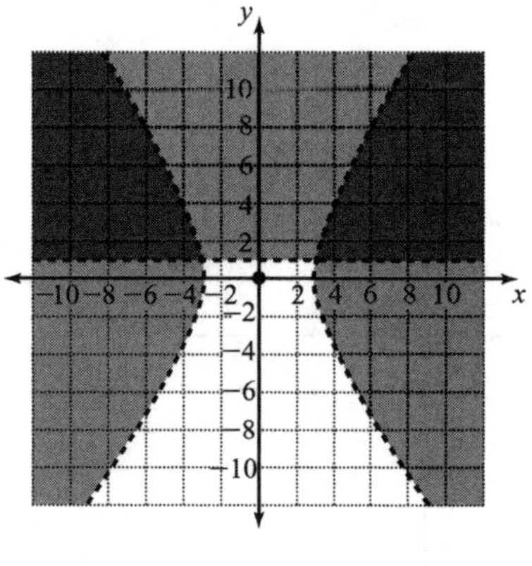

Name: Date:
Instructor: Section:

ADDITIONAL EXERCISES

Objective 1 Graph nonlinear inequalities.

For extra help, see Example 1 on pages 1209-1210 of your text and the Section 18.4 lecture video.

Graph each inequality.

1. $y \geq x^2 - 5$

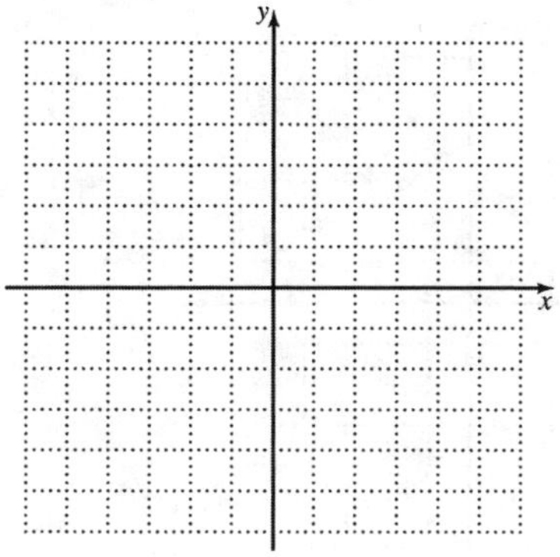

2. $\frac{x^2}{9} + \frac{y^2}{4} < 1$

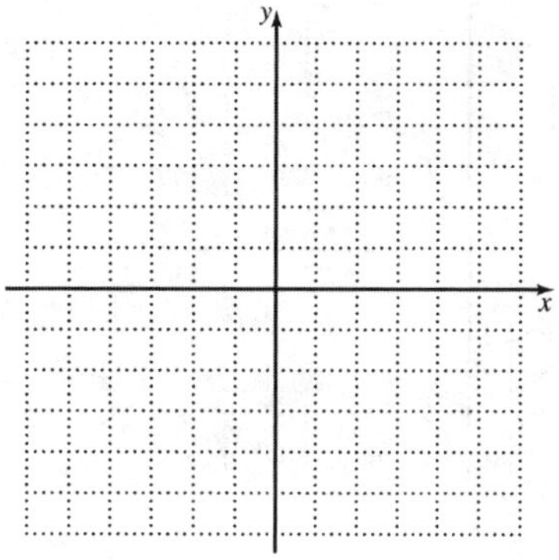

3. $y^2 - \frac{x^2}{4} \geq 1$

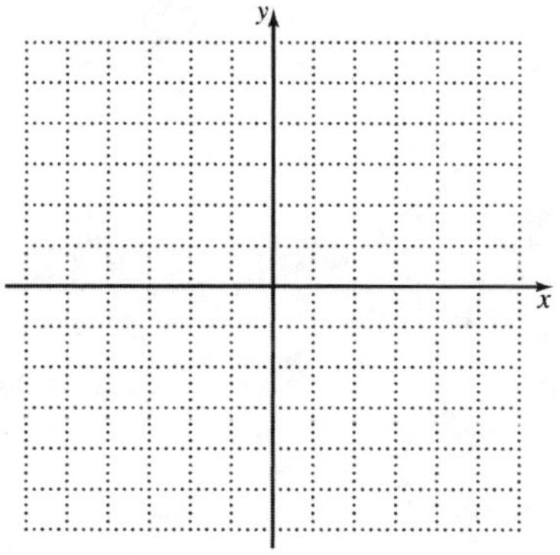

Name: Date:
Instructor: Section:

Objective 2 Graph the solution set of a system of nonlinear inequalities.
For extra help, see Example 2 on pages 1210-1211 of your text and the Section 18.4 lecture video.
Graph the solution set of each system of inequalities.

4. $\begin{cases} -x-3y>3 \\ x^2+y^2<25 \end{cases}$

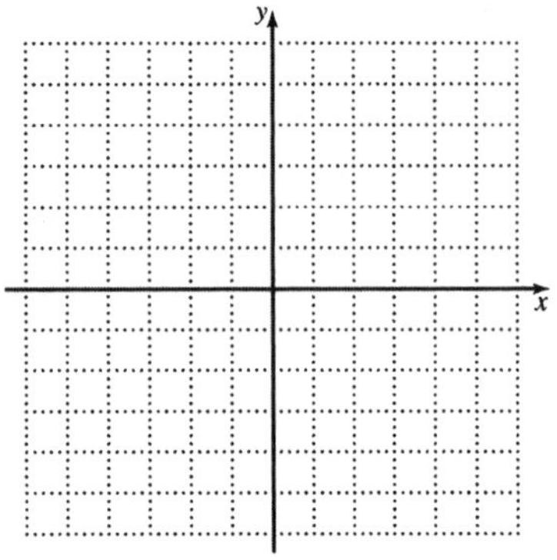

5. $\begin{cases} x^2+y^2\geq 16 \\ x^2+y^2<36 \end{cases}$

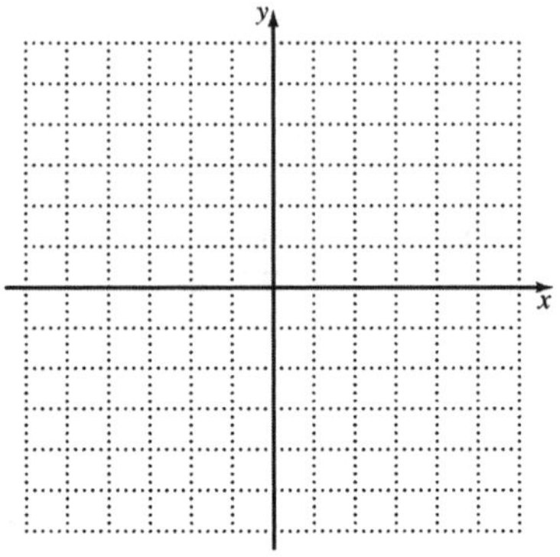

6. $\begin{cases} y\geq x^2-4 \\ y\leq -x^2+2 \end{cases}$

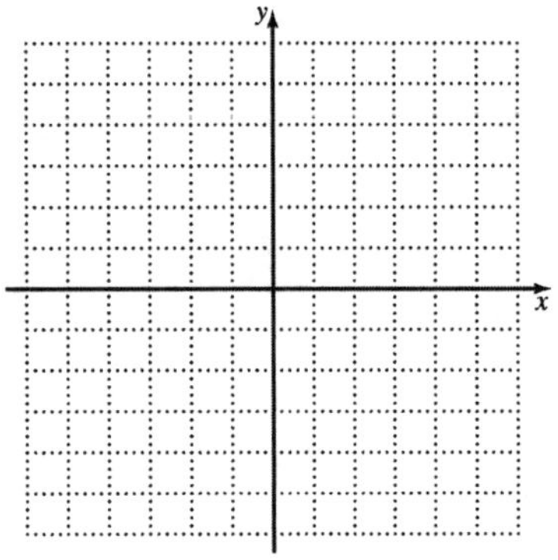

Complete Answers to MyWorkBook

Chapter 1

Section 1.1

Key Terms

1. whole numbers
2. set
3. statistic
4. natural numbers
5. inequality
6. subset
7. equation

Practice

1. 8
2. billions
3. 3 ten millions + 5 millions + 6 hundred thousands + 9 thousands + 9 hundreds + 7 tens + 2 ones
4. 90,238
5. four hundred sixteen thousand, nine hundred ninety-five
6. 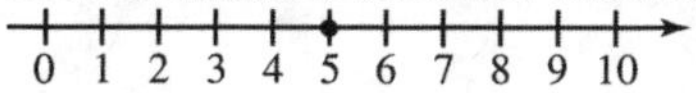
7. <
8. 456,583,100
9. 300 votes

Objective 1

1. 0
2. 9
3. ones
4. ten thousands

Objective 2

5. 1 ten thousands + 4 thousands + 3 hundreds +2 tens +1 ones
6. 7 hundred thousands + 2 ten thousands +5 hundreds +1 tens + 6 ones
7. 90,238
8. 503,572

Objective 3

9. three hundred fifteen thousands, four hundred ninety
10. 2 million, five hundred sixty eight thousand, one hundred fifty-seven

Objective 4

11. 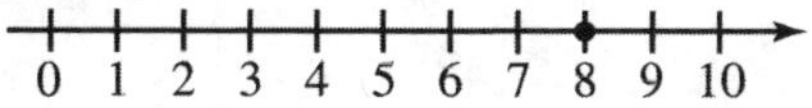
12. 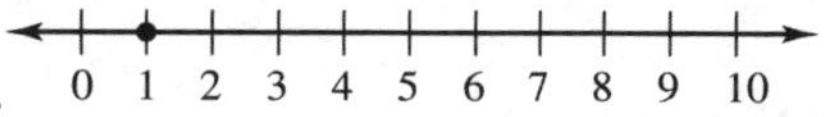

Objective 5

13. =
14. >

Objective 6

15. 19,000,000
16. 18,921,000

Objective 7

17. $6730; $6145
18. 2010; the cost of tuition keeps increasing

Section 1.2

Key Terms

1. solution
2. perimeter
3. variable
4. related equation
5. subtraction
6. addition
7. constant

Practice

1. 71,439
2. estimate: 90,000; actual: 89,778
3. $27,150
4. 50 ft
5. 455
6. $r = 13$
7. 1567 people
8. 30 GB

Objective 1

1. 483
2. 165,752
3. 96
4. 8788

Objective 2

5. estimate: 112,000; actual: 112,079
6. estimate: 11,780; actual: 11,782

Objective 3

7. $121
8. 62 ft.

Objective 4

9. 5377
10. 73,377

Objective 5

11. $t = 56$
12. $x = 138$

Objective 6

13. 118 squares
14. $65

Objective 7

15. $10,559
16. 67 degrees

Section 1.3

Key Terms
1. area
2. exponential form
3. rectangular array
4. base
5. multiplication
6. distributive property
7. formula
8. exponent
9. square unit

Practice
1. 50
2. estimate: 16,000; actual: 15,516
3. 1060 mi.
4. 10,000
5. 3^{11}
6. $6\times10^7+2\times10^6+8\times10^5+1\times10^3+4\times10+3\times1$
7. 17,576,000 tags

Objective 1
1. 62,624
2. 2,848,272

Objective 2
3. estimate: 320,000; actual: 358,224
4. estimate: 72,000; actual: 69,996

Objective 3
5. 840 cupcakes
6. 630 nails

Objective 4
7. 1
8. 81
9. 343
10. 64

Objective 5
11. 21^4
12. 1^7
13. 15^3
14. 7^9
15. $7\times10^5+8\times10^4+5\times10^3+4\times10^2+6\times10+3\times1$
16. $5\times10^4+9\times10^3+2\times10^2+1\times1$

Objective 6
17. 378
18. 600

Section 1.4

Key Terms
1. divisor
2. division
3. dividend
4. remainder

Practice

1. 0; because $8 \cdot 0 = 0$
2. no
3. yes
4. 986 r2
5. $x = 11$
6. 70 bags can be filled; 17 kg will be left over.

Objective 1

1. 27 because $27 \times 1 = 27$
2. undefined because there is no number that can make $? \times 0 = 15$
3. no
4. yes
5. no
6. yes
7. yes
8. no
9. 146 r4
10. 401 r31
11. 320
12. 125 r3

Objective 2

13. $t = 0$
14. $n = 3$
15. $x = 9$
16. no solution

Objective 3

17. 4 feet
18. $37.25

Section 1.5

Key Terms

1. mean (or average)
2. average
3. mode
4. median

Practice

1. 37
2. 24
3. mean: 54; median 54; mode: 54

Objective 1

1. 2
2. 5
3. 640
4. 36
5. 76
6. 7
7. 46
8. 1
9. 3060
10. 1

Objective 2

11. mean: 79; median: 79; mode: 78 and 80
12. mean: 155; median: 162; mode: 187
13. mean: 555; median: 562; mode: none
14. mean: 515; median: 487; mode: 487

Section 1.6

Key Terms
1. right angle
2. volume
3. parallel lines
4. cubic unit
5. parallelogram

Practice
1. 48 in.
2. $1040\ m^2$
3. $450\ in.^2$
4. 12 ft.
5. $2744\ ft.^2$

Objective 1
1. 624 ft.
2. 206 cm

Objective 2
3. $465\ cm^2$
4. $1400\ ft^2$

Objective 3
5. $910\ cm^3$
6. $560\ in.^3$

Objective 4
7. 21 cm
8. 15 feet
9. 2 cm
10. 2 cm

Objective 5
11. $499\ ft.^2$ b. 6 gal. c. $138
12. $880

Chapter 2

Section 2.1

Key Terms
1. absolute value
2. negative
3. integers
4. additive inverse
5. positive

Practice
1. −21
2. +1350
3. 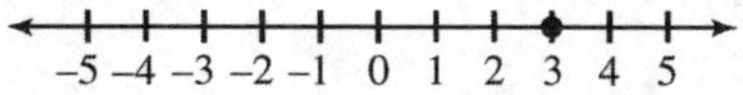

4. 18° F
5. <
6. 59
7. 33
8. 46
9. 17

Answers to MyWorkBook

Objective 1

1. +416
2. +562
3. −1
4. −3142

Objective 2

5. 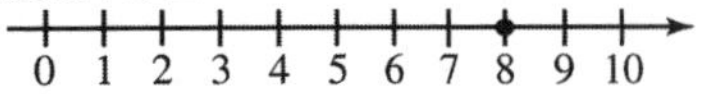

6. 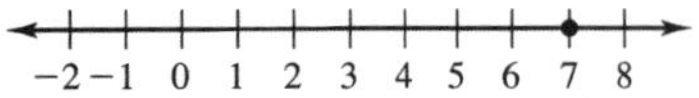

7. 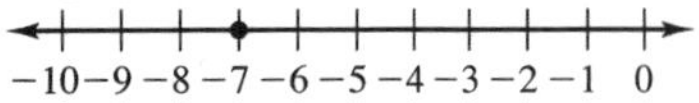

8. 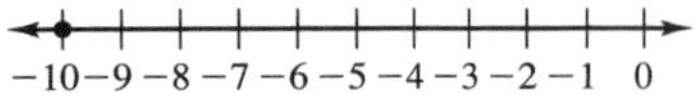

Objective 3

9. $<$
10. $<$
11. $>$
12. $<$

Objective 4

13. 26
14. 41
15. 175
16. 0

Section 2.2

Practice

1. -32
2. -10
3. A debt with a credit, so subtract, and because the debt, -63, is more than the credit, the result is a debt/negative; -22.
4. 11
5. 83
6. Their new altitude is -92 m.

Objective 1

1. 27
2. −13
3. −11
4. 30
5. −39
6. −94

Objective 2

7. −9
6. 0
9. −12
10. −11

Objective 3

11. −12
12. −55
13. 4
14. −2

Objective 4

15. 74
16. −8
17. −4
18. 513

Objective 5

19. $358
20. 32,260 feet
21. 65 feet
22. $4

Section 2.3

Key Terms

1. cost
2. net
3. loss
4. revenue

Practice

1. $15+(-37)=-22$
2. $-14+16=2$
3. $x=-16$
4. $d=30$
5. The submarine ascended 1043 ft.

Objective 1

1. $8+(-4)=-4$
2. $-15+(-31)=-46$
3. $21+(-16)=5$
4. $-316+(-84)=-400$

Objective

5. 12
6. −16
7. 83
8. 1

Objective 3

9. Profit of $1,000,500
10. 2100 feet

Section 2.4

Key Terms

1. even
2. negative
3. positive
4. same
5. positive
6. negative

Practice

1. −40
2. 50
3. −15
4. −8
5. −25
6. 7
7. $f=6$
8. ±13
9. −11
10. Kevin pays a total of $40,500 per year in tuition.

Objective 1

1. −32
2. −32
3. 0
4. 18
5. −240
6. 0

Objective 2

7. 16
8. −32
9. −25
10. 121

Objective 3

11. −9
12. 17
13. undefined
14. 3

Objective 4

15. $y=3$
16. $f=-9$
17. $r=-18$
18. $m=13$

Objective 5

19. 12 and −12
20. 8 and −8
21. 5
22. −7

Objective 6

23. $136

Section 2.5

Practice

1. −14
2. 5
3. 149
4. −12
5. 4

Objective 1

1. −12
2. −326
3. −2
4. 14
5. −66
6. −114
7. 33
8. 13
9. 4
10. 8
11. 12
12. −12
13. −19
14. 41
15. 0
16. undefined

Section 2.6

Practice

1. $5006; profit
2. −495 V
3. The plane travels 453 km.

Objective 1

1. profit of \$125
2. loss of \$16

Objective 2

3. -64 amps
4. -234 volts

Objective 3

5. 52 mph
6. 81 miles

Chapter 3

Section 3.1

Key Terms

1. evaluate
2. equation
3. indeterminate
4. expression
5. undefined

Practice

1. equation
2. expression
3. $5n^2$
4. $\frac{x-y}{x+4}$
5. 18
6. 1

Objective 1

1. equation
2. expression

Objective 2

3. $-3(n-4)$
4. $4-\frac{2}{x}$

Objective 3

5. 0
6. 24
7. 33
8. 0
9. -15
10. 12
11. 4
12. 1
13. -27
14. 3

15.

x	$-3x+5$
−2	11
−1	8
0	5
1	2
2	-1

16.

y	$\lvert y^2-4\rvert$
−4	12
−2	0
0	4
2	0
4	12

17. $x=4$

18. $x=-2,3$

Section 3.2

Key Terms

1. terms
2. coefficient
3. like terms

Practice

1. $4c+28$
2. $-15d$
3. $24x+2$; 262 ft.
4. 54 ft.^{2}

Objective 1

1. $27m+18$
2. $-2b+10$

Objective 2

3. 8
4. −1
5. $-42x$
6. $-17p+47$
7. $4x+2y-129$
8. $15r+19s$

Objective 3

9. $24x+8$; 80 yd.
10. $17x+13$; 64 m

Objective 4

11. 324 cm^2
12. 232 in^2

Section 3.3

Key Terms

1. equation
2. solve
3. expression
4. solution

Practice

1. equation
2. expression
3. 21 is a solution.
4. 2 is not a solution.

Objective 1

1. equation
2. expression
3. expression
4. equation

Objective 2

5. -6 is not a solution.
6. 5 is a solution.
7. 7 is a solution.
8. 17 is a solution.
9. 10 is not a solution.
10. 5 is not a solution
11. 9 is not a solution
12. 8 is a solution.
13. 8 is not a solution.
14. 5 is a solution.
15. 8 is a solution.
16. 9 is a solution.
17. 9 is not a solution.
18. 5 is not a solution.

Section 3.4

Key Terms

1. solution
2. nonlinear
3. add
4. distributive

Practice

1. linear
2. $y=-20$
3. $h=9$
4. $346=168+148+x$; \$30

Objective 1

1. linear
2. nonlinear

Objective 2

3. $n=12$
4. $q=34$
5. $t=6$
6. $m=-1$
7. $w=30$
8. $x=5$

Objective 3

9. $35+33+l=86$; 18 mm
10. $125+t=180$; 55

Section 3.5

Practice

1. $y = 2$
2. $m = -1$
3. $c = -2$
4. $v = -5$
5. The travel time will be 5 hr.

Objective 1

1. $a = -5$
2. $g = -5$
3. $r = 3$
4. $y = -4$

Objective 2

5. $s = 2$
6. $y = -15$
7. $r = -8$
8. $n = -3$
9. $q = 2$
10. $y = -2$
11. $p = -2$
12. $h = 1$
13. $x = 2$
14. $m = 4$

Objective 3

15. The width is 20 cm.
16. The resistance is 24 ohms.

Section 3.6

Practice

1. $n - 7 = 14$; $n = 21$
2. $-2y - 14 = 16$; $y = -15$
3. $6n + 4 = 4(n + 12)$; $n = 22$

Objective 1

1. $n - 14 = 10$; $n = 24$
2. $n + 8 = -12$; $n = -20$
3. $n + 11 = -19$; $n = -30$
4. $n - 6 = -20$; $n = -14$
5. $-5n = 35$; $n = -7$
6. $-1n = 26$; $n = -26$
7. $8n = -16$; $n = -2$
8. $-3n = -18$; $n = 6$
9. $4t + 11 = 6t - 13$; $t = 12$
10. $6x + 3(x - 20) = 62 - (x + 2)$; $x = 12$
11. $-10(y - 15) - 20 = -4y - (y - 5)$; $y = 25$
12. $4(b + 8) = -48$; $b = -20$
13. $7(m + 3) - 6m = 3(m + 3)$; $m = 6$

Chapter 4

Section 4.1

Key Terms

1. prime
2. prime
3. greatest
4. composite

Practice

1. not prime
2. prime
3. $200 = 2^3 \cdot 5^2$
4. 1, 2, 3, 6, 9, 18, 27, 54
5. $\text{GCF} = 1$
6. $\text{GCF} = 1$

Objective 1

1. prime
2. not prime
3. not prime
4. not prime

Objective 2

5. $512 = 2^9$
6. $675 = 3^3 \cdot 5^2$
7. $2205 = 3^2 \cdot 5 \cdot 7^2$
8. $138 = 2 \cdot 3 \cdot 23$

Objective 3

9. 1, 3, 5, 15, 25, 75, 125, 375
10. 1, 3, 5, 9, 15, 25, 45, 75, 225

Objective 4

11. $\text{GCF} = 1$
12. $\text{GCF} = 3$

Objective 5

13. $\text{GCF} = 20$
14. $\text{GCF} = 4$

Section 4.2

Key Terms

1. fewest
2. numerator
3. multiple
4. rational
5. mixed

Practice

1. $\frac{7}{10}$
2. $\frac{31}{61}$
3. 0
4. $5\frac{3}{4}$

5. $-8\frac{5}{7}$

6. $-\frac{32}{3}$

7. 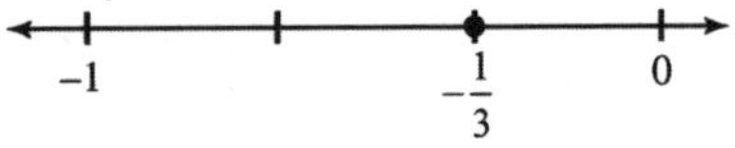

8. 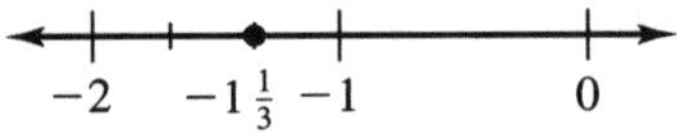

9. −28

10. 3

11. <

Objective 1

1. $\frac{1}{2}$

2. $\frac{5}{8}$

3. $\frac{1}{12}$

4. $\frac{9}{13}$

Objective 2

5. −17

6. undefined

7. $7\frac{5}{6}$

8. $-9\frac{3}{8}$

9. $12\frac{11}{14}$

10. $-14\frac{3}{4}$

Objective 3

11. $-\frac{17}{7}$

12. $\frac{33}{5}$

Objective 4

13. 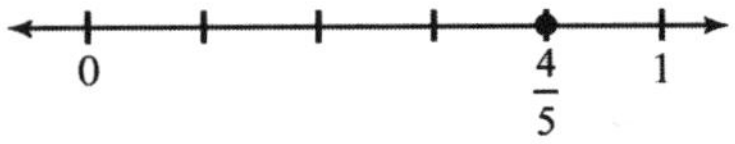

14. 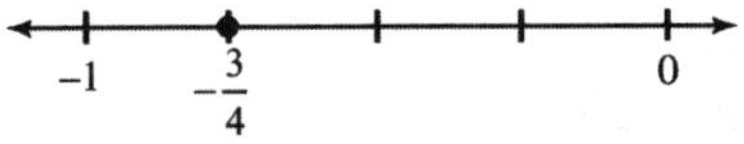

15. 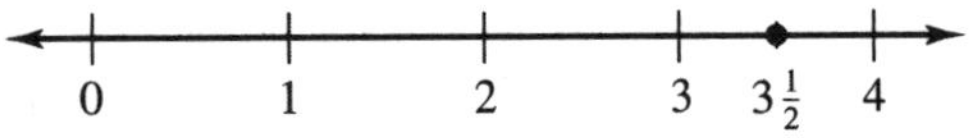

16. 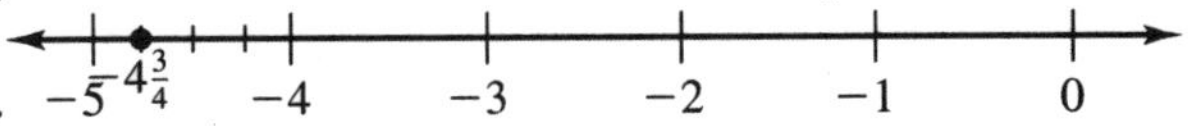

Objective 5

17. 40

18. 7

Objective 6

19. <

20. =

21. >

22. =

Section 4.3

Key Terms

1. rational

2. lowest

3. greatest

4. numerator; denominator

Practice

1. $\frac{3}{5}$
2. $-\frac{4}{5}$
3. $\frac{5}{8}$
4. $3\frac{1}{3}$
5. $4\frac{3}{4}$

Objective 1

1. $\frac{3}{5}$
2. $\frac{1}{3}$
3. $-\frac{7}{9}$
4. $\frac{4}{11}$
5. $\frac{5}{9}$
6. $\frac{5}{7}$
7. $\frac{17}{20}$
8. $\frac{3}{5}$
9. $\frac{10}{11}$
10. 670 kg; $\frac{16}{67}$

Objective 2

11. $1\frac{7}{11}$
12. $-1\frac{3}{14}$
13. $-1\frac{5}{13}$
14. $-12\frac{1}{3}$

Section 4.4

Key Terms

1. irrational
2. circle
3. circumference
4. diameter

Practice

1. $\frac{9}{28}$
2. $-\frac{8}{35}$
3. $\frac{21}{40}$
4. $\frac{1}{12}$
5. $\frac{2}{11}$
6. 10; $9\frac{3}{4}$
7. 5; $6\frac{1}{4}$
8. $11\frac{1}{2}$

9. $\frac{16}{6561}$

10. $-\frac{125}{216}$

11. 217 votes

12. $\frac{3}{8}$

13. $9\frac{3}{8}$ in.^{2}

14. $9\frac{2}{3}$ in.

15. $1\frac{4}{7}$ m

16. $16\frac{1}{2}$ ft.

Objective 1

1. $-\frac{1}{48}$

2. $\frac{30}{91}$

3. $\frac{9}{1000}$

4. $\frac{14}{81}$

Objective 2

5. $\frac{10}{21}$

6. $\frac{7}{26}$

7. $-\frac{5}{52}$

8. $\frac{8}{85}$

9. $\frac{1}{6}$

10. $\frac{1}{3}$

Objective 3

11. estimate: 18; actual: $19\frac{11}{20}$

12. estimate: 45; actual: 36

13. $3\frac{55}{64}$

14. 14

Objective 4

15. $\frac{25}{16}$

16. $-\frac{8}{343}$

Objective 5

17. Cheryl will earn \$12 for working $\frac{3}{8}$ of a day.

18. $\frac{6}{49}$ cup of granola is needed.

Objective 6

19. 324 mm^2

20. 45 m^2

Objective 7

21. $1\frac{1}{4}$ in.

22. $7\frac{1}{2}$ cm

Objective 8

23. $4\frac{2}{5}$ in.

24. 9 m

Section 4.5

Key Terms

1. complex
2. reciprocal
3. denominator
4. improper

Practice

1. $\frac{1}{4}$
2. $-\frac{3}{22}$
3. 14; $16\frac{1}{2}$
4. $\frac{4}{9}$
5. 4
6. $y = 2\frac{4}{5}$
7. $k = 31\frac{1}{3}$
8. $1\frac{5}{8}$

Objective 1

1. $\frac{9}{4}$
2. 8
3. $-\frac{1}{11}$
4. $-\frac{7}{4}$
5. 3
6. $\frac{7}{9}$
7. $-\frac{3}{13}$
8. $2\frac{34}{35}$
9. $5\frac{7}{9}$
10. 35

Objective 2

11. estimate: $\frac{3}{2}$; actual: $1\frac{11}{34}$
12. estimate: $\frac{4}{3}$; actual: $1\frac{31}{104}$
13. estimate: 2; actual: $2\frac{31}{44}$
14. estimate: $3\frac{1}{3}$; actual: $3\frac{49}{52}$

Objective 3

15. $\frac{5}{6}$
16. $\frac{5}{7}$
17. 3
18. 3

Objective 4

19. 28
20. $-\frac{21}{2}$

Objective 5

21. 10 applications can be made.
22. 15 yd of material will make 9 pairs of soccer shorts.

Chapter 5

Section 5.1

Key Terms

1. multiple
2. least
3. greatest

Practice

1. 105
2. 5040
3. $\frac{25}{30}; \frac{26}{30}$

Objective 1

1. 60
2. 63
3. 225
4. 196

Objective 2

5. 36
6. 196
7. 441
8. 945
9. 378
10. 120

Objective 3

11. $\frac{66}{126}$ and $\frac{49}{126}$
12. $\frac{10}{12}$ and $\frac{9}{12}$
13. $\frac{39}{45}$ and $\frac{41}{45}$
14. $\frac{147}{196}, \frac{70}{196}$ and $\frac{172}{196}$

Section 5.2

Practice

1. $\frac{1}{4}$

2. $\frac{4}{9}$

3. $\frac{11}{30}$

4. 9; $8\frac{11}{12}$

5. 3; $2\frac{23}{24}$

6. $2\frac{3}{8}$

7. $f = \frac{17}{24}$

8. $5\frac{29}{30}$ miles per gallon

Objective 1

1. $\frac{4}{5}$

2. $\frac{13}{15}$

3. $-\frac{11}{14}$

4. $\frac{13}{20}$

Objective 2

5. $\frac{7}{33}$

6. $\frac{13}{36}$

7. $-\frac{1}{8}$

8. $\frac{149}{120}$

Objective 3

9. $14\frac{2}{3}$

10. $5\frac{5}{7}$

11. $6\frac{3}{4}$

12. $3\frac{47}{55}$

13. $6\frac{11}{18}$

14. $-3\frac{1}{4}$

15. $-5\frac{1}{30}$

16. $-13\frac{9}{10}$

Objective 4

17. $3\frac{11}{21}$

18. $-\frac{11}{15}$

19. $-15\frac{13}{18}$

20. $3\frac{1}{15}$

Objective 5

21. $y=-\frac{1}{4}$

22. $m=-\frac{37}{63}$

23. $q=\frac{53}{44}$

24. $p=\frac{15}{26}$

Objective 6

25. The new tire tread is $\frac{1}{32}$ inch deeper.

26. The fourth child inherited $\frac{1}{9}$ of the estate.

27. The distance around the book is $31\frac{3}{25}$ inches.

28. Phil would have biked $1\frac{13}{60}$ fewer miles.

Section 5.3

Practice

1. $\frac{3}{4}$

2. $-\frac{1}{5}$

3a. $6\frac{1}{6}$ minutes

3b. $6\frac{1}{6}$ minutes

3c. $5\frac{1}{3}$ minutes

4. $3\frac{7}{9}$

5. $70\frac{7}{8}$ ft^2

6. $1\frac{43}{56}$ m^2

7. $-\frac{4}{15}n+\frac{1}{4}$

8. $\frac{13}{14}j^3-\frac{7}{10}j^2+\frac{3}{4}$

Objective 1

1. $-4\frac{1}{10}$

2. $-2\frac{7}{20}$

3. $2\frac{7}{50}$

4. $-10\frac{1}{2}$

5. $-11\frac{3}{4}$

6. $-9\frac{38}{63}$

Objective 2

7a. height: $46\frac{5}{8}$ in.; weight: $49\frac{13}{24}$ lbs.

b. height: $46\frac{7}{8}$ in.; weight: $49\frac{7}{8}$ lbs.

7c. height: $47\frac{1}{2}$ in.; weight: $42\frac{1}{4}$ lbs.

8a. songs: $2\frac{161}{180}$ min.; ads: $\frac{67}{96}$ min.

b. songs: $2\frac{179}{180}$ min.; ads: $\frac{5}{8}$ min.

8c. songs: no mode; ads: $\frac{1}{2}$ min and $1\frac{1}{4}$ min.

Objective 3

9. $6\frac{1}{4}$

10. $\frac{8}{45}$

11. $-1\frac{17}{48}$

12. $3\frac{9}{11}$

Objective 4

13. 7 m^2

14. $48\frac{11}{16}$ yd.^{2}

Objective 5

15. $254\frac{4}{7}\text{ cm}^2$

16. $21\frac{21}{32}$ ft.^{2}

Objective 6

17. $\frac{5}{2}n-\frac{10}{21}$

18. $-\frac{7}{18}y-\frac{7}{3}$

19. $-\frac{11}{12}y^3+\frac{5}{8}y+\frac{27}{4}$

20. $\frac{5}{6}n^3+\frac{5}{8}n^2-\frac{17}{20}n-\frac{1}{2}$

Section 5.4

Practice

1. $k=\frac{7}{8}$

2. $3\frac{1}{3}n+\frac{3}{8}=\frac{7}{9}n-1\frac{1}{6}$; $n=\frac{111}{184}$

3. $a=2\frac{1}{4}$ ft

4. You should do 38 problems.

Objective 1

1. $t=\frac{1}{14}$

2. $x=-\frac{1}{2}$

3. $x=-\frac{3}{5}$

4. $x=\frac{2}{5}$

5. $n=-5\frac{2}{5}$

6. $x=-1\frac{5}{6}$

7. $y=1\frac{13}{14}$

8. $f=63\frac{1}{3}$

Objective 2

9. $\frac{1}{3}n=25\frac{2}{3};\ n=77$

10. $5\frac{1}{4}n=1\frac{1}{2};\ n=\frac{2}{7}$

11. $h-2\frac{7}{10}=-4\frac{3}{5};\ h=-1\frac{9}{10}$

12. $6\frac{1}{5}+2k=-\frac{1}{3};\ k=-3\frac{4}{15}$

13. $\frac{1}{6}(a+3)=1\frac{1}{9}+a;\ a=-\frac{11}{15}$

14. $\frac{2}{5}z+3=\frac{1}{2}(5z-3);\ z=2\frac{1}{7}$

Objective 3

15. $3\frac{1}{2}$ ft.

16. Warren must score 100 on the fifth test to earn a B.

Chapter 6

Section 6.1

Key Terms

1. fractions
2. ten-thousandths
3. left; right
4. 4

Practice

1. $\frac{7}{20}$

2. one hundred five thousandths

3. one hundred twenty-five and thirty-seven hundredths

4. 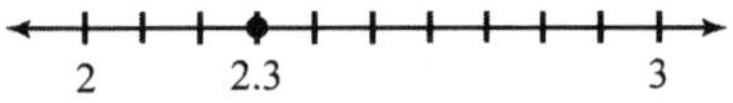

5. >

6. 21.063

Objective 1

1. $\frac{7}{8}$
2. $\frac{3}{50}$
3. $12\frac{1}{5}$
4. $-\frac{3}{500}$

Objective 2

5. seven hundred eighty-nine hundred-thousandths
6. fifty and one hundred seventeen ten-thousandths
7. negative four and eighty-three hundredths

8. negative five hundred twenty-six and three thousand two hundred forty-seven ten-thousandths

Objective 3

9. 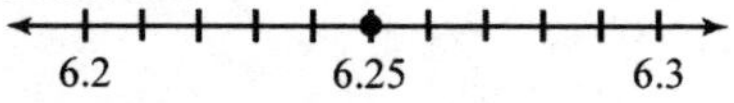

10. 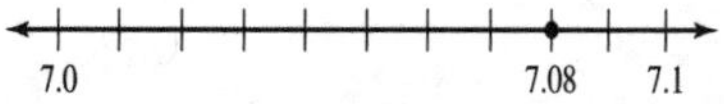

11. 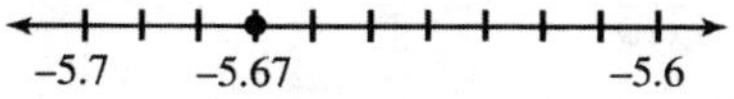

12. 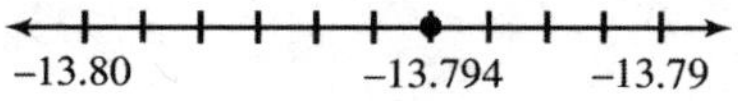

Objective 4

13. > 14. < 15. < 16. <

Objective 5

17. 21.0628 18. 21.06 19. 21 20. 20

Section 6.2

Practice

1. 130.7493
2. 24.09
3. −48.506
4. $12.4d+5.4$
5. $x=-15.69$
6. The missing side length is 21.2 mm.

Objective 1

1. 77.818 2.120.2054 3. 167.6661 4. 188.836

Objective 2

5. 78.21 6. 43.5363 7. 3.471 8. 26.9959

Objective 3

9. −134.3 10 49.7 11. 13.46 12. 7.6

Objective 4

13. $4.3n+3.2$
14. $-3.7y^2+12.3y+41.4$

Objective 5

15. $a=-109.162$
16. $b=-42.56$

Objective 6

17. $757.23
18. 4.6 cm

Section 6.3

Practice

1. 2.0657
2. −16
3. −0.001
4. 267,000,000,000; two hundred sixty-seven billion
5. 7.89545×10^9
6. $42.55

Objective 1

1. 4289.4785 2. 5124.8 3. 3.96534 4. 683.9592

Objective 2

5. −0.429 6. 0.3744 7. −37.03 8. 3141.5

Objective 3

9. 3.375 10. 29.8116 11. 0.0016 12. 94.4784

Objective 4

13. 58,970,000; fifty-eight million, nine hundred seventy thousand
14. 73,814; seventy-three thousand, eight hundred fourteen

Objective 5

15. 3.015715×10^{11} 16. 3.6×10^{6}

Objective 6

25. 14.44 m^2 26. 11.18 $yd.^2$

Section 6.4

Key Terms

1. integers
2. rational numbers; irrational numbers
3. whole numbers
4. decimal numbers

Practice

1. 0.015
2. 20
3. $0.8\overline{3}$
4. 18.71
5. $b = -8.24$
6. You need 3.454 ft. of tape.

Objective 1

1. 0.408 2. 2100 3. 3240
4. 901 5. 2500 6. −621.4

Objective 2

7. 14.5 8. $0.\overline{6}$ 9. −5.4 10. $-4.8\overline{3}$

Objective 3

11. 0.25 12. 7.14

Objective 4

13. $x = 7$
14. $y = .8$
15. $z = -6.85$
16. $t = 5.245$

Objective 6

17. \$451.61
18. ≈ 14.8 in.

Section 6.5

Practice

1. 1.99
2. $0.5\overline{3}$
3. Jonathan's employer will reimburse him \$580.33.
4. 2.73
5. 4.23×10^{21}
6. 10.5 cm^2
7. 10 $in.^2$
8. 34.1946 ft^2
9. 23.55 m^3
10. 5.635 m^3
11. 60.3508 yd^3
12. 2224.96632 mm^3 (however rounded)
13. 17.625 cm^2

Objective 1

1. −339.95
2. −96
3. −11.18
4. −1.61

Objective 2

5. 3
6. $-0.29\overline{3}$
7. 26.2
8. -5.95

Objective 3

9. The mean is 85.8.
10. \$775.95

Objective 4

11. 3.25
12. $3.1\overline{3}$

Objective 5

13. 480
14. 2.16×10^{18}

Objective 6

15. 30.26 cm^2
16. 176.625 ft^2
17. 46.55 $in.^2$
18. 3155.3546 ft^2

Objective 7

19. 321.536 mi^3
20. 109.52 km^3
21. 2906.855 m^3
22. 381.51 mi^3

Objective 8

23. 83.4923 m^2
24. 96.555 ft^3

Section 6.6

Key Terms
1. hypotenuse
2. sum

Practice
1. $y = 1$
2. $x = -0.\overline{3}$
3. $h = 1.25$
4. It is about 93.338 ft. from home to second base.

Objective 1
1. $y = -1$
2. $x = 0.2$
3. $x = 1.5375$
4. $y = 4.2$

Objective 2
5. $m = 0.05625$
6. $x = 2.2\overline{7}$

Objective 3
7. $n = 42$
8. 655 minutes

Objective 4
9. Holly can use her phone 152 feet into the backyard.
10. Around 38.75 m from home to second base

Section 6.7

Key Terms
1. perpendicular
2. acute
3. ray
4. parallel
5. straight
6. complementary
7. congruent
8. interior

Practice
1. plane N
2. straight
3. congruent; They are vertical angles
4. 50°, 130°, 50°, 130°
5. right triangle; The measure of one angle is 90°.

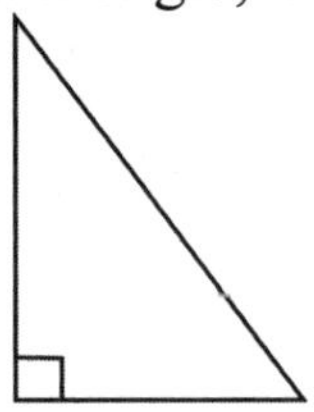

6. 38°, 41°, 101°
7. Not congruent
8. 50°, 38°, 92°

Objective 1

1. point O
2. ray $\overrightarrow{RS}$
3. line segment $\overline{XY}$
4. parallel lines c and d, or $c \parallel d$

Objective 2

5. Right
6. obtuse

Objective 3

7. congruent; They are alternate interior angles.
8. congruent; They are corresponding angles.
9. 116°
10. 52°
11. 55°
12. 121°

13. 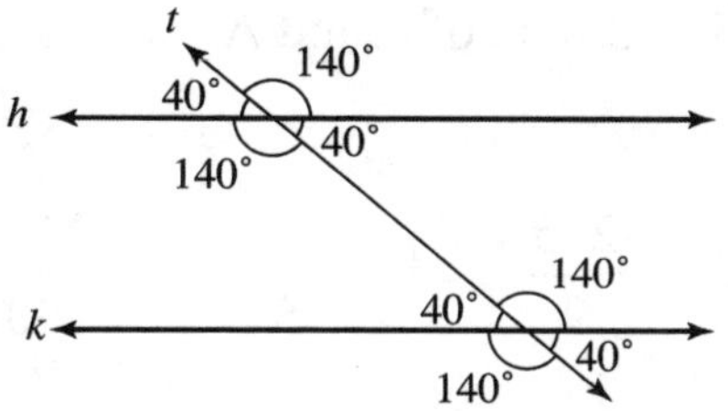

14. 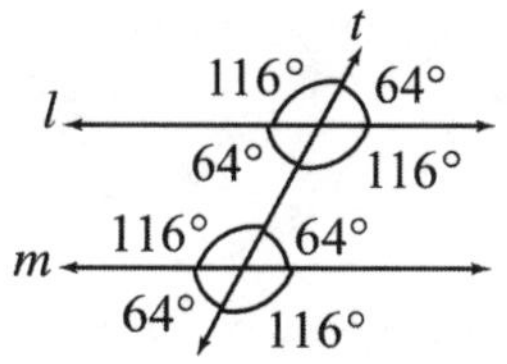

Objective 4

15 acute triangle; The measure of each angle is less than 90°.

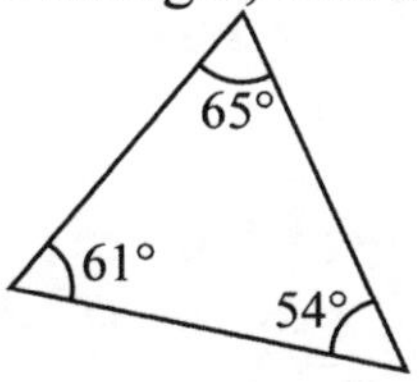

16. Obtuse triangle. The measure of one angle is greater than 90°.

17. 72°, 52°, 56°
18. 24°, 125°, 31°

Objective 5

19. Not congruent
20. Congruent by SAS
21. $18\frac{2}{3}^\circ, 80\frac{2}{3}^\circ, 80\frac{2}{3}^\circ$
22. 47°, 54°, 79°

Chapter 7

Section 7.1

Key Terms

1. rate
2. unit
3. favorable; possible
4. quotient

Practice

1a. $\frac{7}{3}$ b. $\frac{18}{143}$

2. $\frac{1}{4}$

3. ≈ 17.37; The stock is selling at \$17.37 for every \$1 of annual earning.

4. \$1.77/lb.

5. 3 bars of brand A

Objective 1

1. $\frac{25}{42}$

2a. $\frac{275}{9}$ b. $\frac{4}{5}$ c. $\frac{1}{190}$

Objective 2

3. $\frac{2}{13}$

4. $\frac{1}{26}$

5. $\frac{1}{2}$

6a. $\frac{1}{10}$ b. 0

Objective 3

7. ≈ 0.42; They owe about \$0.42 for every \$1 of income.

8. ≈ 16.87; There are about 17 students for every faculty member.

Objective 4

9. $\approx \$0.25$ / ounce

10. 58 mi./hr.

Objective 5

11. half a pound of brand A deli meat

12. three 15–ounce cans of peas

Section 7.2

Key Terms

1. congruent
2. proportional
3. cross products
4. similar

Practice

1. yes
2. $q = 35$
3. 52 apples (will have 50 cents remaining)
4. 150 m

Objective 1

1. no
2. no
3. yes
4. no

Objective 2

5. $n = 18.\overline{6}$
6. $c = 14$
7. $x = 25$
8. $y = 1$

Objective 3

9. 1.2 gal.
10. $2517.50
11. $24.26
12. 46.7 mi.

Objective 4

13. $b = 7.2$ ft; $c = 9.6$ ft

13. 195 meters

Section 7.3

Practice

1. 48 in.
2. 657 ft.^{2}
3. 24 pt.
4. 64 oz.
5. 5 hr.
6. 1,381,320 mi./hr.

Objective 1

1. 5 ft.
2. $5\frac{1}{3}$ yd.
3. 39,600 ft.
4. 0.8 mi.

Objective 2

5. 0.11 yd.^{2}
6. 3 ft.^{2}

Objective 3

7. $9\frac{3}{8}$ pt.
8. 76.8 oz.

Objective 4

9. $277\frac{1}{3}$ oz.
10. 3 T

Objective 5

11. 0.164 yr.
12. 7,884,000 or 7,889,400 min.

Objective 6

13. $95.\overline{3}$ feet per second
14. $550.2\overline{27}$ miles per hour

Section 7.4

Practice

1. 730 mm
2. 780 cl
3. 800 g
4. 0.000038 km^2

Objective 1

1. 2300 m
2. 0.125 dam
3. 0.73 hm
4. 125 mm
5. 40,000 dm
6. 0.00538 km

Objective 2

7. 0.0001 kl
8. 611 ml
9. 6000 L
10. 419 L
11. 15 ml
12. 0.00707 L

Objective 3

13. 890 mg
14. 0.000052 t
15. 25 cg
16. 7.2 kg
17. 60 dg
18. 3660 kg

Objective 4

19. 1.7 cm^2
20. 0.0084 m^2
21. 130 dam^2
22. 472 km^2

Section 7.5

Practice

1. $\approx 42{,}179$ m
2. ≈ 33.3 L
3. ≈ 121 lb.
4. 38.3° C

Objective 1

1. 215.9 mm
2. ≈ 454.01 yd.
3. ≈ 780.19 mi.
4. $\approx 4{,}743{,}256.4$ ft.

Objective 2

5. ≈ 177.4 ml
6. ≈ 1.057 pt
7. ≈ 70.4 L
8. ≈ 0.5 oz

Objective 3

9. ≈ 12.0 oz.
10. ≈ 0.0141 oz.
11. ≈ 18.2 kg
12. ≈ 28.4 g

Objective 4

13. −28.3° C
14. 38.4° C
15. 176.7° C
16. 71.6° F

Section 7.6

Key Terms

1. interest
2. back-end

Practice

1a. 0.19 b. 0.30 c. yes
2. $1344
3. $1767.10
4. 5 drops/min.

Objective 1

1a. 0.25 b. 0.44 c. no
2a. 0.23 b. 0.35 c. yes

Objective 2

3. $1334
4. $1176

Objective 3

5. $1533.60
6. $1795.80

Objective 4

7. 2 tablets
8. 12 drops

Chapter 8

Section 8.1

Key Terms

1. left
2. right

Practice

1. $\frac{2}{5}$
2. 0.95
3. $77.\overline{7}\%$
4. 0.4%

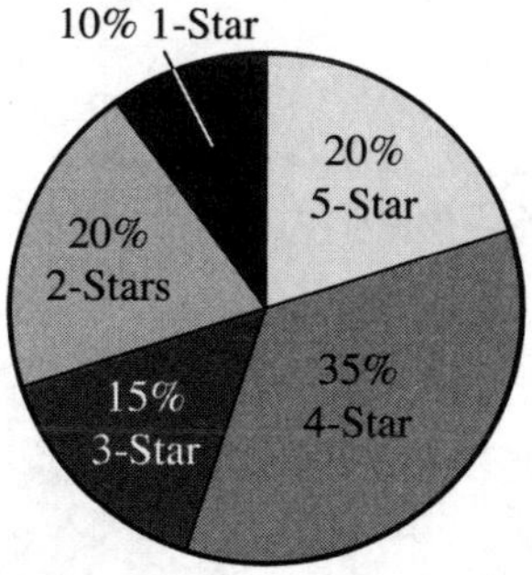

5.

Objective 1

1. $\frac{3}{4}$
2. $\frac{273}{1000}$
3. 0.048
4. 1.26

Objective 2

5. 56%

6. $58\frac{1}{3}\%$

7. 83%

8. 0.1%

Objective 3

9. 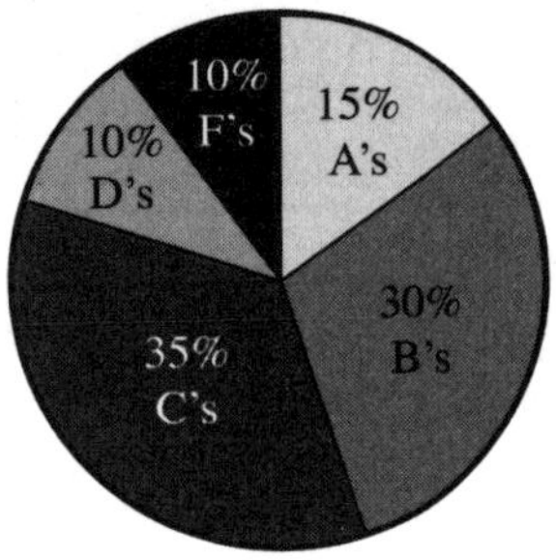

10. 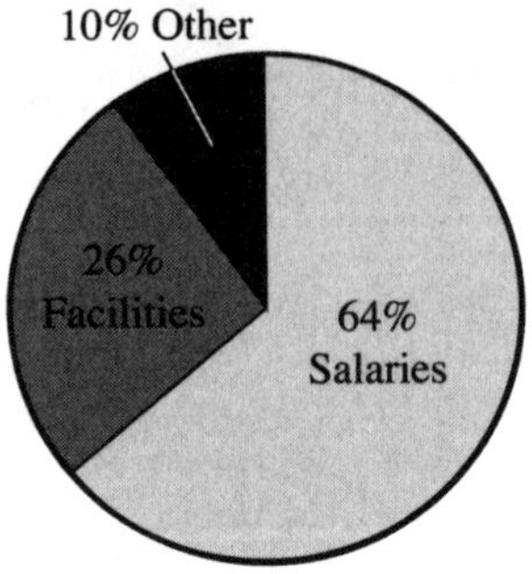

Section 8.2

Key Terms

1. is

2. what

3. %

4. of

Practice

1. whole

2. $0.9 \cdot 72 = n$; 64.8

3. $36\% \cdot n = 23.4$; 65

4. $n \cdot 40 = 16$; 40%

Objective 1

1. part

2. whole

3. percent

4. percent

Objective 2

5. $0.45 \cdot 120 = x$; 54

6. $1.35 \cdot 16 = w$; 21.6

Objective 3

7. $0.4 \cdot w = 62$; 155

8. $64 = 0.96 \cdot x$; $66\frac{2}{3}$

Objective 4

9. $n \cdot 81.5 = 32.6$; 40%

10. $n \cdot 21 = 14$; $66\frac{2}{3}\%$

Section 8.3

Practice

1. $0.10 \cdot 220 = n$; 22

2. $28.50 = 0.68 \cdot n$; \$42.00

3. $32 = n \cdot 33$; $96.\overline{96}\%$

Objective 1

1. $0.028 \cdot 79{,}000{,}000 = n$; \$2,212,000

2. $0.60 \cdot 425 = n$; 255 girls

Objective 2

3. $0.30 \cdot n = 424$; \$1413.33
4. $0.25 \cdot n = 10$; 40 employees

Objective 3

5. $n \cdot 42 = 14$; $33\frac{1}{3}\%$
6. $n \cdot (2{,}300{,}000 + 7{,}800{,}000) = 7{,}800{,}000$; 77.2%

Section 8.4

Key Terms

1. new amount, original amount

Practice

1. Tax: \$71.13; Total price: \$1209.13
2. Amount of discount: \$17.40, Final Price: \$40.60
3. \$16,000
4. 29.4%
5. Rate of discount: 20%; sale price: \$560

Objective 1

1. \$27.00
2. Tax: \$19.30; Total price: \$246.30

Objective 2

3. Amount of discount: \$22; final price: \$198
4. Amount of deduction: \$95.16; final pay: \$1490.84

Objective 3

5. \$26.75
6. \$65.00

Objective 4

7. 3%
8. 14%

Section 8.5

Key Terms

1. simple interest
2. compound interest
3. amortization

Practice

1. \$18
2. Interest: \$3000; Total: \$13,000
3. \$3896.76
4. \$3810.60

Objective 1

	Principal	Rate of interest	Time	Simple interest
1.	\$1200	3%	1 year	\$36
2.	\$500	4.8%	$\frac{1}{2}$ year	\$12
3.	\$25,000	$5\frac{3}{8}\%$	$\frac{1}{4}$ year	\$335.94

4. Interest: \$35.51
5. Total Amount: \$2435.51

Objective 2

6. \$1025.26
7. \$3514.98
8. \$10,934.44
9. \$5016.46

Objective 3

10. \$834.74
11. \$1042.07
12. \$4822.84
13. \$3431.19

Chapter 9

Section 9.1

Key Terms

1. conditional linear equation in one variable
2. contradiction
3. solution
4. solution set
5. identity
6. linear equation in one variable

Practice

1. 4
2. no solution
3. $t = \dfrac{Q}{\pi}$

Objective 1

1. −7
2. 7
3. $-\dfrac{5}{13}$
4. 15

Objective 2

5. −4
6. 0.6
7. $\dfrac{40}{13}$
8. all real numbers

Objective 3

9. $y = \dfrac{-10x+3}{9}$
10. $C = \dfrac{rs}{19} - t$

11. $s = \dfrac{g - jh - 45z}{9}$

12. $d = \dfrac{315 - 7c}{9}$

13. $n = \dfrac{S - Q}{by}$

Section 9.2

Practice

1. 6188 square feet
2. 2

Objective 1

1. 3137.4 mi
2. 60 min
3. 350 gal

Objective 2

4. 140
5. −3

Section 9.3

Key Terms

1. linear inequality in one variable

Practice

1. $\{y \mid y < 0\}; (-\infty, 0)$

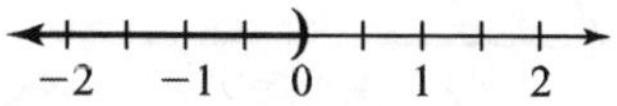

2. $\left\{w \middle| w > -\dfrac{9}{2}\right\}; \left(-\dfrac{9}{2}, \infty\right)$

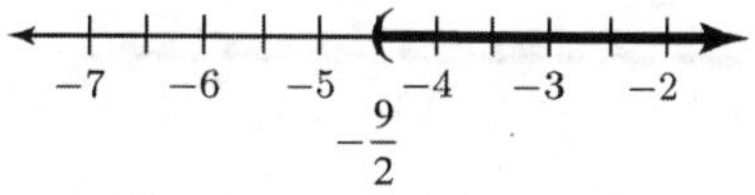

3. 22 cm or less

Objective 1

1. $\{x \mid x \le -1\};\ (-\infty, -1]$

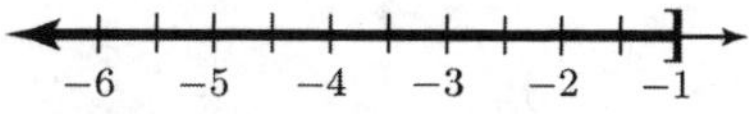

2. $\{y \mid y > 4\};\ (4, \infty)$

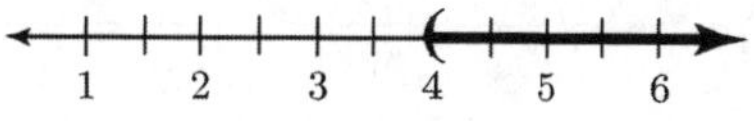

Objective 2

3. $\{x \mid x > -6\}; (-6, \infty)$

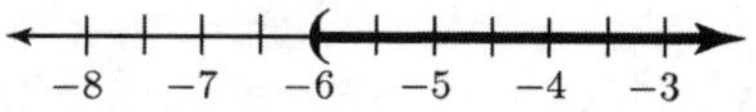

4. $\{k \mid k \ge -8\}; [-8, \infty)$

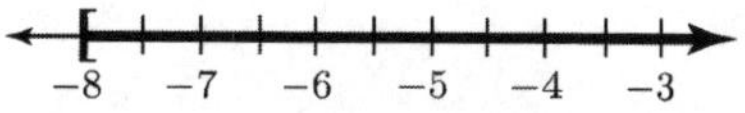

Objective 3

5. $x + 12 > 38;\ x > 26$
6. $3x - 26 \ge 55;\ x \ge 27$
7. $16x - 10 < 5x;\ x < \dfrac{10}{11}$
8. at least 40 units
9. $t < 977.5°\text{C}$

Section 9.4

Key Terms

1. intersection
2. compound inequality
3. union

Practice

1. a) $\{x|-1 < x \le 3\}$; b) $(-1, 3]$;
c) 

2. a) $\{x|-4 \le x < 6\}$; b) $[-4, 6)$;
c) 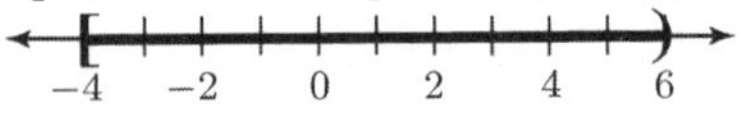

3. a) $\{x|x \ge -6\}$; b) $[-6, \infty)$;
c) 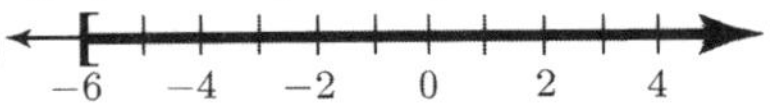

Objective 1

1. a) $\{x|-2 < x < 2\}$; b) $(-2, 2)$
c) 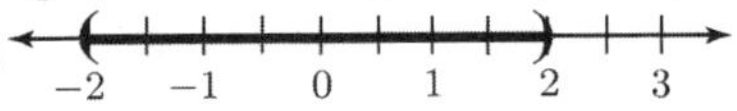

2. a) $\{x|-2 \le x \le 2\}$; b) $[-2, 2]$
c) 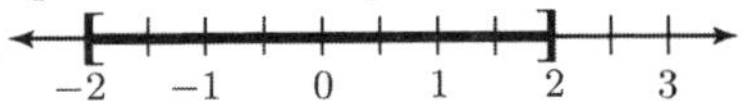

3. a) $\{x|-4 \le x < 3\}$; b) $[-4, 3)$
c) 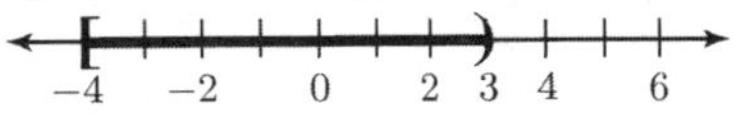

4. a) $\{x|-4 < x < 1\}$; b) $(-4, 1)$
c) 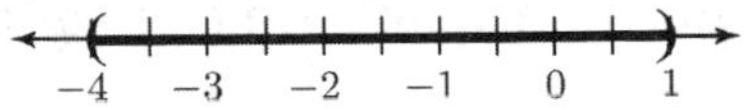

Objective 2

5. a) $\{x|x \text{ is a real number}\}$; b) $\mathbb{R}$;
c) 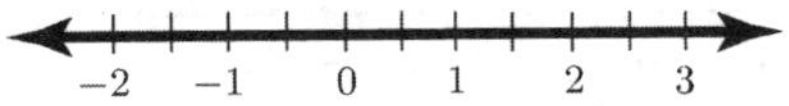

6. a) $\{x|x \text{ is a real number}\}$; b) $\mathbb{R}$;
c) 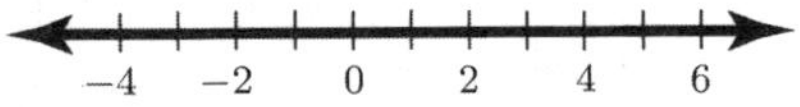

7. a) $\{x|x \ge -9\}$; b) $[-9, \infty)$;
c) 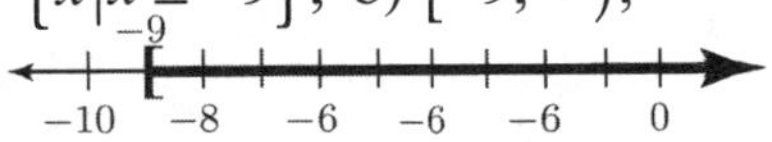

8. a) $\{x|x \ge -6\}$; b) $[-6, \infty)$;
c) 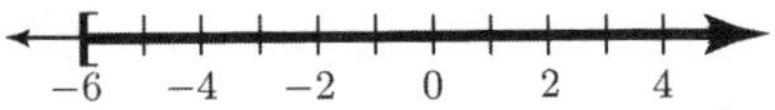

9. a) $\{x|x < -6 \text{ or } x > 4\}$; b) $(-\infty, -6) \cup (4, \infty)$;
c) 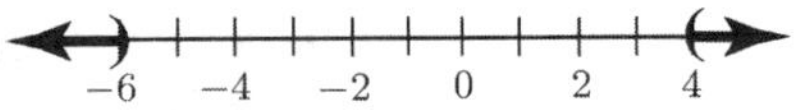

Section 9.5

Practice

1. −9, 9
2. −19, 11
3. $-\frac{19}{2}, \frac{3}{4}$

Objective 1

1. no solution
2. −8, 20

3. 1

4. $\frac{9}{4}, -\frac{3}{4}$

5. −12, 4

6. −6, 6

7. $-\frac{9}{7}, \frac{3}{7}$

8. −15, 1

9. −37, 21

10. all real numbers

11. $-\frac{25}{2}, \frac{9}{4}$

Section 9.6

Practice

1. a) $\{x | -9 \le x \le -1\}$;
   b) b) $[-9, -1]$;
   c) 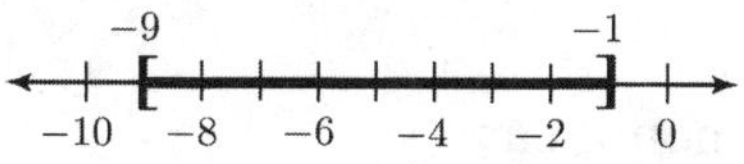

2. a) { } or $\varnothing$;
   b) no interval notation;
   c) 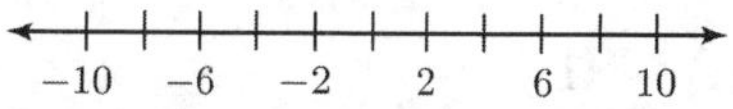

Objective 1

1. a) $\{x | -5 < x < 5\}$; b) $(-5, 5)$;
   c) 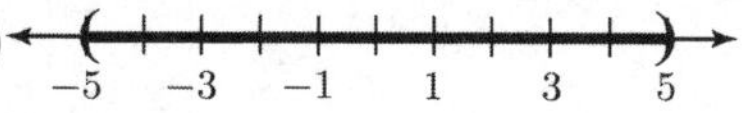

2. a) $\{x | -5 \le x \le 1\}$; b) $[-5, 1]$;
   c) 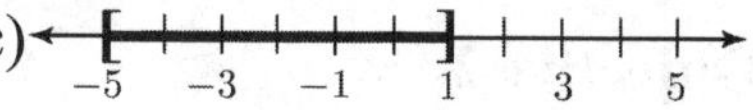

3. a) $\{x | -3 \le x \le 3\}$; b) $[-3, 3]$;
   c) 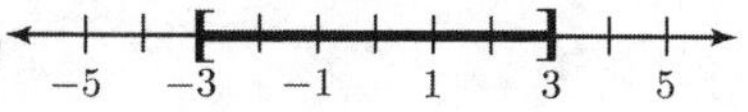

4. a) $\{x | -2 < x < 2\}$; b) $(-2, 2)$;
   c) 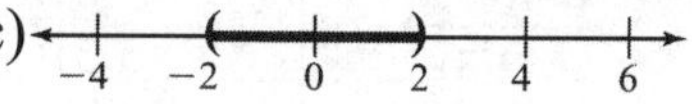

Objective 2

5. a) $\{x | x < -2 \text{ or } x > 2\}$;
   b) $(-\infty, -2) \cup (2, \infty)$;
   c) 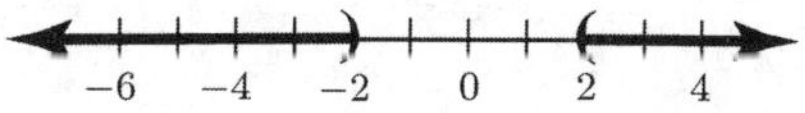

6. a) $\{x | x \text{ is a real number}\}$;
   b) $\mathbb{R}$;
   c) 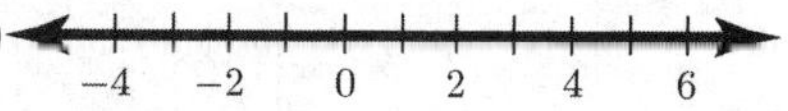

7. a) $\{x | x < -13 \text{ or } x > -1\}$; b) $(-\infty, -13) \cup (-1, \infty)$;
   c) 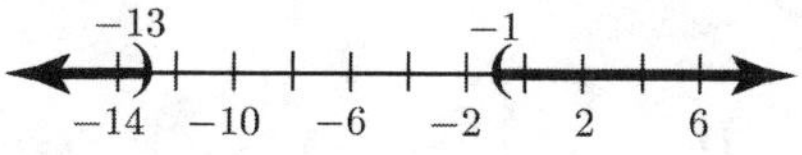

8. a) $\{x | x \le -2 \text{ or } x \ge 2\}$;
   b) $(-\infty, -2] \cup [2, \infty)$;
   c) 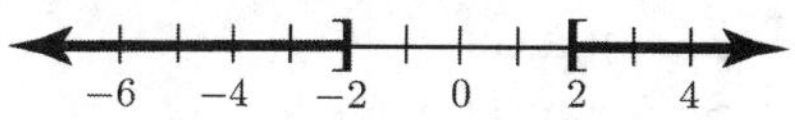

9. a) $\{x | -13 < x < 7\}$; b) $(-13, 7)$

   c) 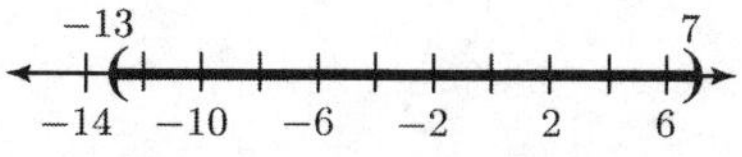

Chapter 10

Section 10.1

Practice

1. $A(-5,1)$, $B(1,2)$, $C(4,-1)$, $D(0,-3)$
2.

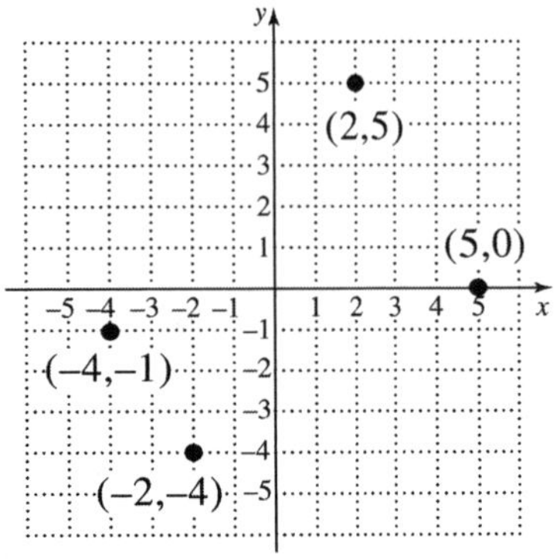

3. II
4. nonlinear

Objective 1

1. $A(-3,0)$, $B(-1,3)$, $C(4,3)$, $D(3,-4)$

Objective 2

2.

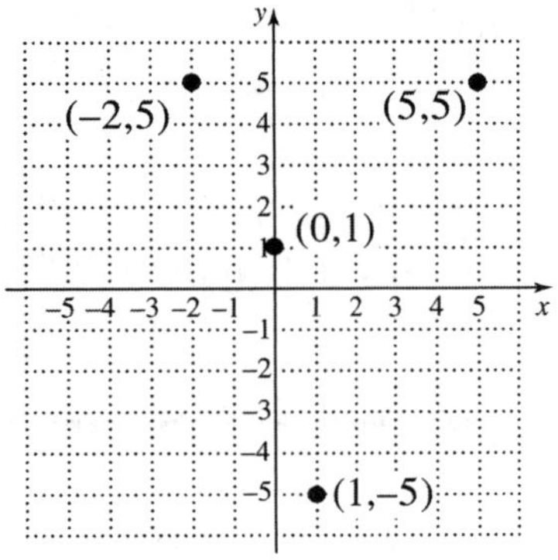

Objective 3

3. y-axis
4. II
5. x-axis
6. I
7. III

Objectivc 4

8. linear
9. nonlinear

Section 10.2

Practice

1. no

2. $(-1, 5)$, $(0, 2)$, $(1, -1)$

3.

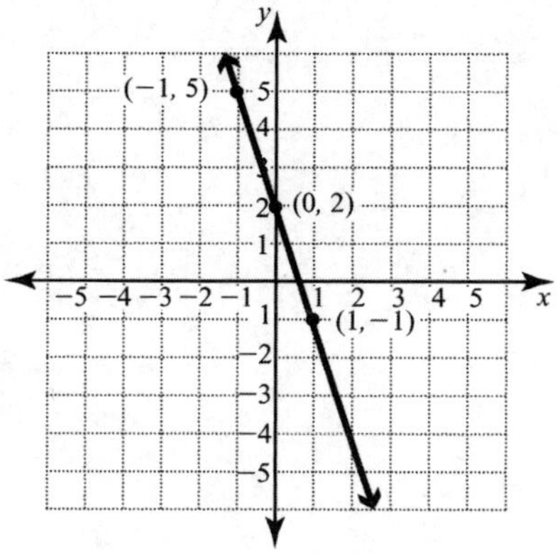

Objective 1

1. no

2. yes

Objective 2

3. $(-1, -4)$, $(0, 0)$, $(1, 4)$

4. $(-4, 4)$, $(0, 3)$, $(4, 2)$

5. $(-5, -8)$, $(0, -4)$, $(5, 0)$

Objective 3

6.

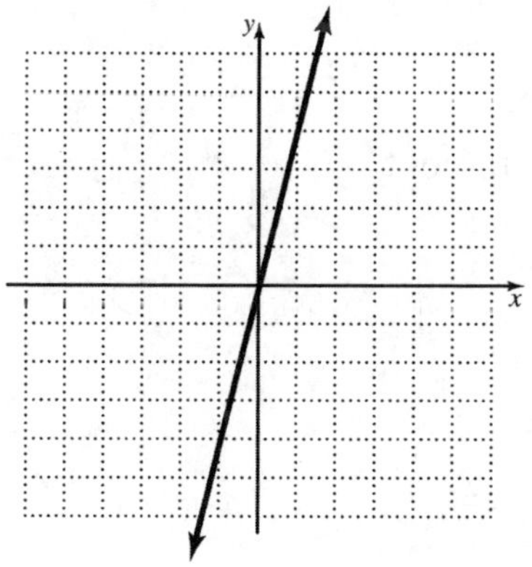

7.

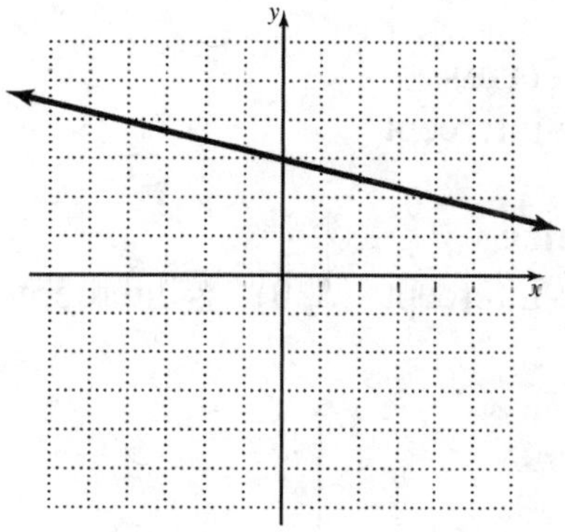

8.

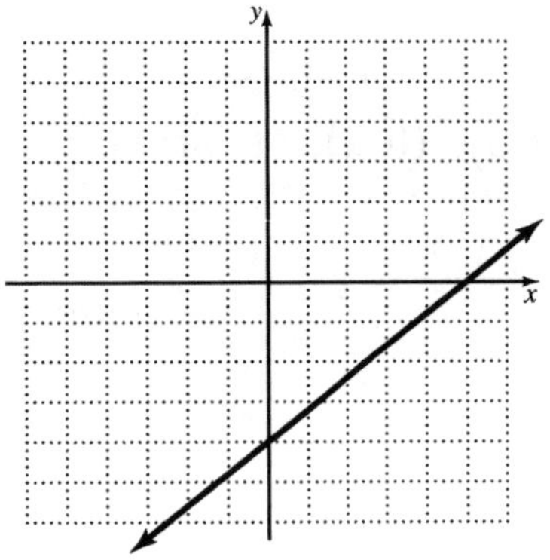

9. a) $1200; b) 4 yr; c) 8 yr
d)

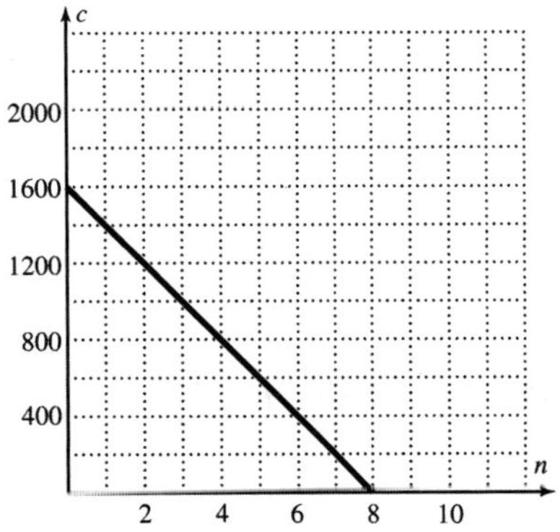

Section 10.3

Key Terms

1. *y*-intercept
2. *x*-intercept

Practice

1. *x*-intercept: $(5, 0)$; *y*-intercept: $(0, 6)$
2.

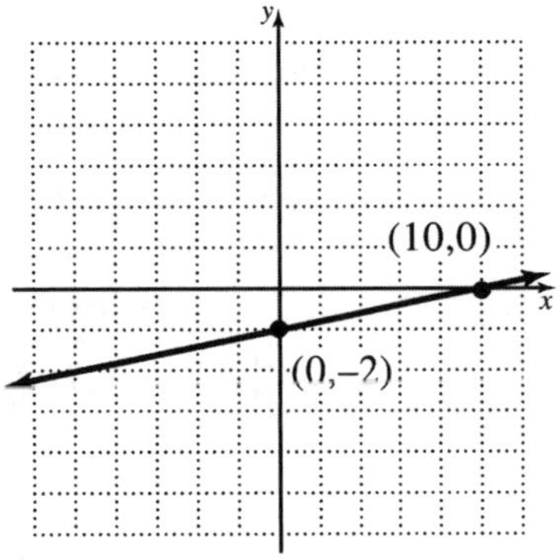

Objective 1

1. $-\frac{5}{2}, 0 \ ,(0,5)$

2. $(4,0)$; no y-intercept

Objective 2

3.

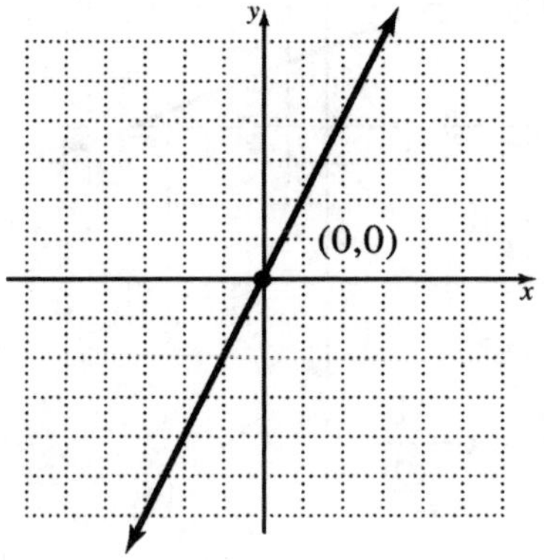

4.

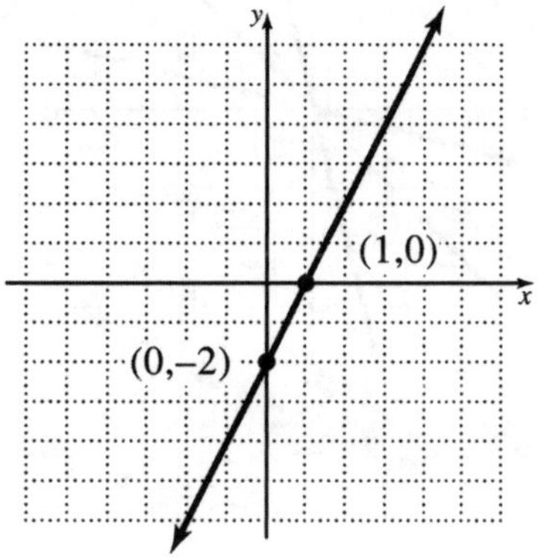

5.

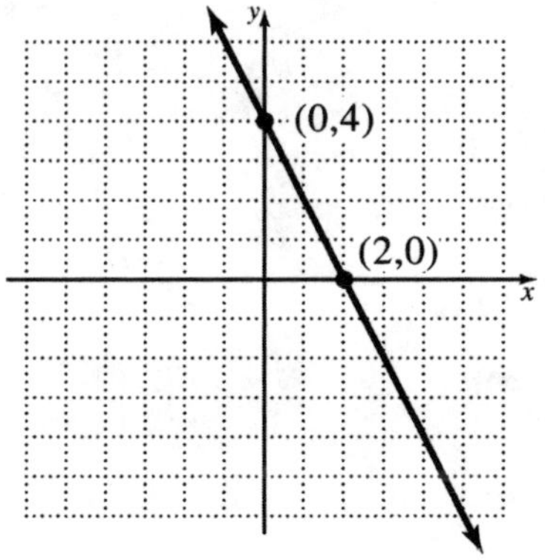

6.

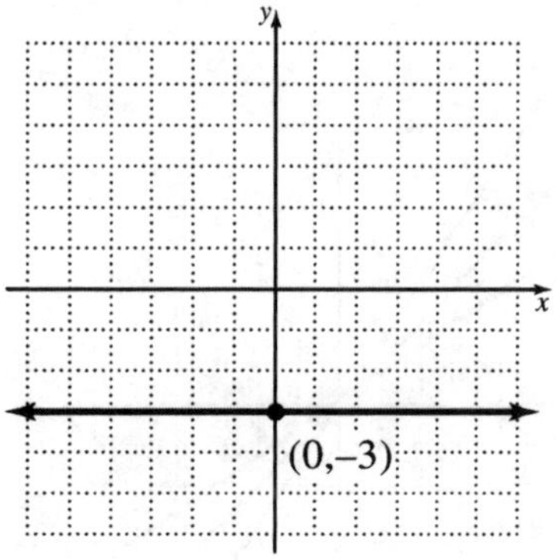

7.

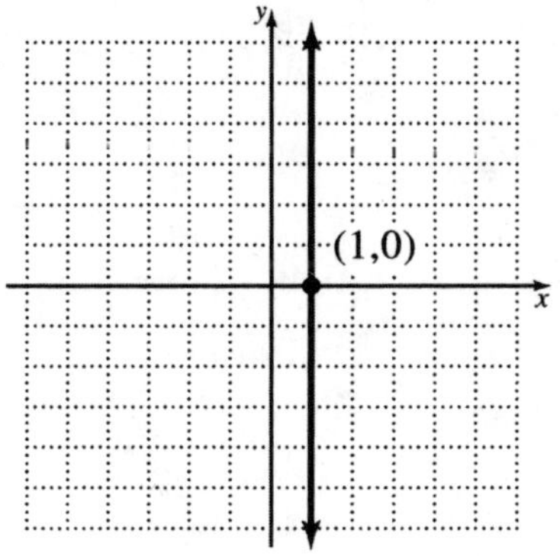

Section 10.4

Key Terms

1. slope

Practice

1.

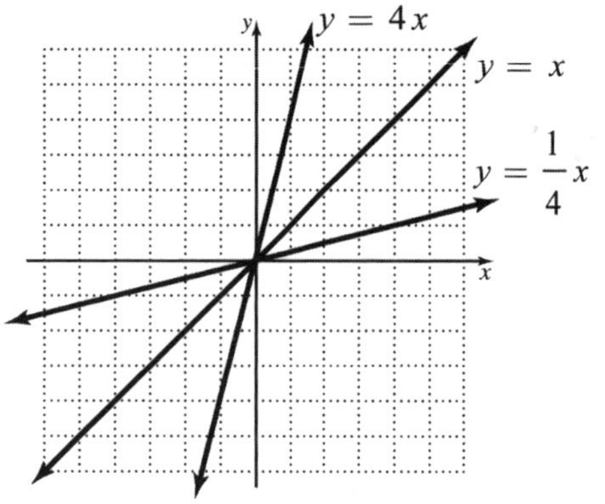

2. $m = -\frac{2}{5}; (0, 2)$

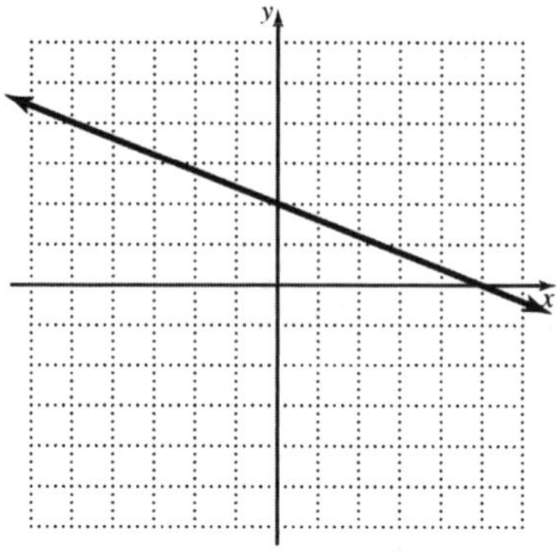

3. $\frac{1}{6}$

4. $y = 4x - \frac{1}{2}$

Objective 1

1.

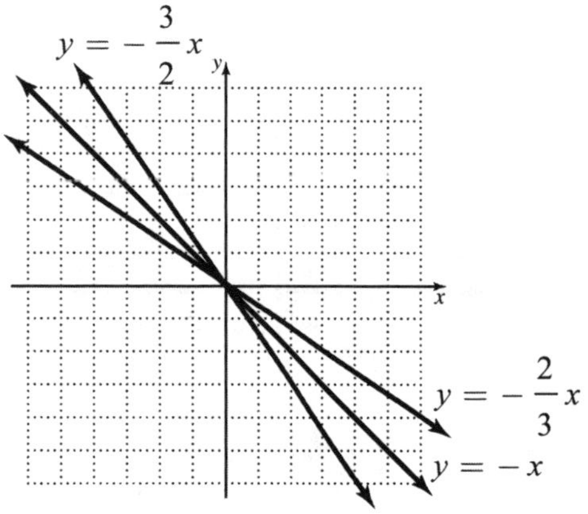

Objective 2

2. $m = -2; (0, 4)$

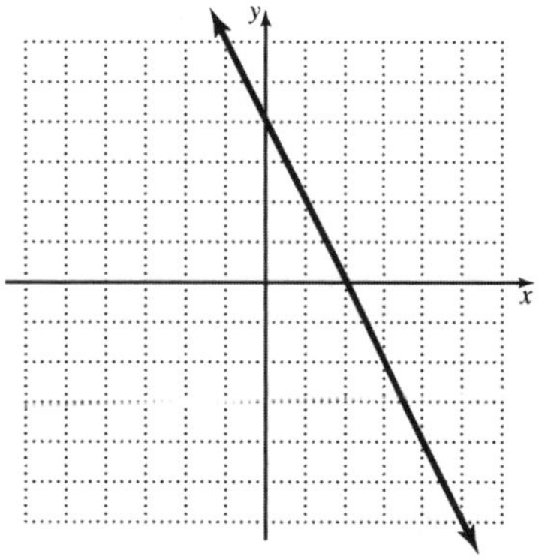

Objective 3

3. $\frac{3}{7}$

4. 1

5. undefined

Objective 4

6. $y = -0.4x + 2.3$

Section 10.5

Practice

1. $y = -8x + 65$

2. a) $y = 5x + 2$; b) $5x - y = -2$

3. a) $y = -2x + 14$; b) $2x + y = 14$

Objective 1

1. $y = \frac{3}{2}x - 7$

2. $y = -\frac{1}{2}x - \frac{21}{2}$

3. $y = \frac{1}{4}x + \frac{7}{2}$

4. $\frac{5}{4}x + y = +5$

5. $2x - y = 5$

6. $7x + 4y = 47$

7. $5x + 4y = -12$

Objective 2

8. a) $y = \frac{4}{7}x - \frac{22}{7}$;; b) $4x - 7y = 22$

Objective 3

9. a) $y = \frac{1}{5}x + \frac{43}{5}$ b) $x - 5y = -43$

10. parallel

11. perpendicular

Section 10.6

Practice

1. yes

2.

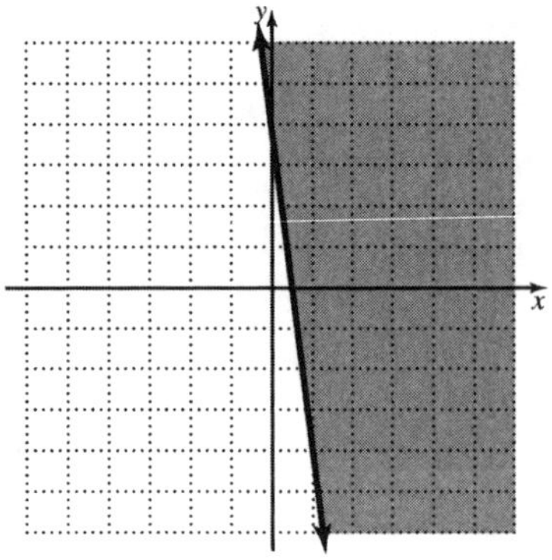

Objective 1

1. yes

2. no

Objective 2

3.

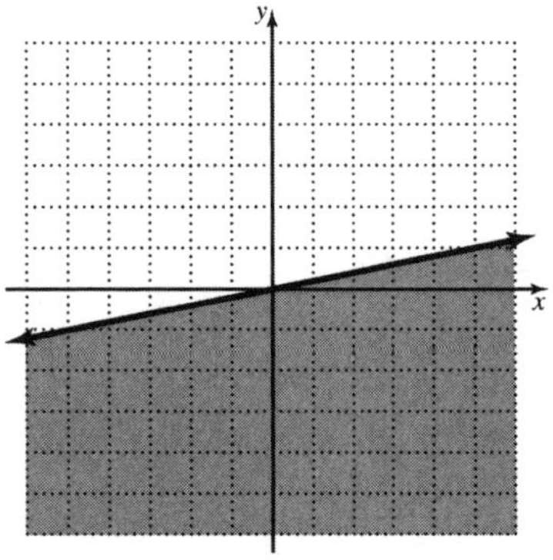

4.

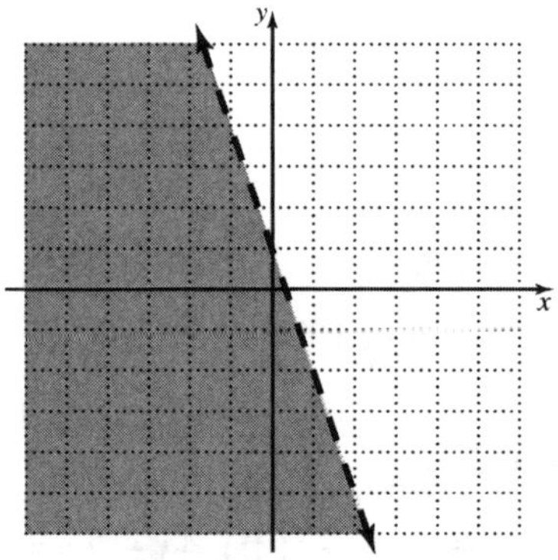

5.

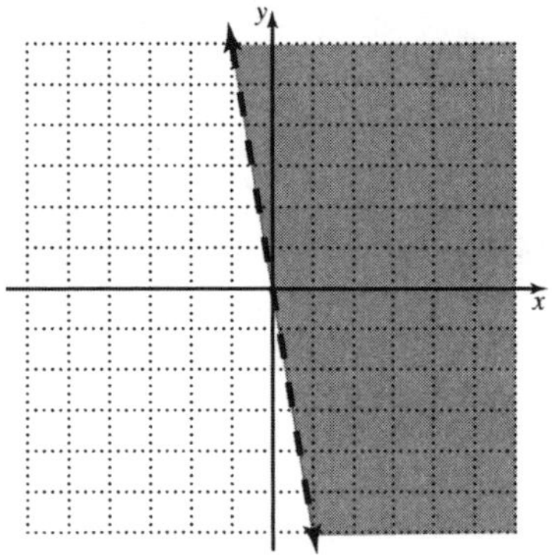

6. $6x + 15y \geq 120$

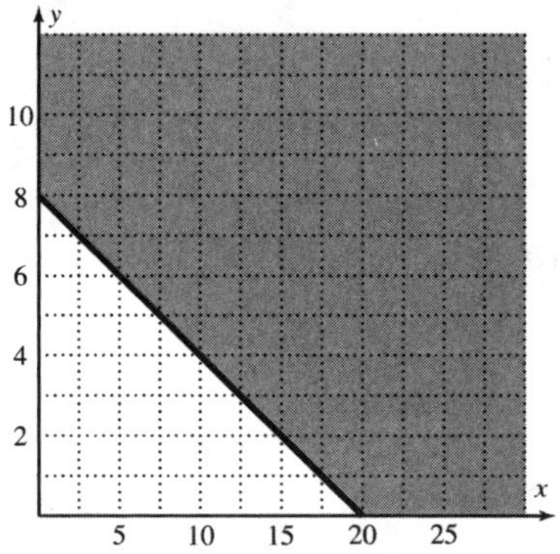

Section 10.7

Key Terms

1. domain
2. relation
3. function
4. range

Practice

1. domain: $\{7, 24, 38, 50\}$; range: $\{2, -9, 8\}$
2. No, an element in the domain is assigned to more than one element in the range. (5 is assigned to both 4 and 3.)
3. a) 10; b) 40; c) 10
4.

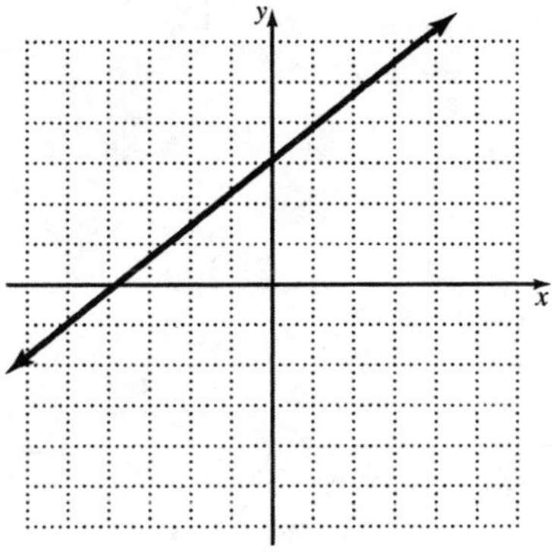

5.

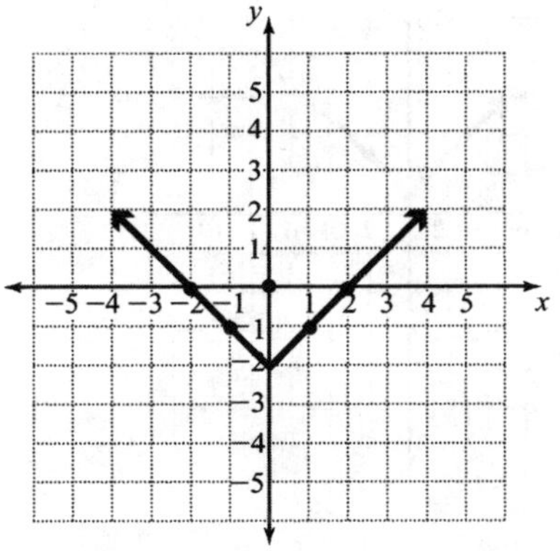

Objective 1

1. domain: $\{6, 16, 35, 59\}$; range: $\{4, -2, 2\}$

Objective 2

2. Yes, every element in the domain is assigned to exactly one element in the range.

Objective 3

3. a) $\sqrt{7}$; b) not a real number; c) $\sqrt{b+7}$ 4. -2

Objective 4

5.

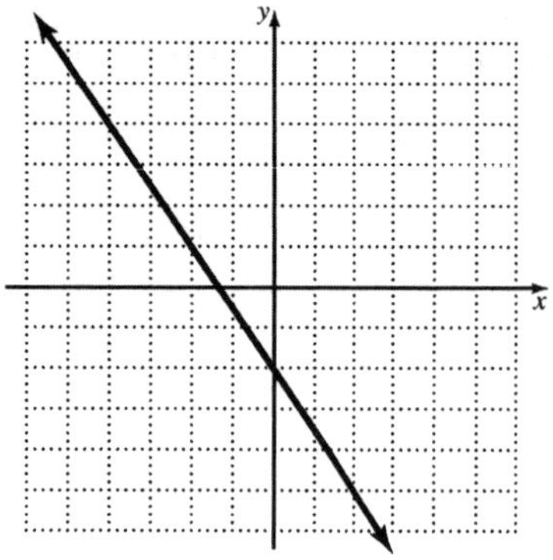

Objective 5

6.

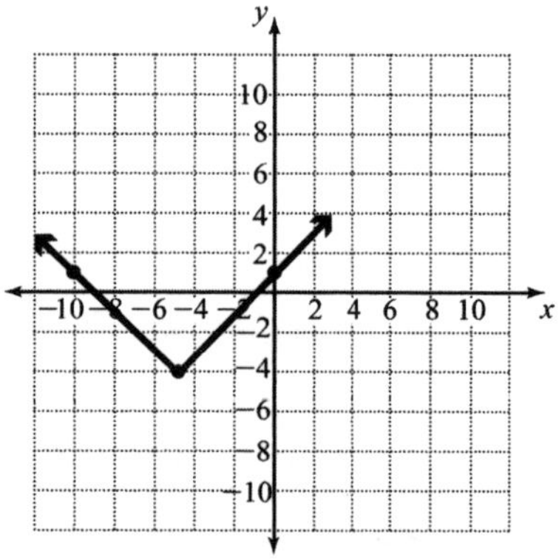

7.

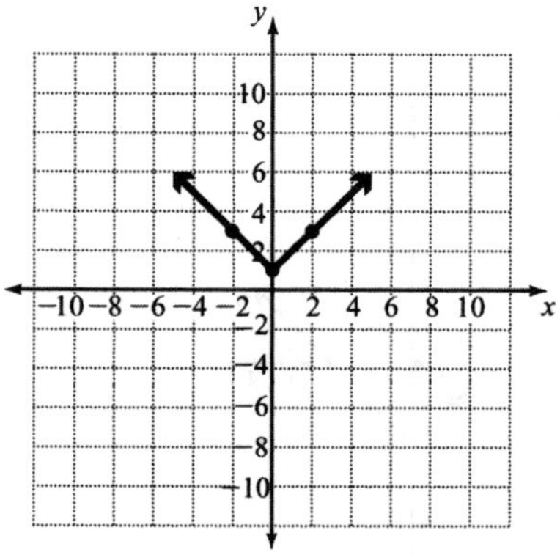

8.

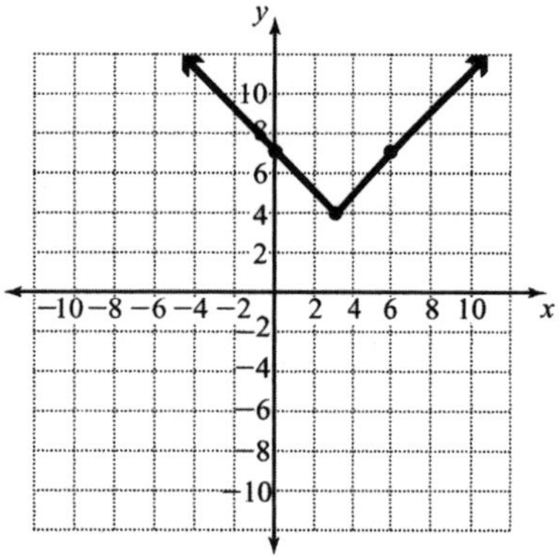

Chapter 11

Section 11.1

Key Terms

1. inconsistent system of equations
2. system of equations
3. solution for a system of equations
4. consistent system of equations

Practice

1. yes
2. $(4, 6)$
3. a) inconsistent; b) no solution

Objective 1

1. no
2. no

Objective 2

3. $(1, 0)$
4. $(1, 0)$
5. no solution
6. all ordered pairs along $4x - 5y = 2$

Objective 3

7. a) consistent with dependent equations; b) infinite number of solutions
8. a) consistent with independent equations; b) one solution

Section 11.2

Practice

1. $(3, 1)$
2. ones: 32; tens: 13

Objective 1

1. $(-7, 4)$
2. $(8, -6)$
3. $(4, 12)$
4. all ordered pairs along $x - 2 = y$
5. $\frac{5}{2}, 0$
6. $(-4, 6)$
7. no solution (inconsistent)
8. $(1, 4)$

Objective 2

9. 22, 13
10. 61°, 119°
11. scientific calculators: 17; graphing calculators: 28
12. 431.25 km

Section 11.3

Practice

1. $(4, -1)$
2. 450 mph

Objective 1

1. $(8, 7)$
2. $(-8, 6)$
3. $(4, 3)$
4 $(11, 17)$

Objective 2

5. 30°, 150°
6. width: 6 ft; length: 8 ft

7. \$20,000 at 6%; \$9000 at 10%
8. \$850 at 4%; \$1700 at 5%
9. 3 L of 25%; 12 L of 10%
10. 392 kg of Kenyan; 196 kg of Turkish
11. \$5000
12. 4200 kg

Section 11.4

Practice

1. yes
2. infinitely many solutions
3. $(-2, -2, 2)$
4. angle A: 21°; angle B: 89°; angle C: 70°

Objective 1

1. no
2. no

Objective 2

3. no solutions

Objective 3

4. $(5, -4, -4)$
5. $(-3, 0, -4)$
6. no solution (inconsistent)
7. infinite number of solutions; (dependent equations)
8. $(-4, 4, -5)$
9. $(6, -6, 15)$
10. $(4, 4, 2)$

Objective 4

11. 1-point foul shots: 12; 2-point field goals: 31; 3-point field goals: 2

Section 11.5

Key Terms

1. augmented matrix
2. row echelon form
3. matrix

Practice

1. $\left[\begin{array}{cc|c} 1 & 6 & 5 \\ 3 & -5 & 7 \end{array}\right]$
2. $(4, -2)$
3. 2850 mi; 3070 mi

Objective 1

1. $\left[\begin{array}{ccc|c} 2 & 8 & 3 & 5 \\ 5 & -4 & 2 & 9 \\ 0 & 4 & -6 & -9 \end{array}\right]$
2. $\left[\begin{array}{ccc|c} 1 & 1 & 0 & 7 \\ 1 & 2 & 5 & 16 \\ 0 & 5 & -3 & -1 \end{array}\right]$

Objective 2

3. $(-31, 7)$

4. $(9, 3, 1)$

5. $\left[\begin{array}{cc|c} 4 & 3 & 1 \\ 0 & 3 & 9 \end{array}\right]$

6. $\left[\begin{array}{ccc|c} 1 & 7 & 4 & 0 \\ 0 & 50 & 23 & 4 \\ 5 & 7 & 0 & 2 \end{array}\right]$

7. $(5, -3)$

8. $(-2, 20, 12)$

9. $(3, -3, 2)$

10. $(5, 4)$

11. $(4, -4, 3)$

12. $(-2, -2, 0)$

Objective 3

13. T-shirt: $8; belt: $6.

Section 11.6

Practice

1.

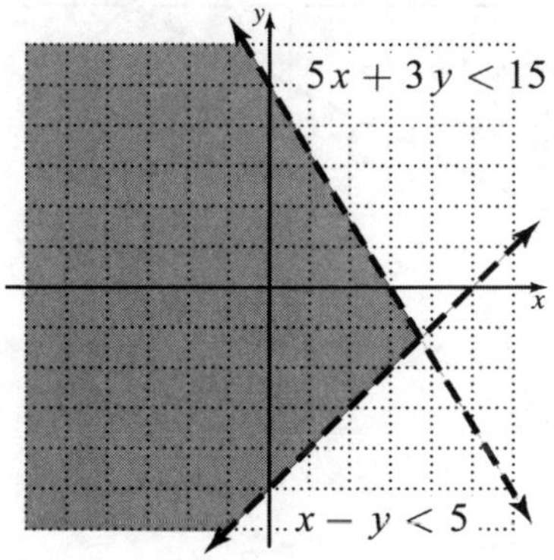

2.

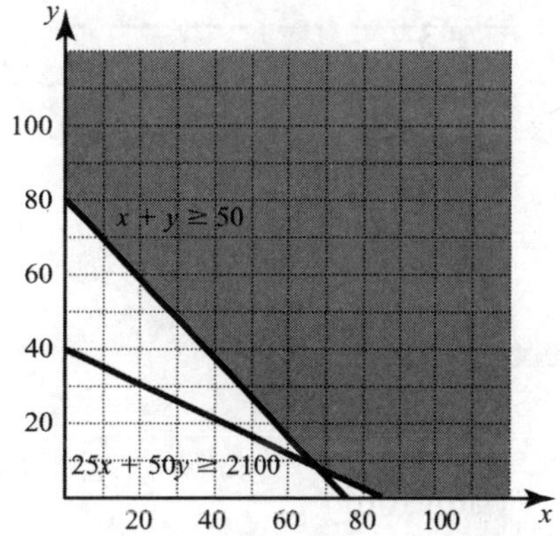

Objective 1

1.

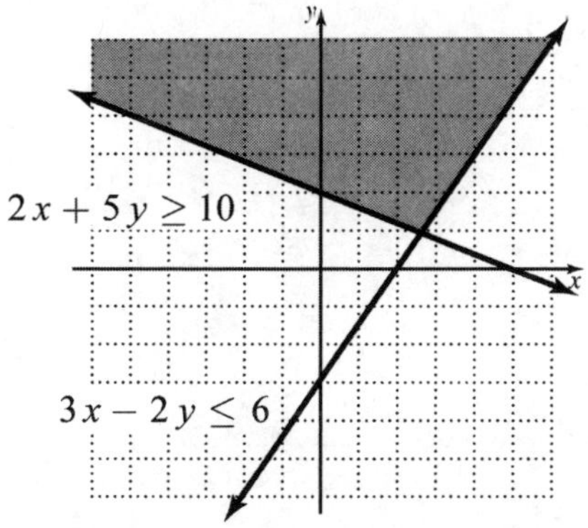

2.

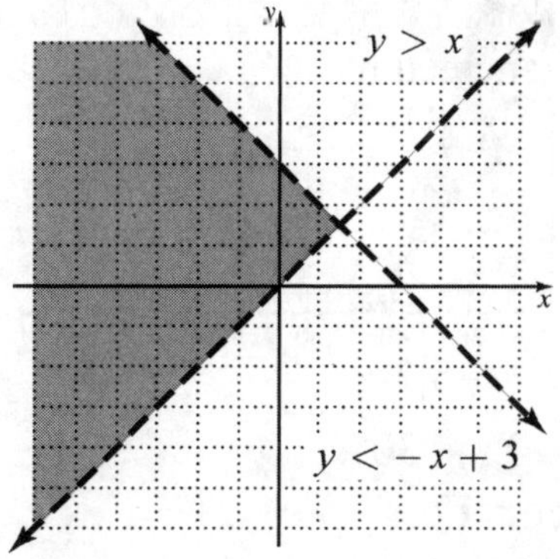

3. no solution

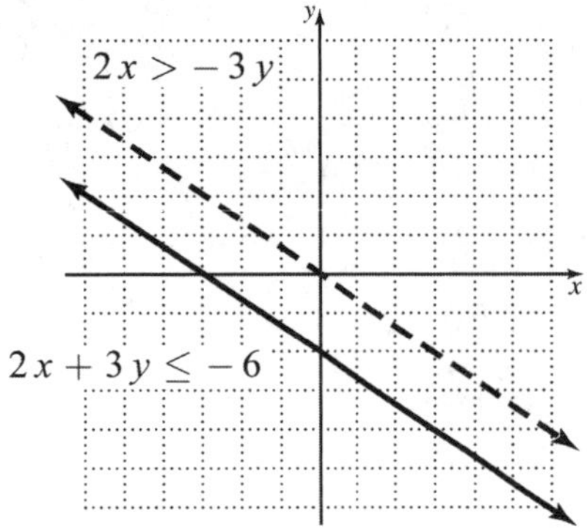

4.

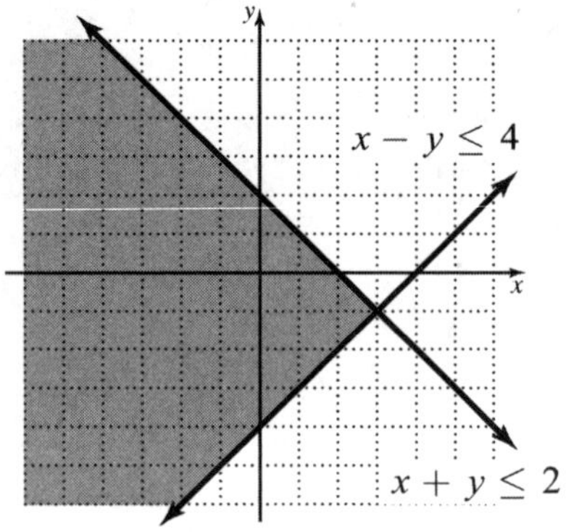

5.

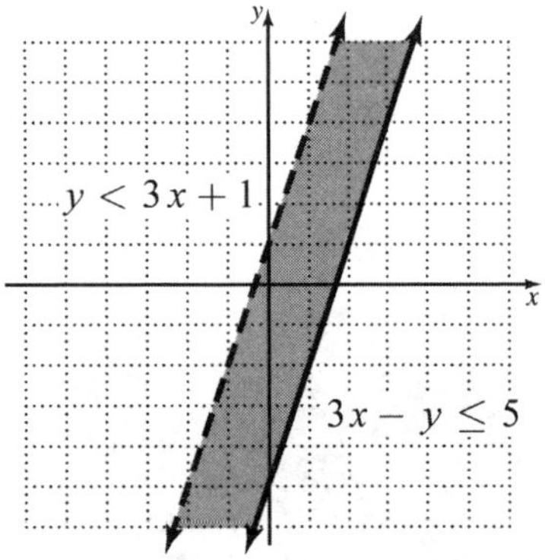

6.

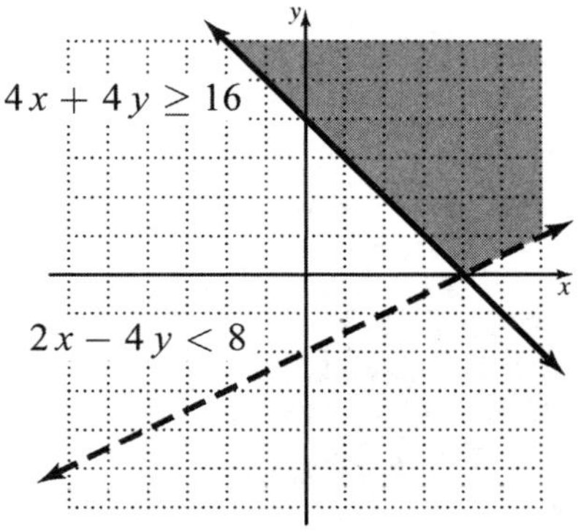

Objective 2

7.

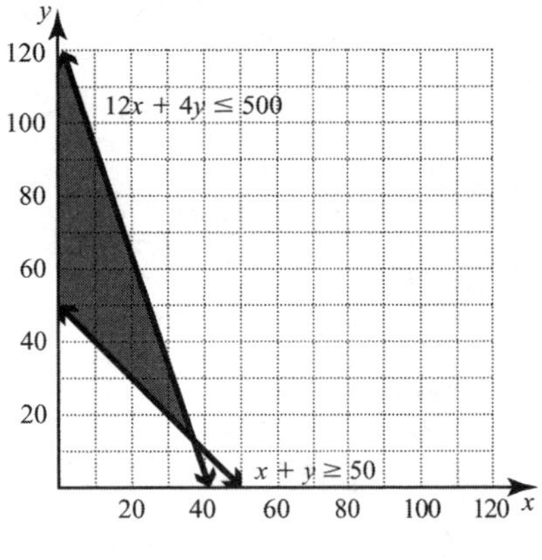

Chapter 12

Section 12.1

Key Terms

1. scientific notation

Practice

1. –49
2. $\frac{1}{36}$
3. 412,000
4. 2.5×10^7

Objective 1

1. $\frac{8}{125}$
2. 1
3. –64
4. $-\frac{1}{36}$
5. $\frac{625}{256}$
6. $-\frac{1}{27}$

Objective 2

7. 7,320,000
8. 0.0000611
9. 0.000000143

Objective 3

10. 7.518×10^9
11. 4.8×10^{-9}

Section 12.2

Key Terms

1. degree of a polynomial
2. coefficient of a monomial
3. trinomial
4. polynomial
5. monomial
6. polynomial in one variable
7. binomial
8. degree of a monomial

Practice

1. monomial
2. coefficient: 6; degree: 7
3. trinomial
4. 7
5. 37
6. $x^5 + 8x^3 + 6x^2 + x + 7$
7. $3b^5 - b^3 - 6b^2$

Objective 1

1. not a monomial
2. is a monomial

Objective 2

3. coefficient: 1; degree: 3
4. coefficient: $\frac{3}{4}$; degree: 6

Objective 3

5. monomial; 0
6. no special polynomial name; 7

Objective 4

7. 5
8. 9

Objective 5

9. 77
10. 89.5 in.^{2}

Objective 6

11. $6z^3 + z^2 + z + 875$
12. $2m^{10} + m^5 + m^3 + 34m$

Objective 7

13. $-22a^2p - 2p^2$
14. $3m^2n + m^2 + 3n^2 + 6mn$

Section 12.3

Practice

1. $-x + 14$
2. $18x + 2$
3. $2x - 4$

Objective 1

1. $-x + 17y$
2. $x^2 - 8x - 2$
3. $16x^2 + 2x - 74$
4. $6x^2 + 11$
5. $-5x^4 + 6x^3 - 2x^2 + 15x - 7$
6. $6x^5 - 4x^4 + 5x^3 + 2x^2 + 7$
7. $14x - 12$

Objective 2

8. $-10v^2 - 2v + 9$
9. $-12a^2 - 4a + 13$
10. $2y^5 + 3y^4 + 5y^2 - 10$
11. $-2w^2 + 16wt - 18t^2$

Section 12.4

Practice

1. v^9
2. $7x^2$
3. 7.44×10^8
4. c^{16}

Objective 1

1. 7^{12}
2. $35r^3$
3. $-20x^{13}y^9$
4. $36q^8r^5$
5. $60x^6y^5$
6. $3a^7b^4$
7. $10x^2$

Objective 2

8. 2.63088×10^4
9. 5.7981×10^2

Objective 3

10. x^{28}

11. $\frac{x^6y^3}{27}$

12. $-r^{22}s^{11}$

13. $225a^7b^3$

Section 12.5

Key Terms

1. conjugates

Practice

1. $-5x^2+15x$
2. $b^2+2b-99$
3. x^3-6x^2-49
4. $a-7$
5. s^2-36

Objective 1

1. $16x^3-36x^2+36x$
2. $-36y^6-16y^5+24y^4$

Objective 2

3. $12x^2+5x-3$
4. $16v^2-4vf-20f^2$

Objective 3

5. $18x^3+28x^2+46x+20$
6. $r^4+8r^3-3r^2-16r+9$

Objective 4

7. $\frac{1}{2}b+2c$
8. $9r^2-25$
9. $25x^2+40xy+16y^2$
10. $y^2-10y+25$

Section 12.6

Practice

1. $\frac{1}{z^7}$
2. -7×10^{-3}
3. $-5a^3b^4$
4. $x+4y$
5. $x+1$
6. m^6

Objective 1

1. 3^3
2. x^5

Objective 2

3. -4×10^{-7}
4. -7.52×10^{-8}

Objective 3

5. $-\dfrac{t^2}{5s^3}$

6. $\dfrac{2a^4}{15b^6}$

Objective 4

7. $3a^4 - 3a^2 + \dfrac{3}{a}$

8. $3z^3 + 2z + \dfrac{1}{z^3}$

Objective 5

9. $x + 6 + \dfrac{-4}{x-7}$

10. $x^2 - 9x + 81$

11. $4y - 21 + \dfrac{315}{3y+8}$

Objective 6

12. $\dfrac{125}{64}$

13. 6

Chapter 13

Section 13.1

Key Terms

1. factored form
2. greatest common factor (GCF)

Practice

1. 1, 2, 3, 6, 9, 18
2. 20
3. $8m^3n^2(n+6)$
4. $(q+z)(s+f)$

Objective 1

1. 1, 5, 25, 125, 625

Objective 2

2. 4

Objective 3

3. $x(x-5)$
4. $5ay(a-3)$
5. $3xy(x^6y^4 + 7x^4y^3 + 4)$
6. $-2z(3z-7)$
7. $(c-4)(x+t)$
8. $(7m-4)(6m-5)$

Objective 4

9. $(s+3)(s+7)$
10. $(z+2)(4z-a)$

11. $(r-5t)(r+2w)$

Section 13.2

Practice

1. $(r-7)(r-10)$
2. $(w+6z)(w+7z)$
3. $2x(x+2)(x+5)$

Objective 1

1. 11
2. 7
3. 11
4. 10
5. $(t+2)(t+4)$
6. $(b-5)(b-9)$
7. $(w+5)(w-6)$
8. prime
9. $6(c-3)(c-7)$
10. $(b+3x)(b-9x)$
11. $(s-3z)(s-5z)$

Objective 2

12. $4x^5(x-4)(x+2)$

Section 13.3

Practice

1. $(t-5)(t-7)$
2. $(4u+5)(u+1)$

Objective 1

1. $(t-2)(t-3)$
2. prime
3. $(4a+1)(a+5)$
4. $(w-5z)(w-4z)$
5. $(a-9f)(a-7f)$
6. $2(2s-1)(s-5)$

Objective 2

7. $(9v-2)(2v-9)$
8. $2(7u-3)(4u-7)$
9. $(7b-1)(4b+3)$
10. $(5c+3)(5c+9)$
11. $(3a+2)(3a+4)$
12. $(5c-6d)(6c+5d)$
13. $5u(6u-7)(u+6)$

Section 13.4

Practice

1. $(5r+7t)^2$
2. $(3a+11g)(3a-11g)$
3. $(r-3)(r^2+3r+9)$
4. $(b+4)(b^2-4b+16)$

Objective 1

1. $(s+6)^2$
2. $(v-8)^2$
3. $(5a-9)^2$
4. $(3c-4g)^2$

Objective 2

5. $(r+6)(r-6)$
6. $(2c+11)(2c-11)$
7. $5m+\frac{6}{5}\quad 5m-\frac{6}{5}$
8. $5(3a+7)(3a-7)$

Objective 3

9. $(3x-4b)(9x^2+12bx+16b^2)$
10. $(y+z-2)(y^2+2yz+z^2+2y+2z+4)$

Objective 4

11. $(3x+4g)(9x^2-12xg+16g^2)$

Section 13.5

Practice

1. $(r+2)(r+8)$
2. $(w+x)(v+k)$
3. $(v+3)(v^2-3v+9)$

Objective 1

1. $(c-5)(a+z)$
2. $6(2b+7)(2b-7)$
3. $(t+6)(t-5)$
4. $(t+5)(t-6)$
5. $(b+5)(b-5)$
6. $(5x-6a)(25x^2+30ax+36a^2)$
7. $3(9u-2)(2u-9)$
8. $(s^2+9)(s+3)(s-3)$
9. $(x+5)(x-9)$
10. $(2v-9)^2$
11. $x(x-6)$
12. $r(3+10r)(3-10r)$

Section 13.6

Key Terms

1. quadratic equation in one variable

Practice

1. 0, -4
2. 5
3. 5, 7
4. 13

Objective 1

1. 25, −64
2. $-\frac{9}{4}$, −8

Objective 2

3. 1, 24
4. 3
5. −3, −4
6. 4, −4
7. 6, $-\frac{4}{5}$
8. −4, 9

Objective 3

9. 8 m by 13 m
10. 2.5 sec and 3.5 sec

Objective 4

11. 5

Section 13.7

Key Terms

1. vertex of a parabola
2. quadratic equation in two variables
3. axis of symmetry

Practice

1.

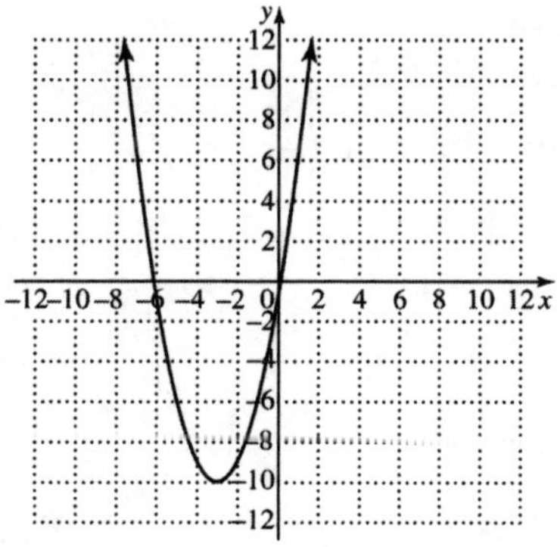

2.

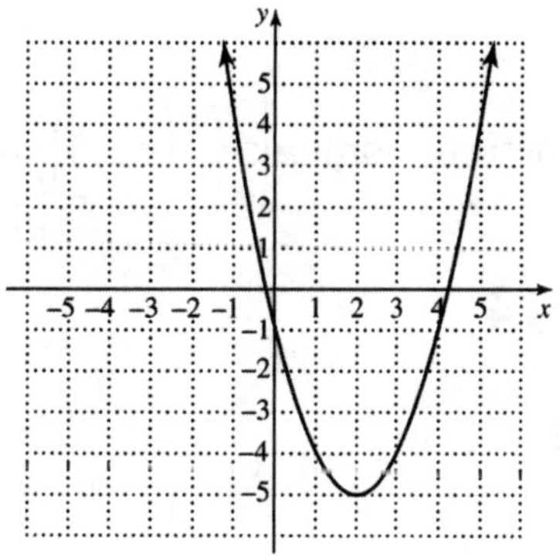

Objective 1

1.

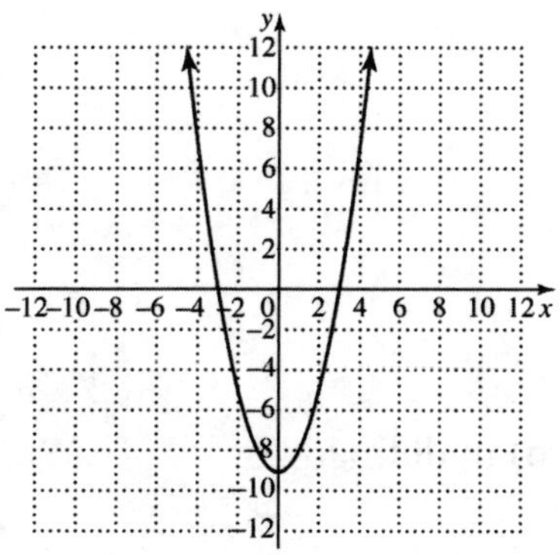

2.

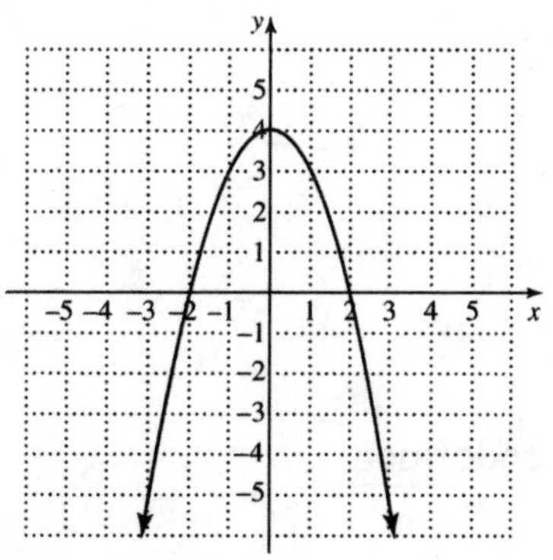

3.

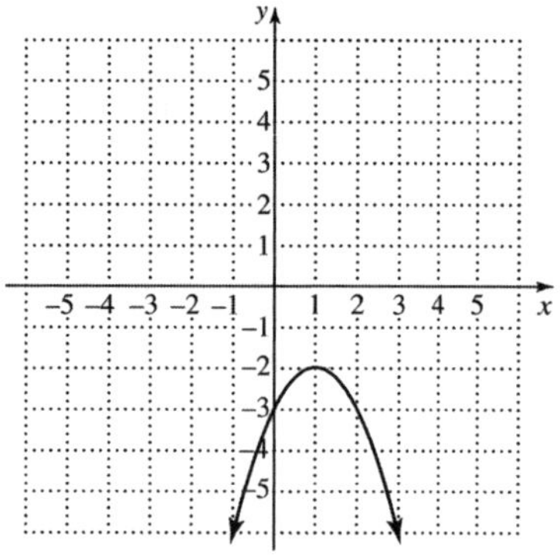

Objective 2

4.

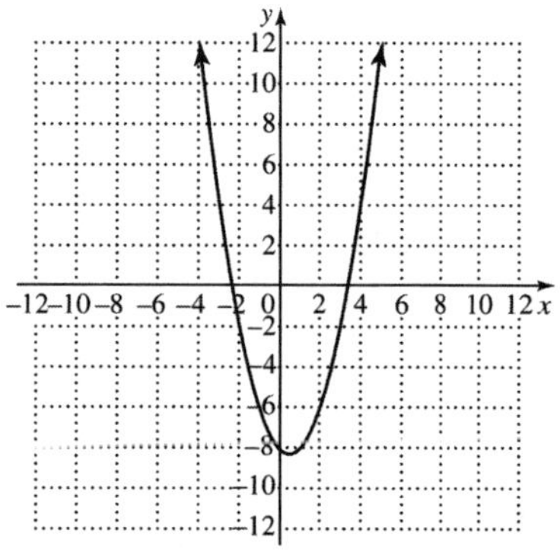

5.

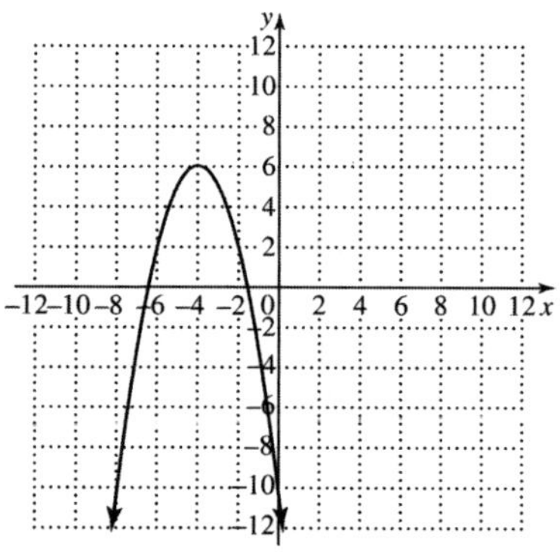

6. not a function; domain: $\{x|x \le 4\}$ or $(-\infty, 4]$; range: all real numbers or $(-\infty, \infty)$

Chapter 14

Section 14.1

Key Terms

1. rational expression

Practice

1. a) 1; b) $\frac{34}{9}$
2. 5
3. $\frac{125u^4z^8}{2}$
4. $\frac{8}{7}$

Objective 1

1. a) 2; b) undefined
2. a) $\frac{8}{25}$; b) undefined

Objective 2

3. $4, \; -4$

4. $3, \; -1$

Objective 3

5. $20u^2z$

Objective 4

6. $\dfrac{a-d}{a+d}$

7. -6

8. $\dfrac{a^2+ac+c^2}{a+c}$

9. $\dfrac{r+7}{r-7}$

10. $\dfrac{w^2+1}{w+1}$

11. $\dfrac{15}{6}$

Section 14.2

Practice

1. m^4

2. $\dfrac{1}{9}$

3. 180 in.

Objective 1

1. $\dfrac{8}{7}$

2. $-\dfrac{4}{7}$

3. $\dfrac{x+2}{2x(x-2)}$

4. $\dfrac{x-2}{(x+2)(x-5)}$

5. $\dfrac{c-d}{c+d}$

Objective 2

6. $\dfrac{5x^2}{2y^3}$

7. $\dfrac{w(w-4)(w+4)}{x}$

8. $\dfrac{4}{3}$

9. $\dfrac{v+4}{(v+3)(5v-3)}$

10. $\dfrac{3}{2}$

Objective 3

11. 80,000 ounces

12. 132 feet per second

Section 14.3

Practice

1. $h^2 - 2$
2. $\dfrac{1}{n^2}$

Objective 1

1. $\dfrac{x}{13}$
2. y
3. $\dfrac{1}{m+7}$
4. 2
5. $\dfrac{21}{w+1}$
6. $z+3$
7. $\dfrac{c+7}{y}$
8. $\dfrac{y-2}{y}$
9. z
10. 2
11. $\dfrac{z+5}{c}$
12. $x+15$
13. $\dfrac{3y-1}{3y(y-2)}$

Section 14.4

Practice

1. $33a^6$
2. $\dfrac{2x^2-2x}{(x+4)(x-1)}, \dfrac{7x+28}{(x+4)(x-1)}$
3. $\dfrac{4z-42}{(z+3)(z-3)}$

Objective 1

1. $12(x-2)$
2. $(p-2)(p+2)$

Objective 2

3. $\dfrac{4x^2+10x+4}{(x+3)(x-3)(x+2)}, \dfrac{2x^2-5x-3}{(x+3)(x-3)(x+2)}$
4. $\dfrac{25y}{5(x-3)^2}, \dfrac{4x-12}{5(x-3)^2}$

Objective 3

5. $\dfrac{49}{5z}$
6. $\dfrac{y+c}{y-c}$
7. $\dfrac{2}{y-5}$
8. $\dfrac{9y-56}{(y-7)(y+5)}$

9. $\dfrac{2x^2+54}{(x+5)(x-8)}$

Section 14.5

Key Terms

1. complex rational expression

Practice

1. -7
2. -7

Objective 1

1. $\dfrac{1}{ut}$
2. $\dfrac{19}{14}$
3. $\dfrac{1+3y}{1-6y}$
4. $\dfrac{29}{14}$
5. $\dfrac{t^2-1}{t^2+1}$
6. $\dfrac{y}{y-7}$
7. 14
8. $-\dfrac{1}{y(y+b)}$
9. $y-1$

Section 14.6

Key Terms

1. extraneous solution

Practice

1. ycs
2. $-\dfrac{3}{8}$

Objective 1

1. yes
2. no
3. no solution (6 is extraneous)
4. $\dfrac{63}{11}$
5. 9
6. 18
7. 45
8. $\dfrac{39}{7}$
9. $-\dfrac{157}{6}$
10. $\dfrac{8}{5}$
11. $\dfrac{37}{3}$

Section 14.7

Key Terms

1. inverse variation
2. direct variation
3. joint variation

Practice

1. 2 hr
2. 40 kg
3. 4.75
4. 16.5
5. 607.64

Objective 1

1. $2\frac{2}{9}$ hr
2. freighter: $\frac{75}{4}$ km per hr; steamboat: $\frac{123}{4}$ km per hr
3. 10 mph

Objective 2

4. 6.9
5. 32

Objective 3

6. 420 m
7. 26

Objective 4

8. 1440 cubic inches
9. 40

Objective 5

10. 18
11. 9

Chapter 15

Section 15.1

Key Terms

1. *n*th root
2. radical function.

Practice

1. a) $\sqrt[4]{-16}$ is not a real number; b) 6; c) −3
2. 2.962
3. $6x^8y^{17}$
4. 11
5. $x \mid x \geq -\frac{9}{5}$ or $-\frac{9}{5}, \infty$
6. 26 ft

Objective 1

1. 9
2. −2
3. 7

Objective 2

4. 7.874
5. 3.849

Objective 3

6. $4c^2$
7. $|x-6|$
8. $(x+6)^6$
9. $(y-2)^3$

Objective 4

10. $\sqrt{17}$
11. 8

Objective 5

12. $x \, \Big| \, x \le \frac{5}{2}$ or $\infty, \frac{5}{2}$
13. $\{x | x \ge -10\}$ or $[-10, \infty)$

Objective 6

14. 28 mph

Section 15.2

Key Terms

1. radical exponent

Practice

1. $-\sqrt{25} = -5$
2. $\sqrt[6]{(3y+x)^5}$
3. $19^{1/3}$
4. $16x^8 y^{24}$
5. $\sqrt[3]{a^2}$
6. $\sqrt[14]{y^9}$
7. $\sqrt[15]{z}$

Objective 1

1. $\sqrt[5]{x}$
2. $\left(\sqrt[3]{216}\right)^4 = 1296$
3. $\frac{1}{\left(\sqrt[4]{256}\right)^5} = \frac{1}{1024}$

Objective 2

4. $n^{5/6}$
5. $5xz^{-3/5}$

Objective 3

6. $a^{5/2}$
7. $6^{1/2}$
8. $-42y^{1/6}$

Objective 4

9. x^3y^4z
10. $\sqrt[4]{m^2n}$
11. $y^{5/6}$
12. $n^{7/4}$

Section 15.3

Practice

1. $10pn^4$
2. $\dfrac{4}{7}$
3. $2\sqrt{10}$
4. $30x^4\sqrt{3}$

Objective 1

1. $\sqrt[3]{85ab}$
2. $\sqrt{\dfrac{7x}{5y}}$

Objective 2

3. 7
4. 3

Objective 3

5. $x^8 \cdot y^5$
6. $3x^2\sqrt[3]{9x^2}$
7. $2\sqrt{35}$
8. $a^4b^3\sqrt{ab}$
9. $5\sqrt{7}$
10. x^3y
11. $18xy^4\sqrt{2xy}$

Section 15.4

Key Terms

1. like radicals

Practice

1. $5\sqrt{5}$
2. $7+\sqrt{70}$
3. $56\sqrt{21}+21\sqrt{14}-32\sqrt{15}-12\sqrt{10}$
4. $20\sqrt{3}$

Objective 1

1. $4b\sqrt[3]{3}$
2. $\sqrt{5}$
3. $-24\sqrt{5}$
4. $-7x^2\sqrt{x}$
5. $(3-x)\sqrt[3]{2x}$

Objective 2

6. $30x - 70x\sqrt{2}$
7. $3\sqrt{5} + 5\sqrt{15} - 5\sqrt{3} - 25$
8. 13
9. $2\sqrt[3]{49} - 3\sqrt[3]{14} - 2\sqrt[3]{4}$
10. $46 + 12\sqrt{10}$
11. -190
12. $3c - d$

Objective 3

13. $72 - 48\sqrt{3}$

Section 15.5

Practice

1. $\frac{7\sqrt{6y}}{6y}$
2. $-18 + 9\sqrt{5}$
3. $\frac{7x}{6\sqrt{7x}}$

Objective 1

1. $\frac{5\sqrt{7}}{14}$
2. $\frac{2\sqrt{5}}{25}$
3. $\frac{x\sqrt{15}}{15}$
4. $\frac{5\sqrt[3]{4}}{2}$
5. $\frac{\sqrt[3]{18}}{3}$
6. $\frac{2\sqrt[3]{9x}}{3x}$
7. $\frac{\sqrt[3]{15z}}{5z}$
8. $-\frac{10 + 12\sqrt{5}}{31}$

Objective 2

9. $\frac{-7 - 6\sqrt{7}}{29}$
10. $\frac{4y - 24\sqrt{y}}{y - 36}$
11. $\frac{x - \sqrt{xy}}{x - y}$

Objective 3

12. $\frac{1}{\sqrt{11} - 2}$

Section 16.6

Key Terms

1. radical equation

Practice

1. 0, 256

Objective 1

1. 49
2. –64
3. 112
4. 444
5. 63
6. 59
7. 11
8. no real-number solution
9. 9
10. no real-number solution
11. 4
12. 5 (–6 is extraneous)
13. 1, 2
14. $-\frac{9}{5}$

Section 16.7

Key Terms

1. complex conjugate
2. complex number
3. imaginary unit
4. imaginary number

Practice

1. $3i\sqrt{5}$
2. $-12-16i$
3. i

Objective 1

1. $11i$
2. $i\sqrt{6}$

Objective 2

3. $9+i$
4. $3+9i$
5. $6+15i$
6. $-40-35i$
7. $89+33i$
8. $-3i$
9. $\frac{20}{13}-\frac{4}{13}i$
10. $\frac{32}{17}-\frac{9}{17}i$

Objective 3

11. 1
12. –1

Chapter 16

Section 16.1

Practice

1. $-25.5, -32.5$

2. $\frac{-10 \pm \sqrt{109}}{3}$

Objective 1

1. ± 1.4
2. $\pm 3\sqrt{3}$
3. $\pm 8i$
4. ± 7
5. ± 6
6. ± 12
7. $\frac{8}{3}, -4$
8. $8 \pm 9i$

Objective 2

9. $-1 \pm 4i$
10. $-3, 1$
11. $-7, 4$
12. $-\frac{5}{3}, \frac{1}{4}$
13. $-\frac{3}{8}, 5$

Section 10.2

Key Terms

1. discriminant

Practice

1. $-\frac{5}{3}, \frac{10}{3}$
2. two nonreal complex solutions
3. x-intercepts: $(-8, 0)$, $(2, 0)$
   y-intercept: $(0, 16)$
4. 5.6 sec

Objective 1

1. $-3, 6$
2. 2
3. $\frac{1 \pm 3i\sqrt{3}}{2}$

Objective 2

4. one rational solution
5. one rational solution
6. two irrational solutions

Objective 3

7. x-intercepts: $(3, 0)$, $(-2, 0)$
   y-intercept: $(0, -6)$

Objective 4

8. $x^2 + 4(x+1) = 81$; 7, 8

9. $w(2w-3) = 527$
   width: 17 in.
   length: 31 in.

Section 16.3

Key Terms

1. quadratic in form

Practice

1. $-\frac{5}{6}, 1$

2. $\frac{16}{81}$ (1 is extraneous)

3. Georgia: 35 mph; Benny: 40 mph

Objective 1

1. $-50, 55$

2. $-2, 3$

3. $-\frac{3}{2}, \frac{3}{2}$

4. 1, 25

5. 10 (5 is extraneous)

6. 2, 3

7. 342 (2 is extraneous)

Objective 2

8. $\pm 4, \pm 5$

9. $\pm\frac{\sqrt{10}}{5}, \pm\frac{\sqrt{2}}{2}$

10. 3, 4

11. -1, 27

Objective 3

12. Jane 7.831 hours; Mike 10.831 hours

Section 16.4

Practice

1. a) downward; b) $(0, 0)$; c) $x = 0$;

2. a) downward; b) $(0, 3)$; c) $x = 0$;

d)

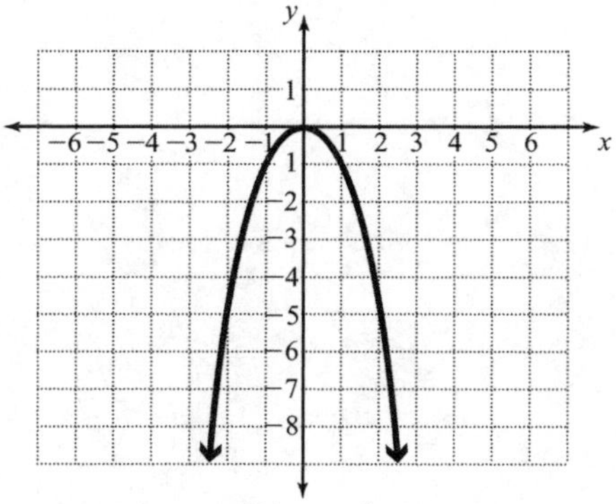

$f(x) = -x^2$

d)

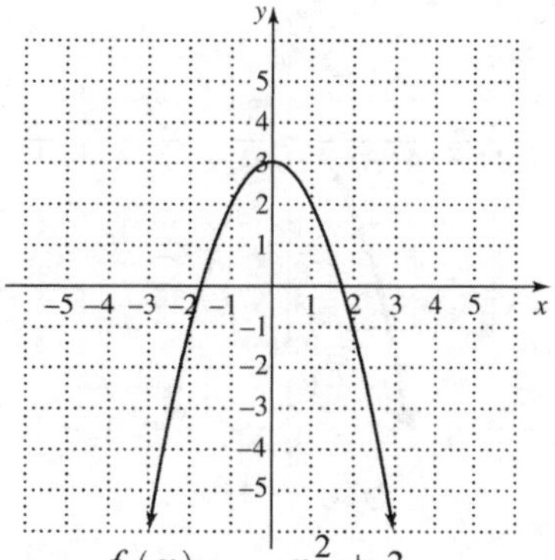

$f(x) = -x^2 + 3$

3. a) downward; b) $(-3, 0)$; c) $x = -3$;
   d)

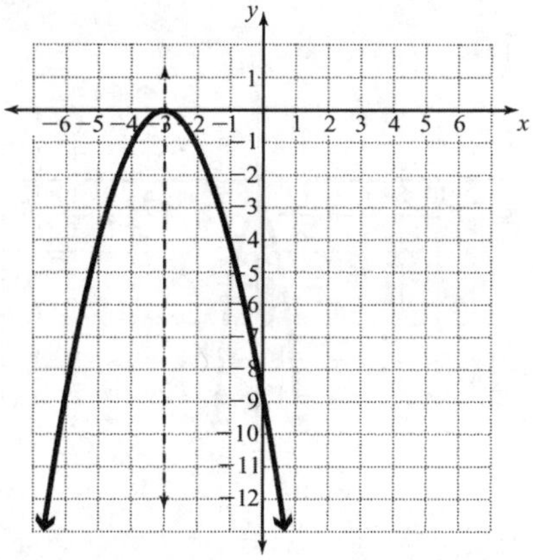

$f(x) = -(x+3)^2$

4. a) downward; b) $(7, 3)$; c) $x = 7$;
   d)

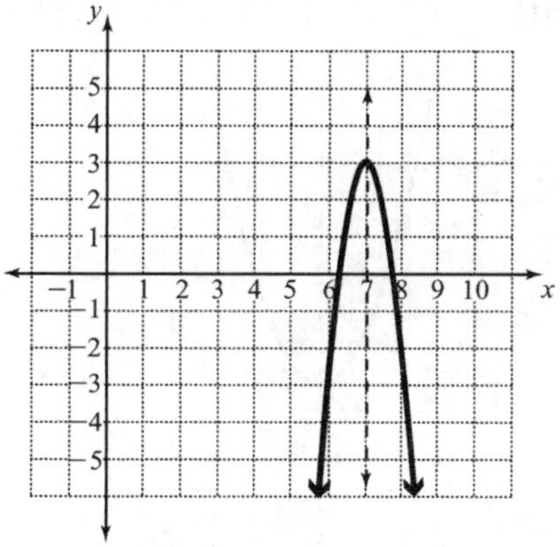

$f(x) = -6(x-7)^2 + 3$

5. a) $(0, 7)$; b) $k(x) = (x+2)^2 + 3$; c) upward; d) $(-2, 3)$; e) $x = -2$;
   f)

$k(x) = x^2 + 4x + 7$

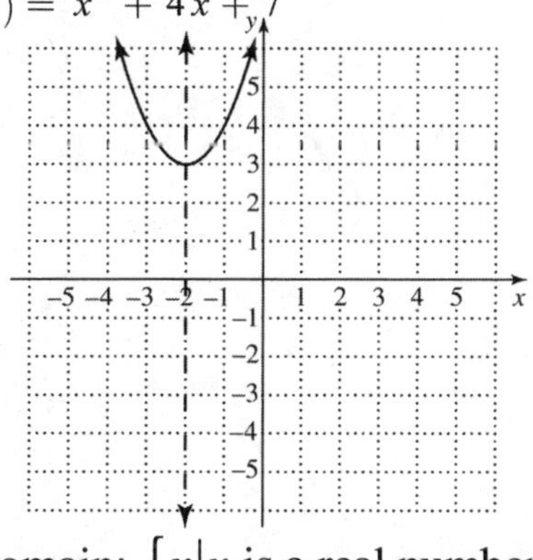

g) domain: $\{x | x \text{ is a real number}\}$ or $(-\infty, \infty)$; range: $\{y | y \geq 3\}$ or $[3, \infty)$

6. 21.875 ft

Objective 1

1. a) downward; b) $(0, 0)$; c) $x = 0$;

d)

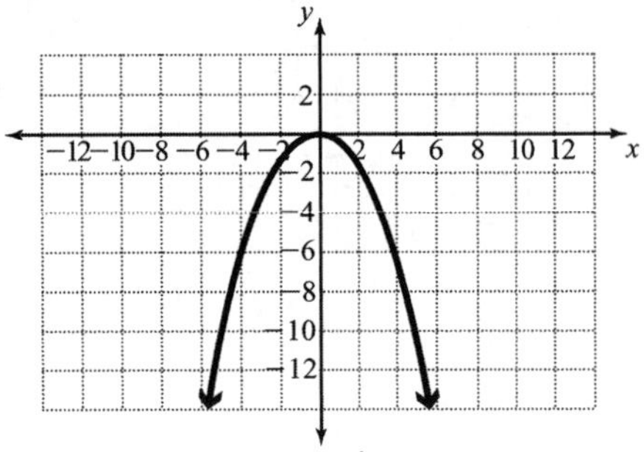

Objective 2

2. a) upward; b) $(0, 5)$; c) $x = 0$;

d)

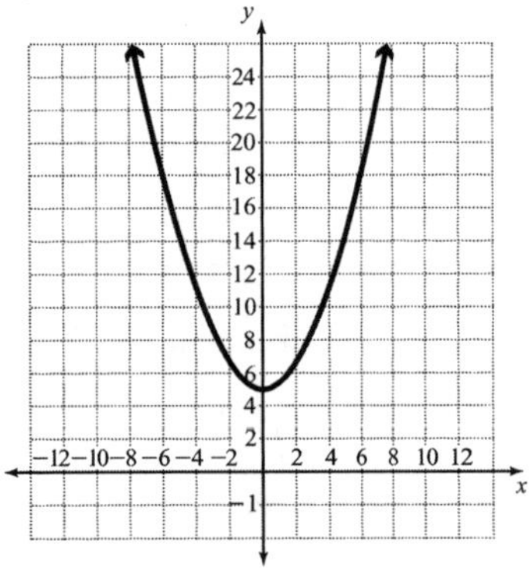

3. a) downward; b) $(0, -2)$; c) $x = 0$;

d)

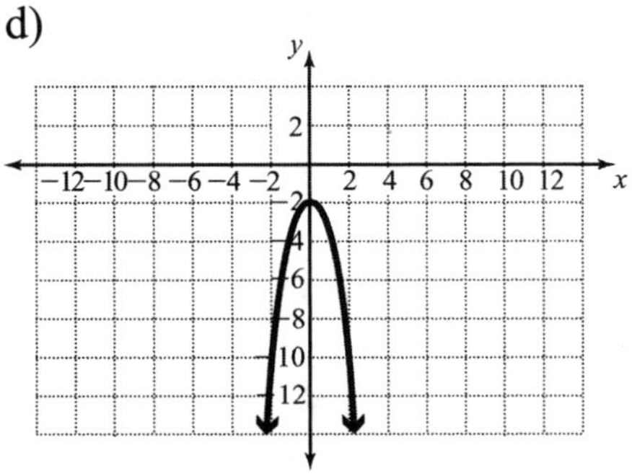

Objective 3

4. a) downward; b) $(-3, 0)$; c) $x = -3$;

d)

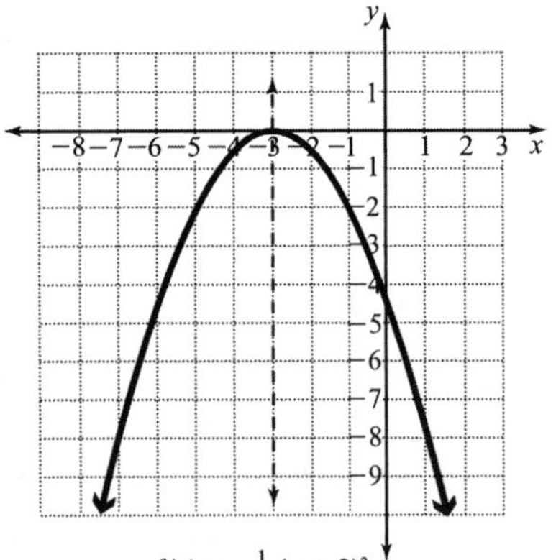

Objective 4

5. a) upward; b) $(-2, 1)$; c) $x = -2$;

d)

$f(x) = \frac{1}{4}(x+2)^2 + 1$

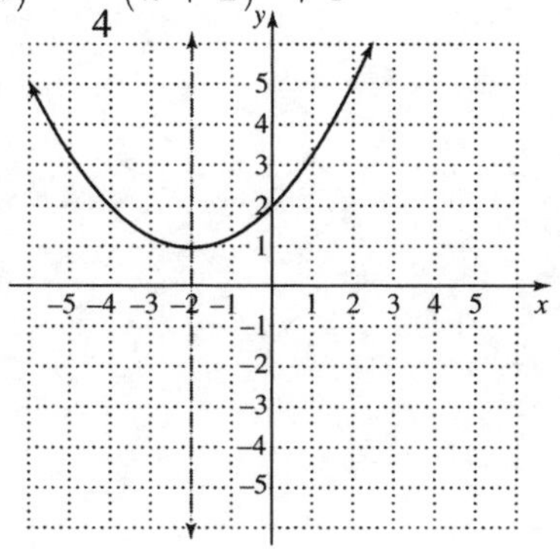

Objective 5

6. a) $(-6, 0)$, $(0, 36)$; b) $h(x) = (x+6)^2$; c) upward; d) $(-6, 0)$; e) $x = -6$;
f)

$h(x) = x^2 + 12x + 36$

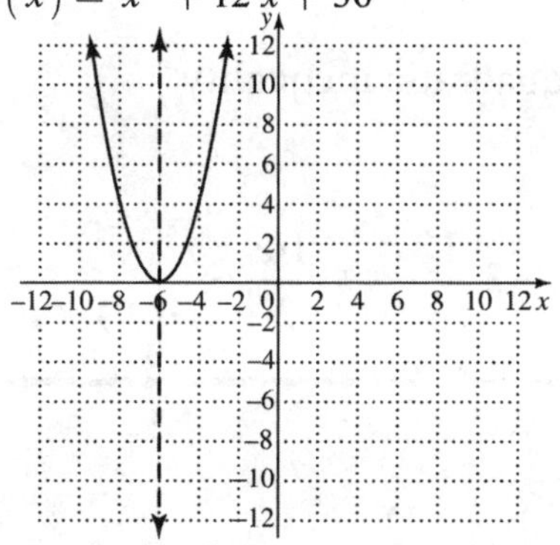

g) domain: $\{x|x \text{ is a real number}\}$ or $(-\infty, \infty)$; range: $\{y|y \geq 0\}$ or $[0, \infty)$

7. a) $(-3, 0)$, $(1, 0)$, $(0, 3)$; b) $g(x) = -(x+1)^2 + 4$; c) downward d) $(-1, 4)$; e) $x = -1$;
f)

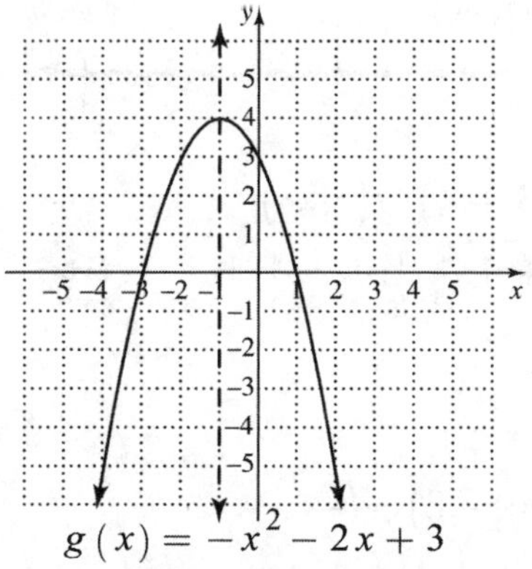

$g(x) = -x^2 - 2x + 3$

g) domain: $\{x|x \text{ is a real number}\}$ or $(-\infty, \infty)$; range: $\{y|y \leq 4\}$ or $[4, -\infty)$

Objective 6

8. a)

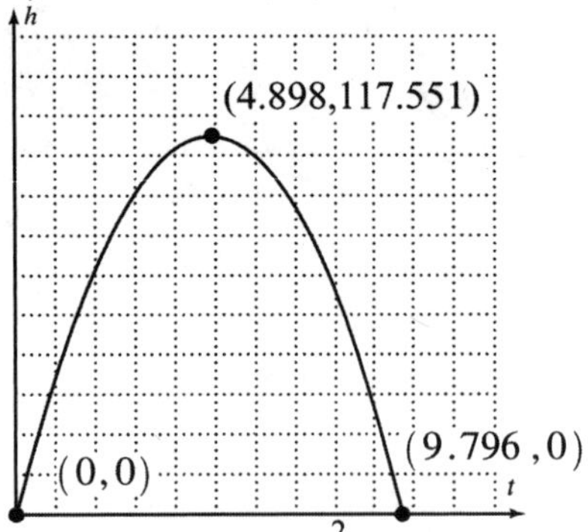

$h = -4.9t^2 + 48t$

b) 117.551 m; c) 9.796 sec.

9. a) length: 84 ft; width: 84 ft; b) 7056 sq. ft

Section 16.5

Key Terms

1. rational inequality
2. quadratic inequality

Practice

1. $(-\infty, -6) \cup (-5, 2)$

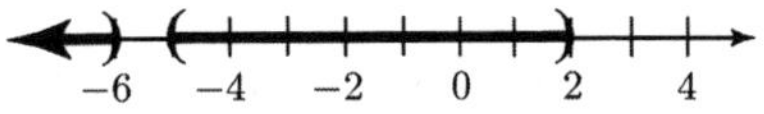

2. $-3, \frac{10}{3} \cup \frac{10}{3}, \infty$

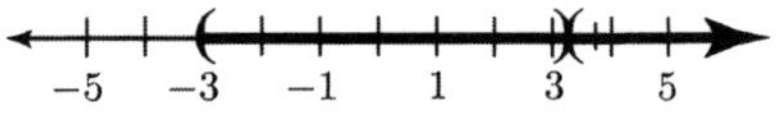

Objective 1

1. $(-3, 4)$

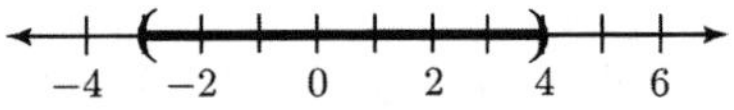

2. $(-\infty, 5) \cup (8, \infty)$

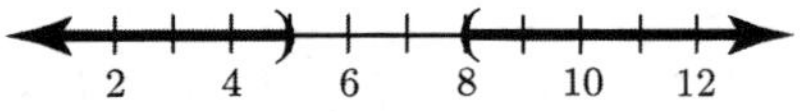

3. $(4, 11)$

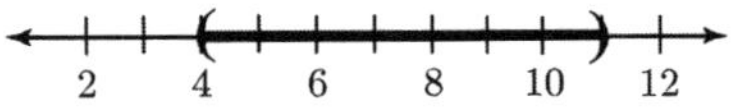

4. $(-\infty, \infty)$

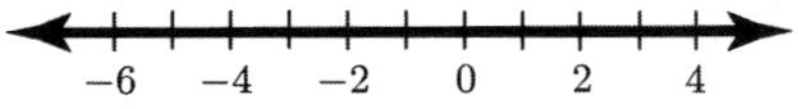

5. No solution or $\varnothing$

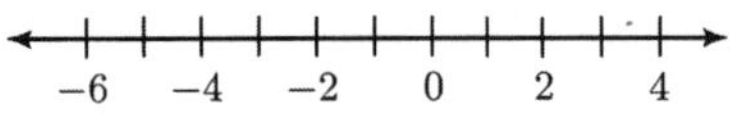

6. $(-5, -4) \cup (5, \infty)$

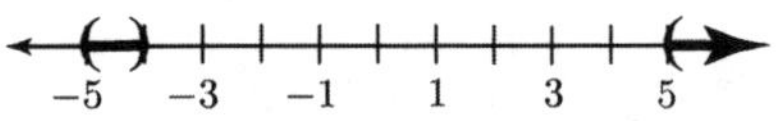

Objective 2

8. $(2, 6]$

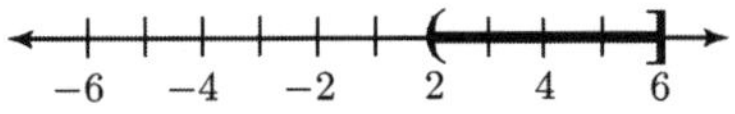

9. $(-\infty, -8) \cup (-4, \infty)$

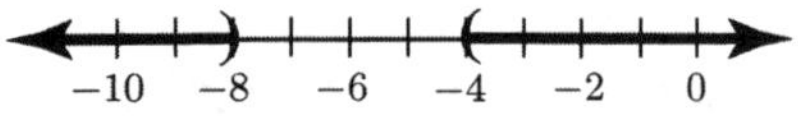

10. $(8, 8.5)$

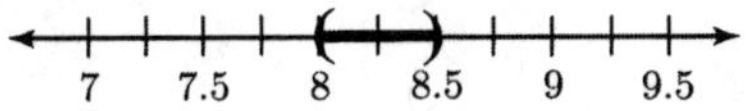

Section 16.6

Practice

1. $3x^3 - x^2 + 5x + 4;\ -x^3 + x^2 - 9x + 4$
2. $x^4 - x^3 - 13x^2 + 31x - 18$
3. $x^2 - 2x,\ x \neq 0$

Objective 1

1. $4x^2 + 8x - 7;\ 4x^2 - 8x + 7$
2. $x^2 - 5x + 5;\ -x^2 - 5x + 1$
3. $2x;\ 4$
4. $x^2 - 1;\ -x^2 + 2x - 1$
5. $x^3 + 2x^2 + 6x + 3;\ -x^3 + 2x^2 - 6x + 3$
6. $3x^2 + 2x + 7;\ -11x^2 + 4x - 3$
7. $4x^2 + 9x - 5;\ 3x + 3$

Objective 2

8. $6x^2 - 25x - 9$
9. $-x^3 + 14x^2 + 18x - 252$

Objective 3

10. $x - 5,\ x \neq 2$
11. $x - 1,\ x \neq -1$

Chapter 17

Section 17.1

Key Terms

1. one-to-one function
2. inverse functions
3. composition of functions

Practice

1. $(f \circ g)(x) = 50x^2 - 60x + 22;\ (g \circ f)(x) = 10x^2 + 17$
2. Yes
3. Yes
4. $f^{-1}(x) = \dfrac{x - 1}{4}$

5.

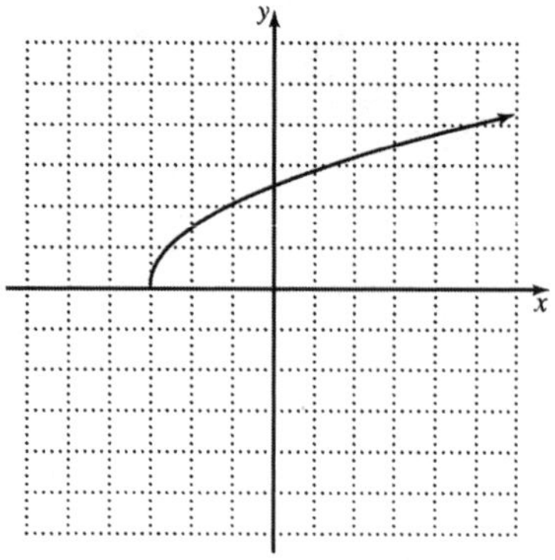

Objective 1

1. $(f \circ g)(x) = 7\sqrt{x+8} + 5$; $(g \circ f)(x) = \sqrt{7x+13}$
2. $(f \circ g)(x) = 8 - x$; $(g \circ f)(x) = \sqrt{4-x^2}$

Objective 2

3. No

Objective 3

4. No
5. Yes

Objective 4

6. $f^{-1}(x) = \dfrac{9x+8}{1-x}$
7. $f^{-1}(x) = \sqrt[3]{\dfrac{x+6}{5}}$

Objective 5

8.

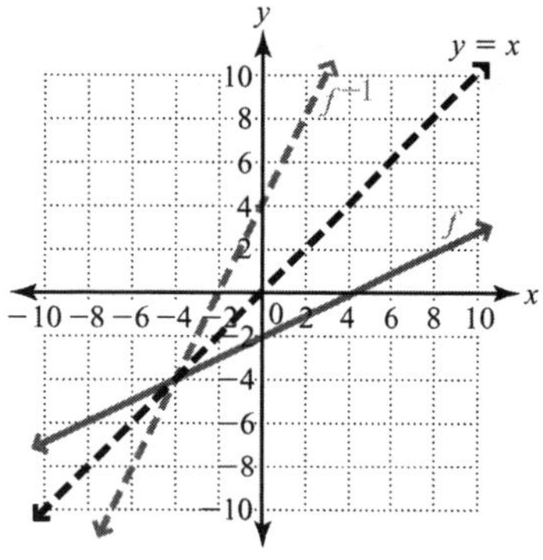

9.

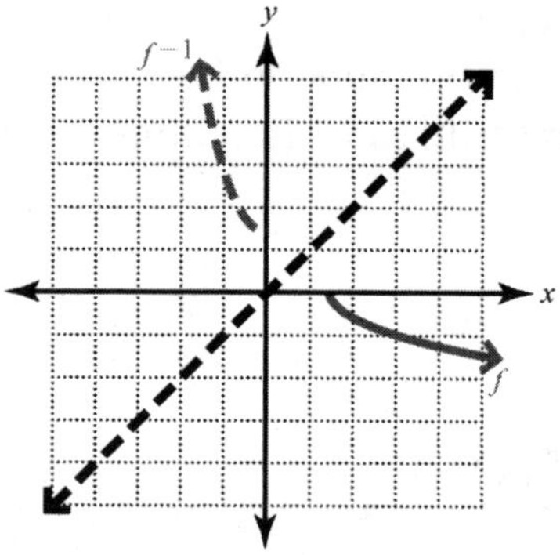

Section 17.2

Key Terms

1. exponential function

Practice

1.

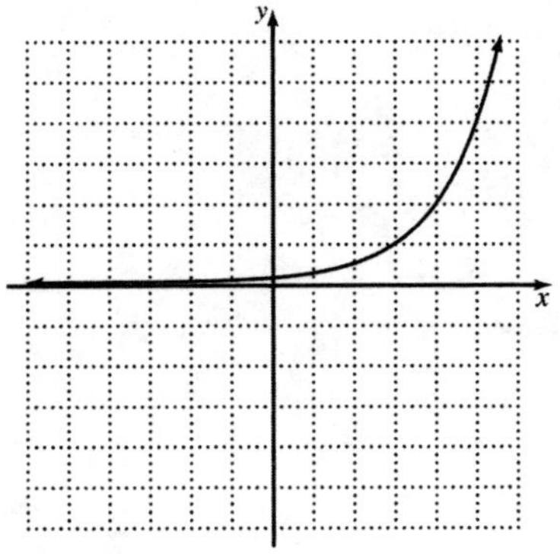

2. $\frac{3}{2}$

3. $28,827.22

Objective 1

1.

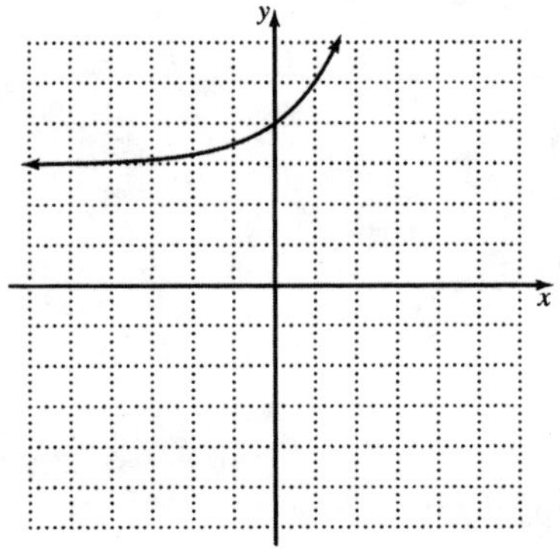

2.

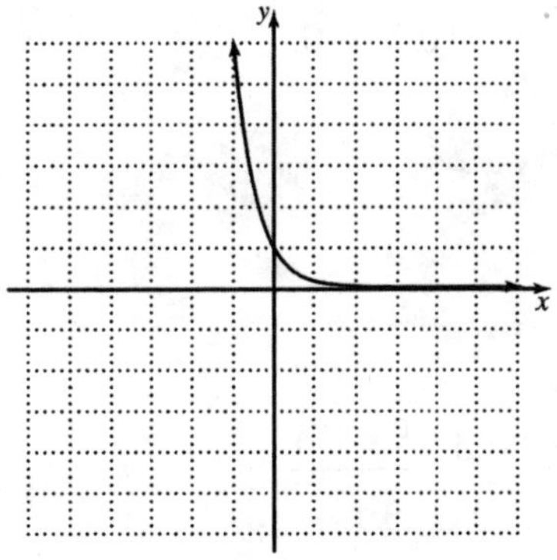

Objective 2

3. –4

4. –3

Objective 3

5. 2828 cells

Section 17.3

Key Terms

1. logarithm

Practice

1. $\log_3 243 = 5$

2. $5^4 = 625$

3. $\frac{1}{2}$

4.

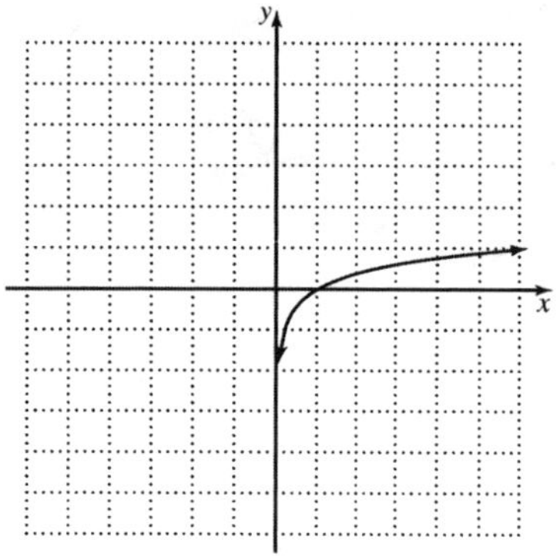

5. 68%

Objective 1

1. $\log_{64} 4 = \frac{1}{3}$

2. $\log_{1/5} 125 = -3$

3. $10^{-5} = \frac{1}{100,000}$

4. $2^{-2} = \frac{1}{4}$

Objective 2

5. 4

6. −3

7. 3

8. 64

9. 4

Objective 3

10.

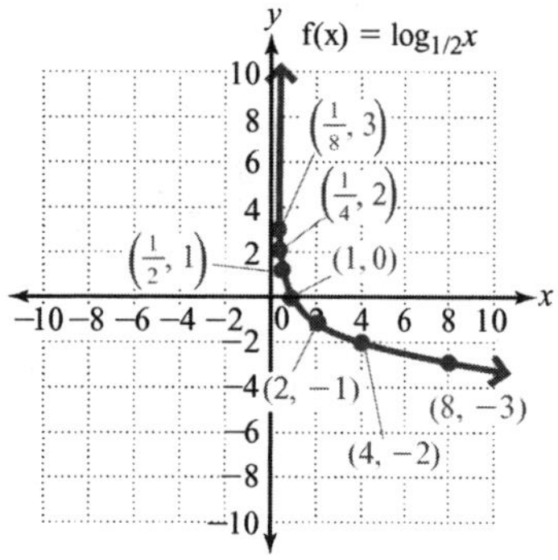

Objective 4

11. 74%

Section 17.4

Practice

1. f

2. 18

3. $\log_d a + \log_d q + \log_d r$

4. $\log_3 HZ$

5. $\log_6 13 - \log_6 7$

6. $\log_c 7$

7. $15\log_N t$

8. $\log_6 7^2$

Objective 1

1. z

2. 22

Objective 2

3. $\log_b y + \log_b (y-4)$

4. $\log_3 (x^2 + 16x + 60)$

5. $1 - \log_d k$

6. $\log_a (x+4)$

7. $\frac{1}{3}\log_7 b$

8. $\log_2 \frac{1}{a^3}$

9. $\log_b x + 7\log_b y - 6\log_b z$

10. $\log_a x^4 z^7$

Section 17.5

Key Terms

1. common logarithms

2. natural logarithms

Practice

1. 1.9685

2. 4.1790

3. 50 dB

4. 16.1

Objective 1

1. −2.0386

2. 4.7412

Objective 2

3. 8.5129

Objective 3

4. 3.4

5. $10^{-3.6}$ moles/L

Objective 4

6. 16.2935

Section 17.6

Practice

1. 2.5850

2. 11

3. 1.5 yr

4. 1.1833

Objective 1

1. 0.7953

2. 561.2160

3. −2.7004

Objective 2

4. 5
5. 2
6. $\dfrac{27}{10}$
7. 7

Objective 3

8. a) \$98,497.0; b)\$99,334.73
9. a) 126,000 b)1998

Objective 4

10. 17

Chapter 18

Section 18.1

Key Terms

1. circle
2. conic section
3. radius

Practice

1. opens right; vertex: $(-6, -3)$;

   axis of symmetry: $y = -3$

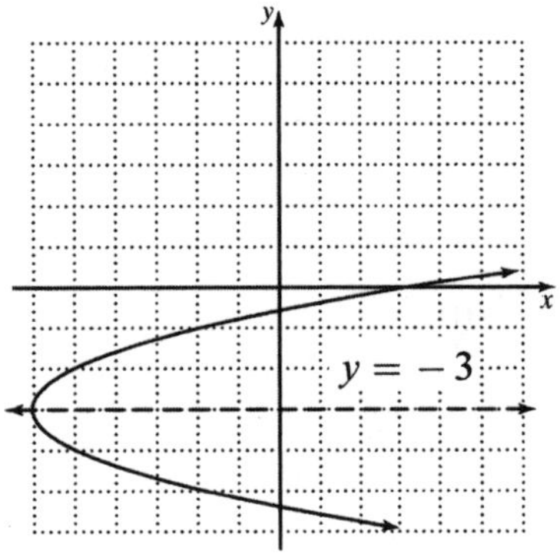

2. 25; $0, -\dfrac{19}{2}$

3. center: $(3, 2)$; radius: 6

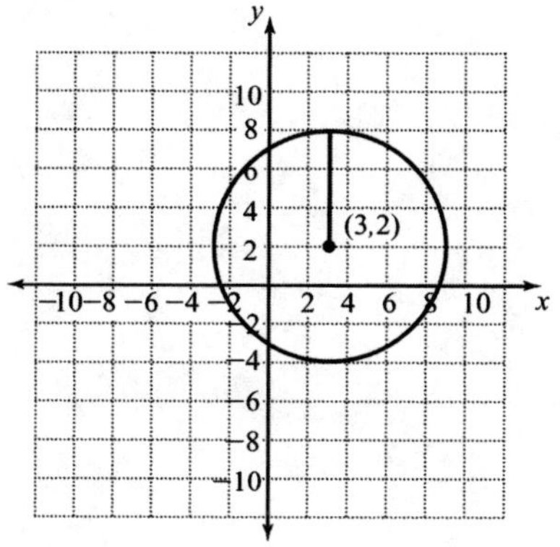

4. $(x-9)^2+(y-4)^2=4$

5. center: $(3, -2)$; radius: 6

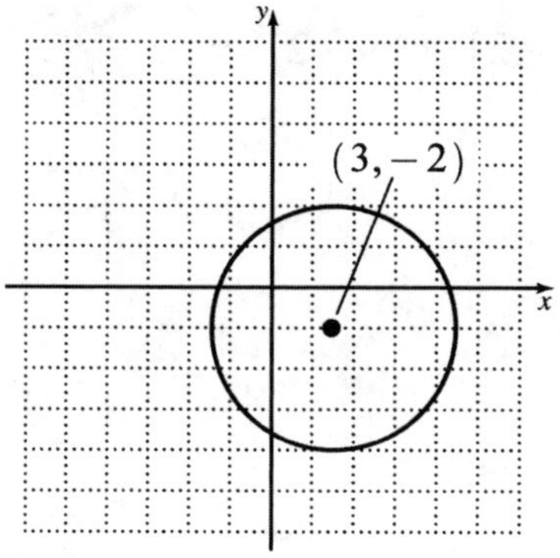

(scl = 2)

Objective 1

1. opens downward; vertex: $(-2, 1)$; axis of symmetry: $x=-2$

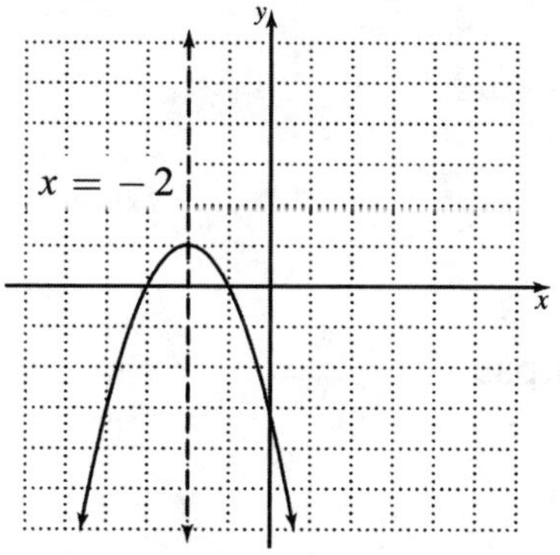

Objective 2

2. $\sqrt{58}$; $\frac{7}{2}, -\frac{5}{2}$

3. 5; $0, -\frac{1}{2}$

Objective 3

4. center: $(-2,-1)$; radius: 3

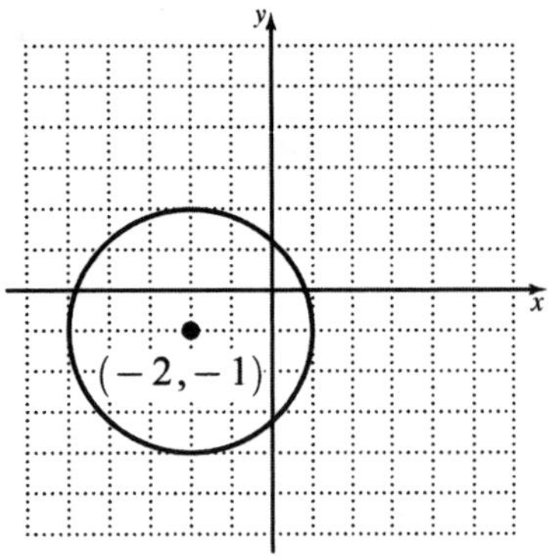

Objective 4

5. $(x+1)^2+(y-3)^2=6$

6. $(x-2)^2+(y-2)^2=51$

Objective 5

7. center: $(4,-2)$; radius: 2

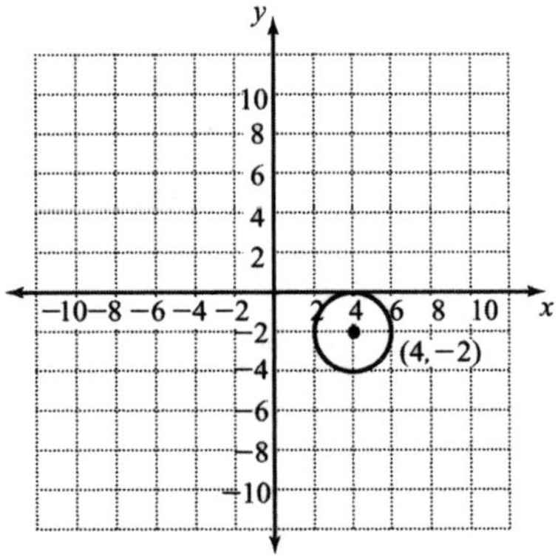

Section 18.2

Key Terms

1. hyperbola

2. ellipse

Practice

1.

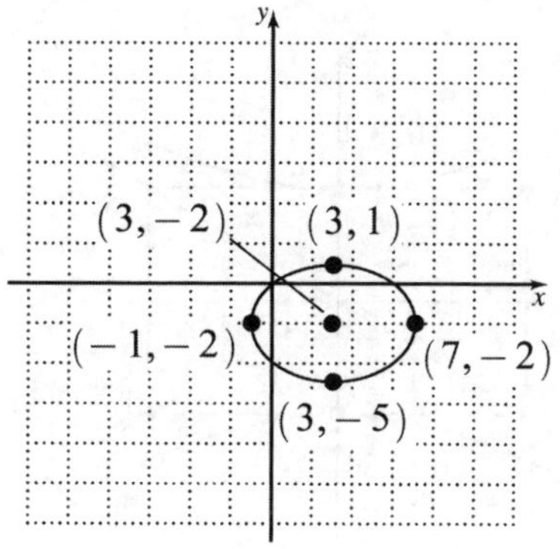

2.

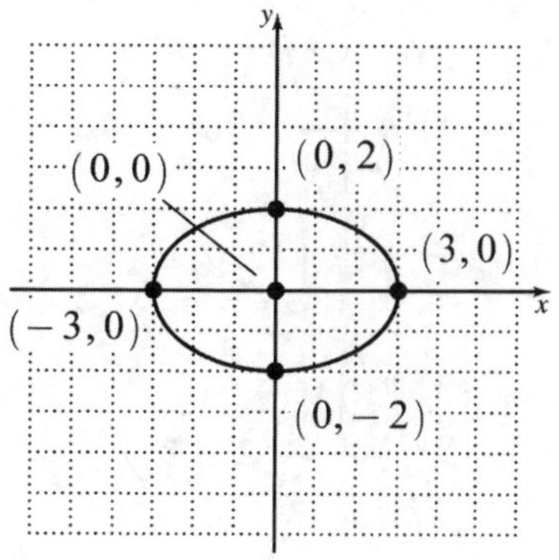

(scl = 2)

3.

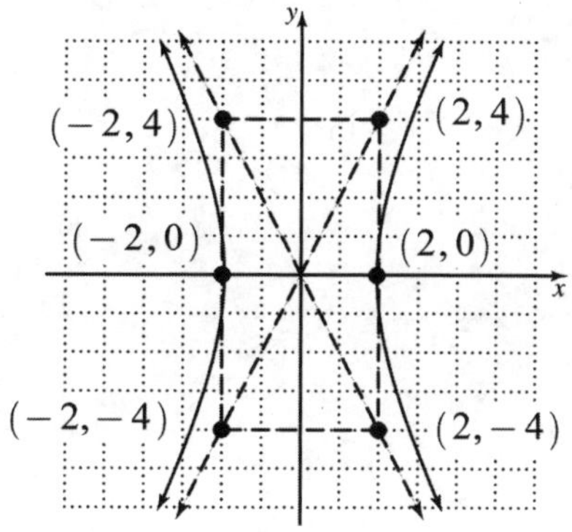

Objective 1

1.

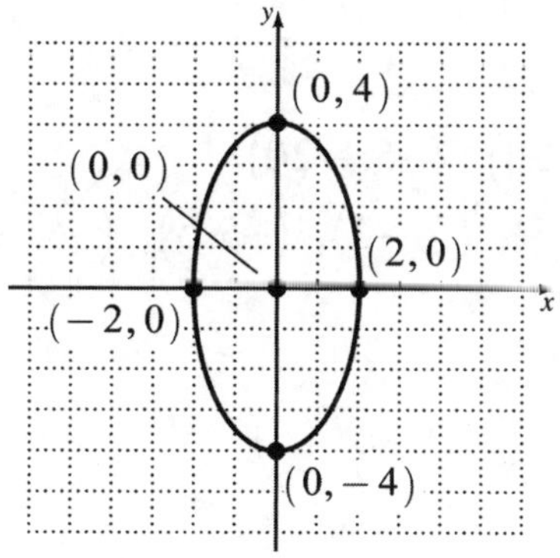

2.

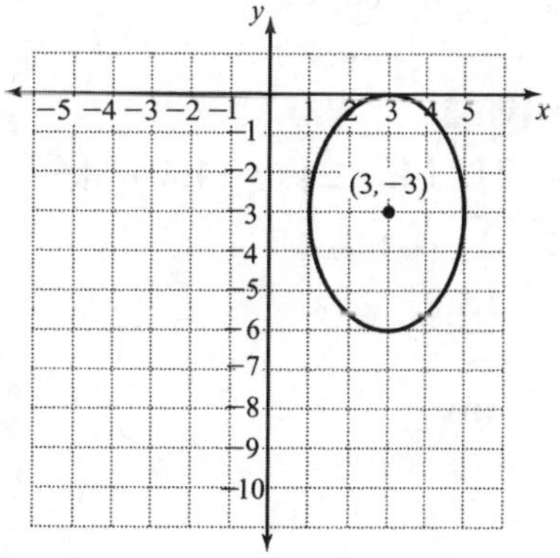

Objective 2

3.

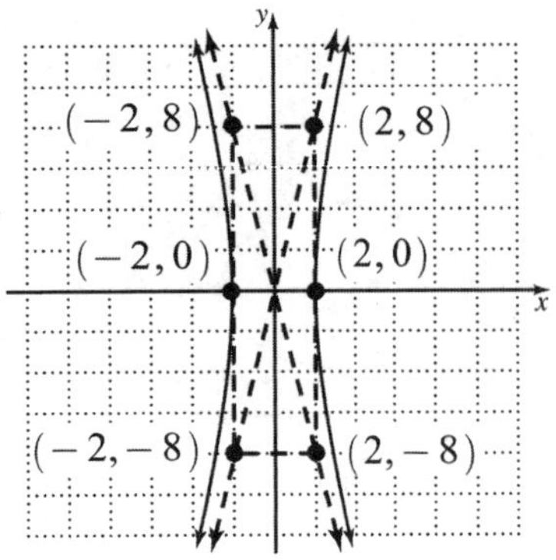

4.

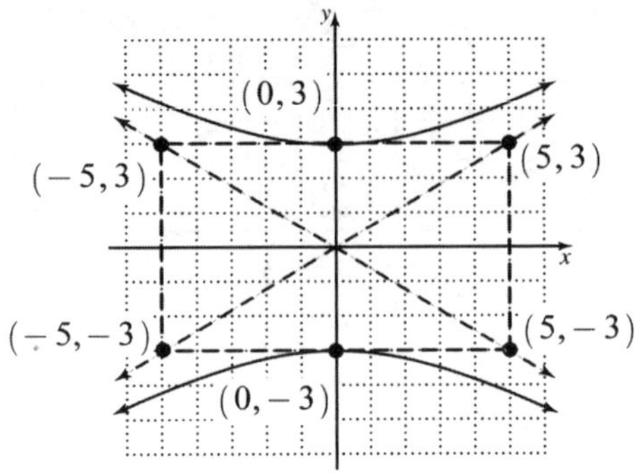

Section 18.3

Key Terms

1. nonlinear systems of equations

Practice

1. $(4,11)$, $(-7,-22)$
2. $(6,8)$, $(6,-8)$, $(-6,8)$, $(-6,-8)$

Objective 1

1. $(-1,7)$, $(-3,9)$
2. $(6,8)$, $(-8,-6)$
3. $(-4,-8)$, $(6,12)$
4. $(5,0)$

Objective 2

5. $(4,2)$, $(4,-2)$, $(-4,2)$, $(-4,-2)$
6. $(5,20)$, $(-5,-20)$
7. $(2,5)$, $(-1,-4)$
8. $(5,5)$, $(5,-5)$, $(-5,5)$, $(-5,-5)$

Section 18.4

Practice

1.

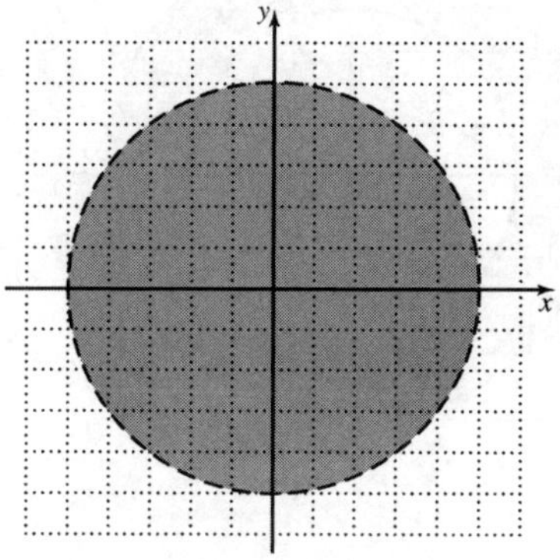

2.

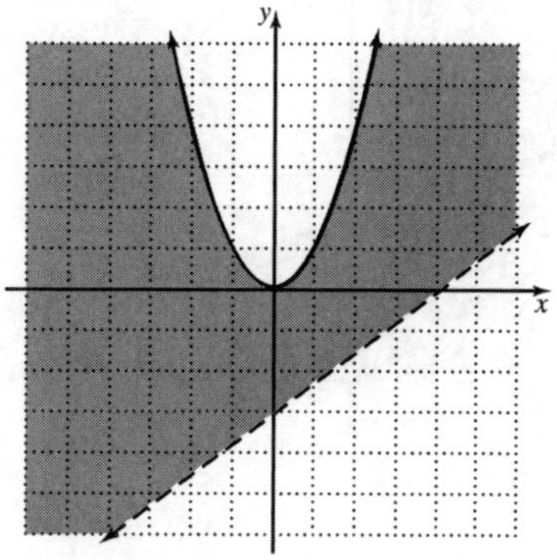

Objective 1

1.

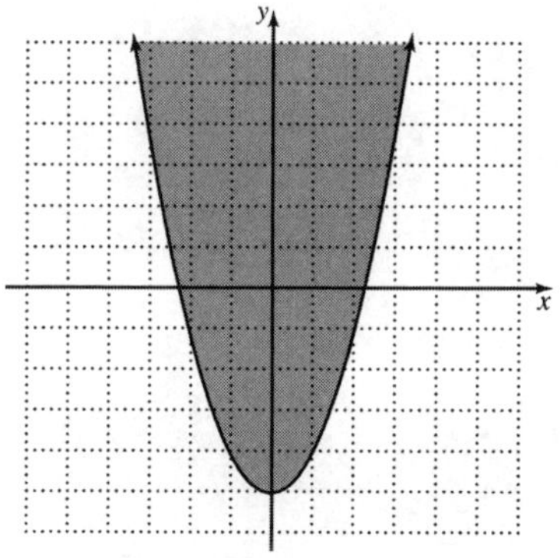

2.

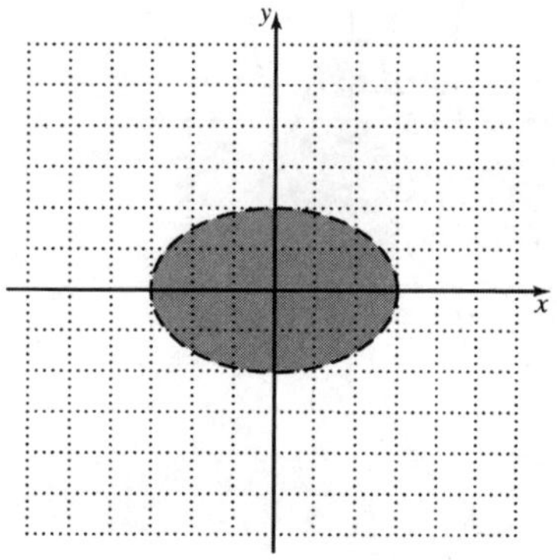

3.

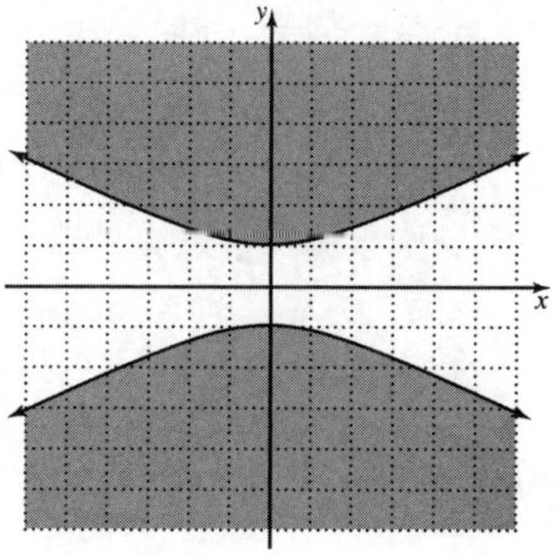

Objective 2

4.

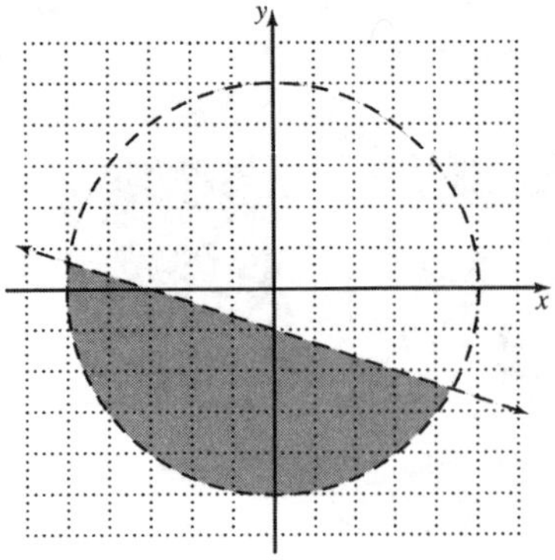

5.

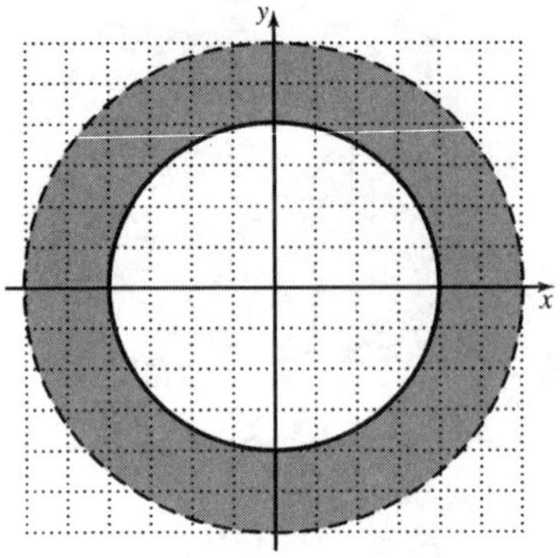

6.

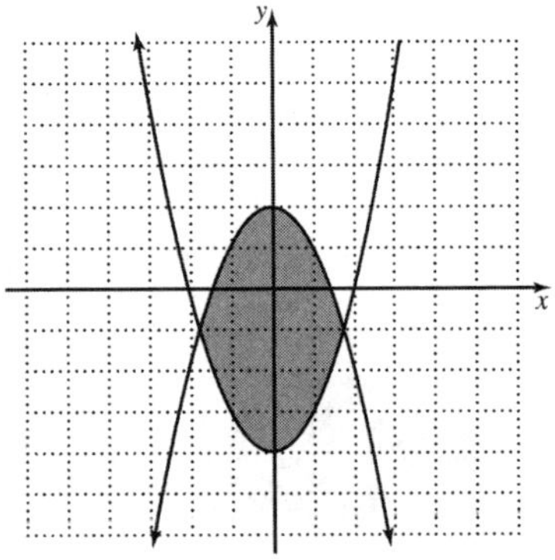